Journal
of the American Ceramic Society

ISSN 0002-7820
Volume 94
Number S1
June 2011

Institutional subscription prices for 2011 are:
Print & Online: US$2,269 (US), US$2,479 (Rest of World), €1,608 (Europe), £1,265 (UK). Prices are exclusive of tax. Asia-Pacific GST, Canadian GST and European VAT will be applied at the appropriate rates. For more information on current tax rates, please go to www.wileyonlinelibrary.com/tax-vat. The price includes online access to the current and all online back files to January 1st 2007, where available. For other pricing options, including access information and terms and conditions, please visit www.wileyonlinelibrary.com/access.

Delivery Terms and Legal Title Where the subscription price includes print issues and delivery is to the recipient's address, delivery terms are Delivered Duty Unpaid (DDU); the recipient is responsible for paying any import duty or taxes. Title to all issues transfers FOB our shipping point, freight prepaid. We will endeavour to fulfil claims for missing or damaged copies within six months of publication, within our reasonable discretion and subject to availability.

Publisher: Journal of the American Ceramic Society is published by Wiley Periodicals Inc., 350 Main St., Malden MA 02148-5020.

Journal Customer Services: For ordering information, claims and any enquiry concerning your journal subscription please go to www.wileycustomerhelp.com/ask or contact your nearest office.

Americas: Email: cs-journals@wiley.com; Tel: +1 781 388 8598 or +1 800 835 6770 (toll free in the USA & Canada).

Europe, Middle East and Africa: Email: cs-journals@wiley.com; Tel: +44 (0) 1865 778315.

Asia Pacific: Email: cs-journals@wiley.com; Tel: +65 6511 8000.

Japan: For Japanese speaking support, Email: cs-japan@wiley.com; Tel: +65 6511 8010 or Tel (toll-free): 005 316 50 480.

Visit our Online Customer Get-Help available in 6 languages at www.wileycustomerhelp.com.

Production Editor: Josh Martino (Email: JACE@wiley.com)

Advertising: Karl Franz (email: kfranz@wiley.com)

Commercial Reprints: Lydia Supple-Pollard (email: lsupple@wiley.com)

Back Issues: Single issues from current and recent volumes are available at the current single issue price from cs-journals @wiley.com. Earlier issues may be obtained from Periodicals Service Company, 11 Main Street, Germantown, NY 12526, USA. Tel: +1 518 537 4700, Fax: +1 518 537 5899, Email: psc@periodicals.com

Abstracting and Indexing Services: The Journal is indexed by Science Citation Index and METADEX. For a complete list of A&I services please visit www.wileyonlinelibrary.com.

This journal is available online at Wiley Online Library. Visit www.wileyonlinelibrary.com to search the articles and register for table of contents e-mail alerts.

ISSN 0002-7820 (Print)
ISSN 1551-2916 (Online)
For submission instructions, subscription and all other information visit: www.wileyonlinelibrary.com

Journal of the American Ceramic Society accepts articles for Open Access publication. Please visit http://olabout.wiley.com/WileyCDA/Section/id-406241.html for further information about OnlineOpen.

Printed in the USA by The Sheridan Group.

Wiley's Corporate Citizenship initiative seeks to address the environmental, social, economic, and ethical challenges faced in our business and which are important to our diverse stakeholder groups. Since launching the initiative, we have focused on sharing our content with those in need, enhancing community philanthropy, reducing our carbon impact, creating global guidelines and best practices for paper use, establishing a vendor code of ethics, and engaging our colleagues and other stakeholders in our efforts. Follow our progress at www.wiley.com/go/citizenship.

Introduction

Feature

Review

Original Articles

Contents

Register Now!

MS&T'11®

Materials Science & Technology
2011 Conference & Exhibition

OCTOBER 16–20
COLUMBUS, OHIO, USA

www.matscitech.org

The leading forum addressing structure, properties, processing and performance across the materials community.

Organized by: ACerS | AIST | ASM | TMS
Co-sponsor: NACE

Plenary Session Announced!

Dr. Subra Suresh, Director of the U.S. National Science Foundation, will discuss the importance of the U.S.'s science & technology work force, followed by several prominent speakers on related topics.

More information coming soon!

J. Am. Ceram. Soc., **94** [S1] S1–S2 (2011)
DOI: 10.1111/j.1551-2916.2011.04592.x
© 2011 The American Ceramic Society

journal _____

A Tribute to Anthony G. Evans:
Materials Scientist and Engineer December; 4, 1942–September 9, 2009

John W. Hutchinson
January 4, 2011

T HIS special issue of the Journal of the American Ceramics Society brings together papers written by some of the many colleagues who have worked with Tony Evans over his career. The issue itself reflects the extraordinary breadth of Evans' scientific interests. This brief Tribute is an attempt to capture in a few words Tony Evans' remarkable influence and contribution to materials science and more broadly to engineering science. The task would be more daunting were it not for the testimony to Tony Evans in celebration of his 65th birthday by A. Heuer (Int. J. Mat. Res. **98** (2007) 1168–1169) and an obituary by N. A. Fleck, which appeared in the November 29, 2009 issue of the Guardian newspaper.

Anthony G. Evans was one of the most influential materials scientists and materials engineers of his generation. He had no rival when it came to the grasp of the underlying fundamentals of material behavior coupled with an extraordinary ability to focus his attention and to inspire and lead collaborative efforts. Evans was born and raised in Porthcawl, Wales to William Glyn and Annie May Evans. He met his wife, Trisha, in their hometown and they were married in 1964. Trisha and their daughters and grandchildren survive Tony. After obtaining BSc (1964) and PhD (1967) degrees in metallurgy at Imperial College, London, Evans began work in 1967 as a ceramist at the Atomic Energy Research

Establishment, Harwell. Following a sabbatical period at UCLA, Evans worked at the National Bureau of Standards from 1971 to 1974 and then served as a group leader at the Rockwell International Science Center from 1974 to 1978. In 1978 he joined the faculty of the Department of Materials Science and Mineral Engineering, University of California, Berkeley, where he remained until 1985. During these years the emphasis of much of Evans' research was on ceramics and he began his long association with the American Ceramics Society. In 1985 he moved to the Santa Barbara campus of the University of California as the Alcoa Professor. Evans was the founding chair of the Materials Department at UCSB (1985–1991) which would become one of the leading materials departments in the world. Those of us who conducted research with Evans during this period were largely unaware of his efforts as department chair—he later attributed this to the fact that he allocated the period from 7:30 to 9 in the morning each day to his departmental duties, finishing before his colleagues had an opportunity to perturb the process. This was also the period that Evans established himself as a research leader par excellence heading major projects on ceramic matrix composites, toughening of ceramics, and thin films and multilayers. Evans made two more academic moves before completing a circle back to UCSB in 2002. From 1994 until 1998 Evans was the Gordon McKay Professor of Materials Engineering in the Division of Engineering and Applied Sciences at Harvard University and from 1998 to 2002 he served as the Gordon Wu Professor in the Department of Mechanical and Aerospace Engineering at Princeton University. The pull of Santa Barbara and UCSB remained strong, however, and he returned in 2002 to a joint appointment in the Departments of Materials and Mechanical Engineering where he focused primarily on teaching and research.

Any tribute to Tony Evans must begin with the impact of his work. Evans is the most highly cited materials scientist with almost 35 000 citations to over 650 published journal papers. His h-index will soon pass 100. A short list of subjects to which Evans has made major contributions includes micro-cracking and transformation toughening of ceramics, ceramic matrix composites and metal matrix composites, thin film mechanics, interface mechanics, thermal barrier coatings, metallic foams, morphing structures, aerospace materials with special thermo-mechanical properties, lightweight lattice materials, and superior blast and ballistic resistant structural materials. On each of these problem areas, Evans brought to bear a fundamental understanding of material behavior at all scales together with innovative experiments in the laboratory. The experimental work he and his collaborators performed more often than not focused on observation of micromechanical behavior and new phenomena rather than on refined measurement of material properties.

The synergy between Evans' grasp of theory and his insightful exploitation of experiment, combined with his love of subject and legendary ability to focus, would have been enough, by themselves, to establish his primacy. However, there is more. Evans' skills at assembling, inspiring and leading interdisciplinary teams of engineers and scientists to tackle challenging technological problems is the additional component of his approach which truly set him apart. Those who have had the experience of working with Tony Evans on one or more of the teams he put together will be aware that he had no match as a technology leader in the arena of structural materials. These include his former students, post-doctoral fellows, and a large cadre of colleagues which he brought together from many academic, government and industrial institutions, here and abroad. Particularly notable for each major project that Evans led were the workshops where ongoing work was reviewed and new work was planned with criticism and input from experts from industry and government labs. These workshops, which were always enlivened by Evans' active participation in every detail of the research, were exceptional in identifying the challenges and moving the research forward. Skepticism about the effectiveness of shifting research funding from smaller projects to relatively fewer large projects, a trend that has taken place over the past several decades in the US and is now spreading around the world, would be far less warranted if more large projects were led by individuals with the abilities of an Evans.

Tony Evans provided leadership in the materials community throughout his career in other ways as well. As already noted, he provided critical leadership in founding the Materials Department at UCSB for nearly six years and then later for the Princeton Materials Institute for four years, in each case without appearing to break step in his own research activities. Starting his service in 1974, Evans became the longest serving member of the Defense Sciences Research Council (formerly the Materials Research Council) of DARPA which played a major role in setting the agenda for research in advanced materials in the US. Simultaneous with all his other activities, during his entire career, Evans was a highly engaged consultant to many companies, not only in the materials industry but also in aerospace and electronics. To those of us who worked closely with Evans, it was clear that much of his research emphasis was motivated by problems that surfaced through his consulting activities. There is little wonder that his work has had, and will continue to have, such major impact.

In his testimony in celebration of Tony Evans' 65th birthday, Arthur Heuer noted that Evans' exceptional generosity to students and colleagues with his ideas and time was one of the keys to his success. Quoting Heuer directly: "It is his incredible ability to focus, his "nose" for important problems to work on, and his generosity in collaborative research, that have led to his revered status in our field." Throughout his career, Evans enjoyed working closely with his many students and post-docs who are now spread far and wide in academia, industry and government. These former junior colleagues, together with all his other collaborators, will keep Evans' legacy alive for years to come.

Tony Evans' contributions have been recognized in many ways—a few of them are listed below. He was a member of both the National Academy of Engineering and the National Academy of Sciences. He was also a Fellow of both the Royal Society (FRS) and the Royal Academy of Engineering in the UK. He was a Distinguished Life Member of the American Ceramics Society and received essentially every major award of this Society. He won the Henry Marion Howe Medal of ASM International, the Turnbull Award of the Materials Research Society, the Griffith Medal and Prize and Mellor Memorial Lectureship of the Institute of Materials, UK, and the Nadai Medal of ASME. He was an Alexander von Humboldt Senior Scientist. In 2008, Evans was especially pleased to receive the highest award granted by his alma mater, Fellow of Imperial College, London. □

J. Am. Ceram. Soc., **94** [S1] S3–S14 (2011)
DOI: 10.1111/j.1551-2916.2011.04559.x
© 2011 The American Ceramic Society

journal

Hybrid Materials to Expand the Boundaries of Material-Property Space

Mike Ashby[†]

Engineering Department, Cambridge University, CB2 1PZ Cambridge, U.K.

The materials we use today for mechanical design are the outcome of at least 3000 years of development, much of it empirical but much the outcome of systematic science. Both approaches have been motivated by the desire for stiffer, stronger, more durable, and lighter structures, progressively populating material property "space". We first examine the extent to which this space is now filled and estimate the ultimate constraints on this filling. Strategies for expanding the filled regions further include *hybrid material design*—making materials by combining two or more monolithic materials in chosen configurations and connectivity. Here we explore two classes of hybrid—lattice materials and sandwich structures—developing ways of comparing their properties with those of conventional monolithic materials. The comparison reveals the potential of hybrids. This paper builds on ideas that have grown from a long and rewarding collaboration with Tony Evans with whom it has been a privilege to have worked.

I. Introduction and Synopsis

HYBRID materials (Fig. 1) are combinations of two or more materials, or of materials and space, assembled in such a way as to have attributes not offered by any one material alone.[1,2] Particulate and fiber composites are examples of one type of hybrid, but there are many others: sandwich structures, foams, lattice structures, segmented structures, zero-expansion material, and more. Here we explore the potential of hybrid materials, emphasizing the choice of the components, their configuration, and their relative volume fraction. The new variables expand design space, allowing the creation of new "materials" with specific property profiles. But how are we to compare a hybrid—a sandwich structure for example—with monolithic materials such as polycarbonate or titanium? To do this we must think of the sandwich (as an example) not only as a hybrid with faces of one material bonded to a core of another, but as a "material" in its own right, with its own set of *effective properties*; it is these that allow the comparison.

The approach adopted here is one of breadth rather than precision. The aim is to assemble methods to allow the properties of alternative hybrids to be scanned and compared with those of monolithic materials, seeking those that best meet a given set of design requirements. Once materials and configuration have been chosen, standard methods—optimization routines, finite-element analyzes—can be used to refine them. But what the standard methods are *not* good at is the quick scan of alternative combinations. That is where the approximate methods developed below, in which material and configuration become the variables, pay off.

In this paper we explore two broad classes of hybrid, each with a number of discrete members.

(1) *Cellular structures* are combinations of material and space giving precise control of density, stiffness, strength, and thermal conductivity.

(2) *Sandwich structures* have outer faces of one material supported by a core of second, low density, material—a configuration that can offer a flexural stiffness per unit weight that is greater than that offered by either component alone.

We use simple continuum and micromechanical models to estimate the *equivalent properties* of each configuration, allowing direct comparison with monolithic engineering materials. The examples of Section IV illustrate how this is done. The methods are developed further in a White Paper.[3]

II. Holes in Material-Property Space

Material properties can be "mapped" as *Material Property Charts* of which Fig. 2 is an example: it is a chart of Young's modulus plotted against density. The small bubbles show the range of these two properties exhibited by a given material type. The larger colored envelopes enclose material families. All the charts have in common that parts of them are populated with materials but other parts are not: there are *holes*. Some parts of the holes are inaccessible for fundamental reasons that relate to the size of atoms and the nature of the forces that bind them together. But others are empty even though, in principle, they could be filled.

Is anything to be gained by developing materials (or material combinations) that lie in these holes? To answer this we need criteria of excellence to assess the merit of any given hybrid. These are provided by the material indices, described fully elsewhere.[4] If a possible hybrid has a value of any one of these that exceed those of existing materials, it has the potential to increase performance.

The axes in Fig. 2 are Young's modulus, E, and density ρ. The material indices E/ρ, $E^{1/2}/\rho$, and $E^{1/3}/\rho$, indicated by the "Guide lines" at the lower right, are criteria of the excellence or material indices for selecting material for light, stiff structures. A

F. Zok—contributing editor

Manuscript No. 29018. Received December 7, 2010; approved March 9, 2010.
A contribution to the celebration of the life of Tony Evans.
[†]Author to whom correspondence should be addressed. e-mail: mfa2@eng.cam.ac.uk

Fig. 1. Hybrid materials combine the properties of two (or more) monolithic materials, or of one material and space. They include fibrous and particulate composites, foams and lattices, sandwiches, and almost all natural materials.

grid of lines of one index—E/ρ—is plotted in the Figure. The orange arrow lies normal to the index lines. If the filled areas can be expanded in the direction of this arrow (i.e. to greater values of E/ρ) the materials so created will enable lighter, stiffer structures. The arrow thus defines a *vector for material development*.

Figure 3 shows a second example. The axes are density ρ and strength σ_f, meaning the yield strength for ductile materials, the modulus of rupture for those that are brittle. Once again small bubbles enclosed the property ranges of individual material types, the larger colored envelopes enclose families. Again some regions are filled and some are empty. The material indices σ_f/ρ, $\sigma_f^{2/3}/\rho$, and $\sigma_f^{1/2}/\rho$, characterizing materials that allow light, strong structures, are indicated by "Guide lines." A grid showing σ_f/ρ with superimposed vector, indicates the desired direction of development for materials for structures that are both light and strong.

So far, so good. But how much of the holes are accessible? Does the underlying physics of stiffness, strength, and density prohibit the existence of materials within the holes? Figure 4 is an attempt to answer this question. The stable element with the highest density is osmium ($E = 560$ GPa, $\rho = 22\,550$ kg/m³). Vegard's law, though approximate, effectively rules out the creation of an alloy or compound with a larger density this. The bulk material with the highest modulus is diamond ($E = 1{,}050$ GPa, $\rho = 3.520$ kg/m³), a consequence of the great stiffness of the covalent carbon–carbon bond and the small size of the carbon atom. It is conceivable that structures with more densely packed, more tightly bound atoms might exist but no examples of bulk materials with such a combination are at present known. Structures with diamond-like bonding but with lower density than diamond could be made by creating lattice structures like those described in the next section of this paper. It is shown there that the modulus of such structures falls linearly with density. Combining these facts allows part of the modulus-density space to be blanked out as inaccessible, as indicated in Fig. 4. That still leaves possible room of expansion by creating materials that lie in the area marked "accessible."

A similar, approximate boundary for the accessible space can be plotted on the strength-density chart (Fig. 5). The idea that

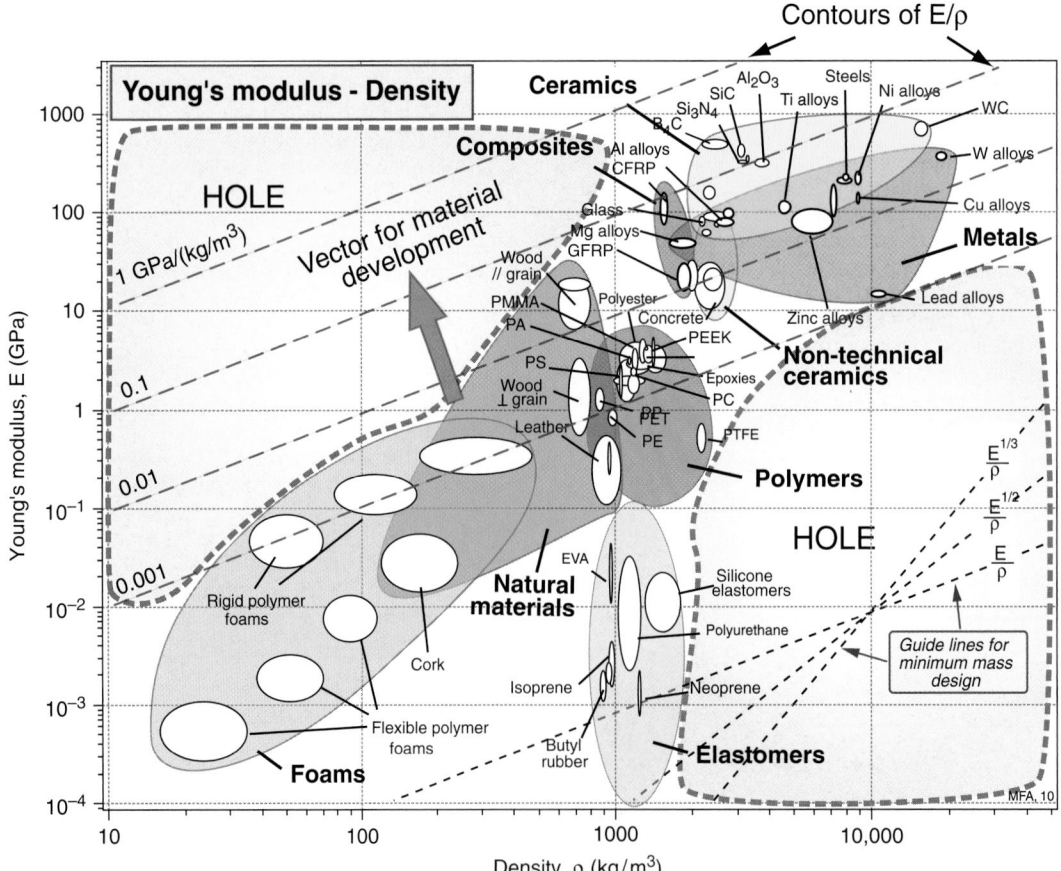

Fig. 2. A material property chart showing modulus-density space, with contours of specific modulus, E/ρ. Part of the space is occupied by materials, part it empty (the "holes"). Material development that extended the occupied territory in the direction of the arrow (the "vector for material development") allows components with greater stiffness-to-weight than is at present possible.

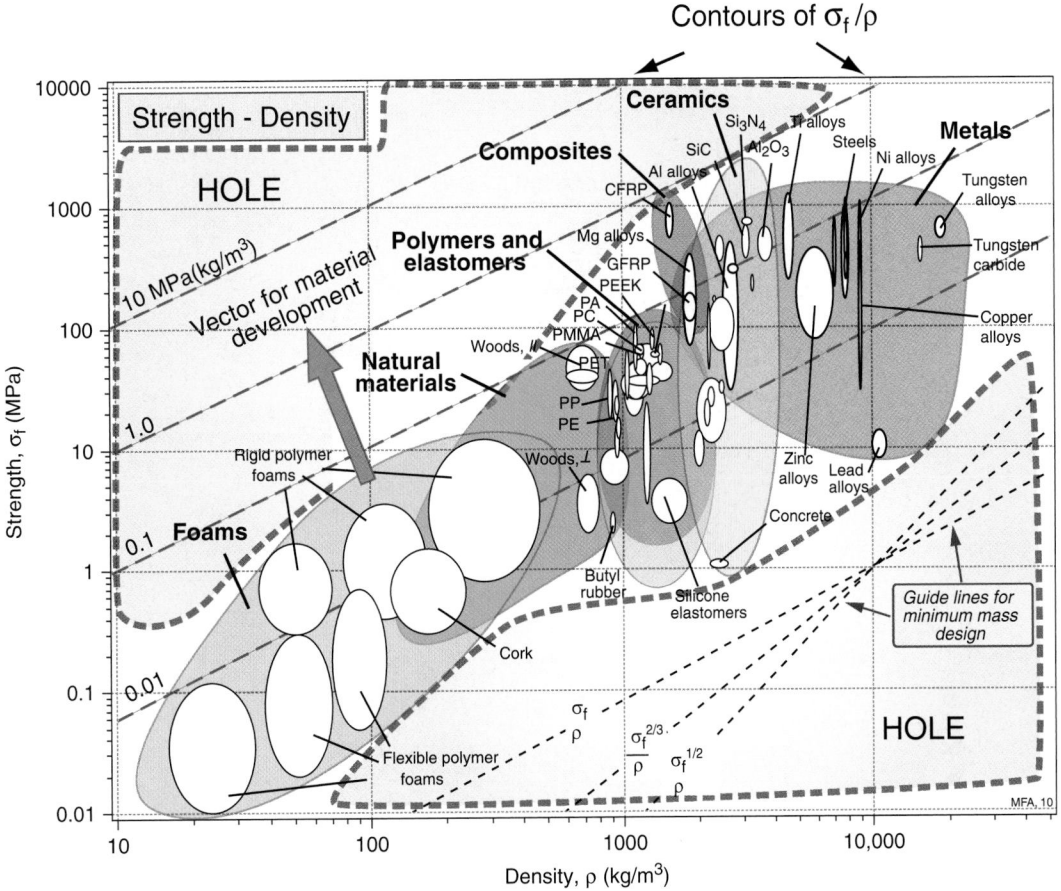

Fig. 3. A material property chart showing strength-density space, with contours of specific strength, $\sigma_{f/\rho}$. Here σ_f means the yield strength metals and polymers, the modulus of rupture of ceramics. Like Fig. 2, it has holes. Material development that extended the occupied territory in the direction of the arrow enables components with greater strength-to-weight than is at present possible.

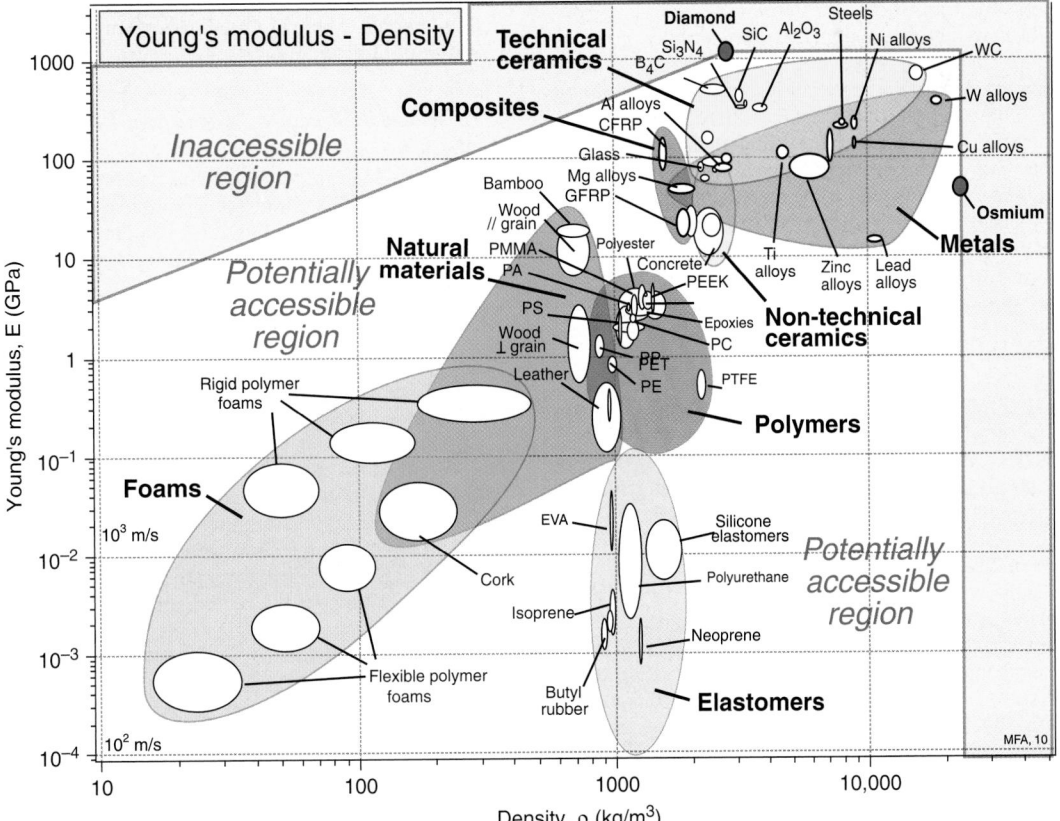

Fig. 4. Some unfilled areas of modulus-density space are inaccessible for fundamental reasons, but other areas are potentially accessible, suggesting that materials might be created to fill them. Two "limiting" solids, osmium and diamond, are indicated.

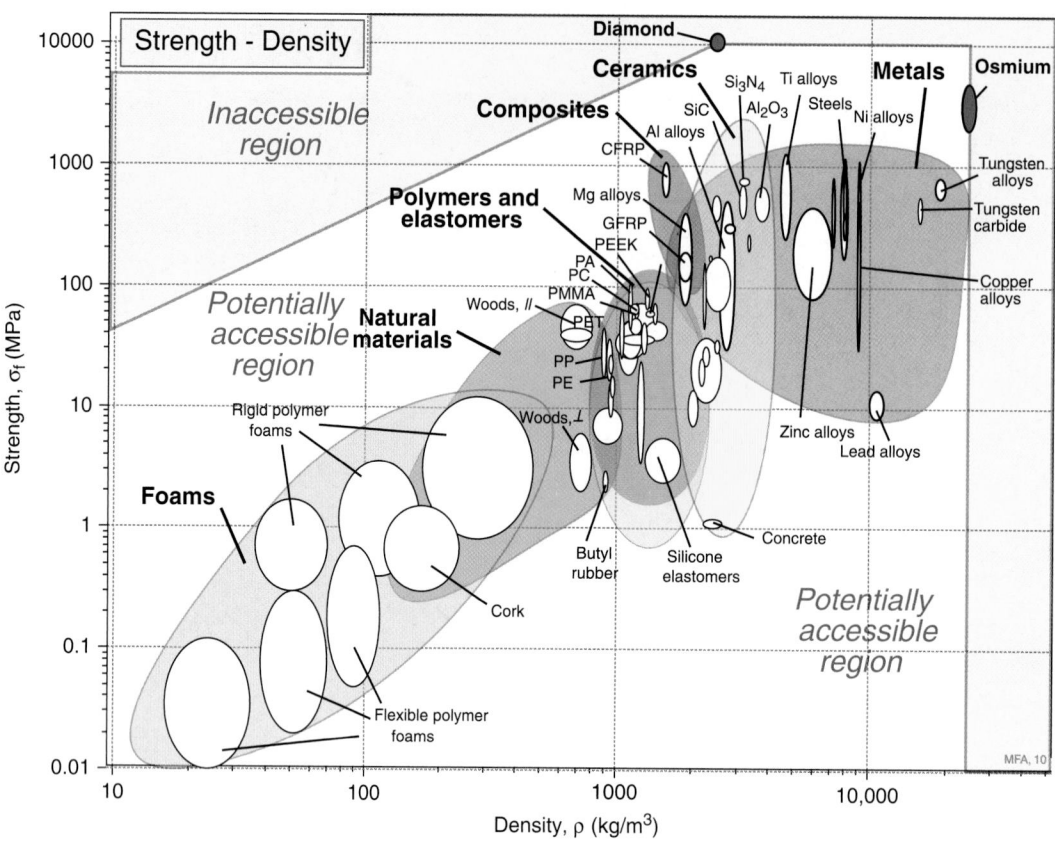

Fig. 5. Like the previous figure, the strength-density space has potentially accessible areas that are not yet filled.

the "ideal" strength of a solid is a fraction of its modulus—$E/30$ is often cited—has a sound physical basis. The upper boundary on the figure is based on the conservative value $\sigma_f = E/100$ Once again it appears that room exists for developing materials with higher strength-to-weight ratios than those we already have.

One approach to filling these holes—the long-established one—is that of developing new metal alloys, new polymer chemistries, and new compositions of glass and ceramic so as to create monolithic materials that expand the populated areas of the property charts. But developing new materials can be expensive and uncertain, and the gains tend to be incremental rather than step like. An alternative is to combine two or more existing materials so as to allow a superposition of their properties—in short, to create hybrids. The great success of carbon and glass–fiber reinforced composites at one extreme, and of foamed materials at another (hybrids of material and space) in filling previously empty areas of the property charts is encouragement to explore ways in which such hybrids can be designed.

When is a hybrid a "material"? There is a certain duality about the way in which hybrids are thought about and discussed. Some, like filled polymers, composites or wood are treated as materials in their own right, each characterized by its own set of macroscopic material properties. Others—like galvanized steel—are seen as one material (steel) to which a coating of a second (zinc) has been applied, even though if could be regarded as a new material with the strength of steel but the surface properties of zinc. Sandwich panels illustrate the duality, sometimes viewed as two sheets of face material separated by a core material, and sometimes—to allow comparison with monolithic materials—as a "material" with their own density, axial and flexural stiffness and strength, thermal conductivity, expansion coefficient, etc. To call any one of these a "material" and characterize it as such is a useful shorthand, allowing designers to use existing methods when designing with them. But if we are to design the hybrid itself, we must deconstruct it, and think of it as a combination of materials (or of material and space) in a chosen configuration.

III. Approximate Models for Hybrid Properties

In this section we assemble approximate equations for the density, moduli, strengths, and toughnesses for two families of hybrid materials. These are used in Section IV to compare hybrids with monolithic engineering materials.

(1) Cellular Structures: Foams and Lattices

Cellular structures—foams and lattices—are hybrids of a solid and a gas. The properties of the gas might at first sight seem irrelevant, but this is not so. The thermal conductivity of low-density foams of the sort used for insulation is determined by the conductivity of the gas contained in its pores; and the dielectric properties, and even the compressibility, can depend on the gas properties.

There are two distinct species of cellular solid. The first, typified by foams, are *bending-dominated structures*; the second, typified by triangulated lattice structures, are *stretch dominated*—a distinction explained more fully below.[5–8]

Foams are cellular solids made by expanding polymers, metals, ceramics, or glasses with a foaming agent. Figure 6, left, shows an idealized cell of a low-density foam. It consists of solid cell walls or edges surrounding a void space containing a gas or fluid. Foams have the characteristic that, when loaded, the cell walls bend. *Lattice-structures* (Fig. 6, right) are configured to suppress bending, so the cell edges have to stretch instead.

(A) Density: Cellular solids are characterized by their *relative density*, which for the structure shown here (with $\ll L$) is

$$\frac{\tilde{\rho}}{\rho_s} = C_1 \left(\frac{t}{L}\right)^2 \tag{1}$$

where $\tilde{\rho}$ is the density of the foam, ρ_s is the density of the solid of which it is made, L is the cell size, t is the thickness of the cell edges, and C_1 is a constant, approximately equal to 1.

Foam cell   Lattice cell

Fig. 6. Left: an idealized cell in a low density foam. Right: a fully triangulated lattice structure.

(B) Elastic Moduli of Bending-Dominated Foams: A remote compressive stress σ exerts a force $F \propto \sigma L^2$ on the cell edges, causing them to bend and leading to a bending deflection δ, as shown in Fig. 7(a). For the open-celled structure shown in the figure, the bending deflection scales as

$$\delta \propto \frac{FL^3}{E_s I} \tag{2}$$

where E_s is the modulus of the solid of which the foam is made and $I = \frac{t^4}{12}$ is the second moment of area of the cell edge of square cross section, $t \times t$. The compressive strain suffered by the cell as a whole is then $\varepsilon = 2\delta/L$. Assembling these results gives the modulus $\tilde{E} = \sigma/\varepsilon$ of the foam as

$$\tilde{E} = C_2 \left(\frac{\tilde{\rho}}{\rho_s}\right)^2 E_s \quad \text{(bending-dominated behavior)} \tag{3}$$

(a)   Elastic deformation   (b)   Plastic collapse

(c)   Cell edge buckling   (d)   Cell edge fracture

Fig. 7. (a) Elastic deformation of a foam cell by cell wall bending. (b) Plastic collapse: the cell edges bend plastically. (c) An elastomeric foam, by contrast, collapses by the elastic buckling of its cell edges. (d) A brittle foam collapses by the successive fracturing of cell edges.

Because $\tilde{E} = E_s$ when $\tilde{\rho} = \rho_s$, we expect the constant of proportionality C_2 to be close to unity—a speculation confirmed both by experiment and numerical simulation. The quadratic dependence means that a small decrease in relative density causes a large drop in modulus. When the cells are equiaxed in shape, the foam properties are isotropic with shear modulus, bulk modulus, and Poisson's ratio given

$$\tilde{G} = \frac{3}{8}\tilde{E} \qquad \tilde{K} = \tilde{E} \qquad v = \frac{1}{3} \tag{4}$$

(C) Elastic Moduli of Stretch-Dominated Lattices: The structure on the right in Fig. 6, is fully triangulated. This means that the cell edges must stretch when the structure is loaded elastically. On average one third of its edges carry tension when the structure is loaded in simple tension, regardless of the loading direction. Thus

$$\tilde{E} = C_3 \left(\frac{\tilde{\rho}}{\rho_s}\right) E_s \quad \text{(strech-dominated behavior)} \tag{5}$$

with $C_3 = 1/3$. The modulus is linear, not quadratic, in density giving a structure that is stiffer than a foam of the same density. The structure is almost isotropic, so we again approximate the shear modulus, bulk modulus, and Poisson's ratio by Eq. (4).

(D) Strength of Bending-Dominated Foams: When the foam structure in Fig. 6 is loaded beyond the elastic limit, its cell walls may yield, buckle elastically, or fracture as shown in Fig. 7. Consider yielding first (Fig. 7(b)). Cell edges yield when the force exerted on them exceeds their fully plastic moment.

$$M_f = \frac{\sigma_{y,s} t^3}{4} \tag{6}$$

where $\sigma_{y,s}$ is the yield strength of the solid of which the foam is made. This moment is related to the remote stress by $M \propto FL \propto \sigma L^3$. Assembling these results gives the compressive failure strength when yield dominates, $\tilde{\sigma}_c$

$$\tilde{\sigma}_c = C_4 \left(\frac{\tilde{\rho}}{\rho_s}\right)^{3/2} \sigma_{y,s} \quad \text{(yield of foams)} \tag{7}$$

where the constant of proportionality, $C_4 \approx 0.3$, has been established both by experiment and by numerical computation.

Elastomeric foams collapse not by yielding but by elastic bucking; brittle foams by cell wall fracture (Figs. 7 (c) and (d)). As with plastic collapse, simple scaling laws describe this behavior well. Collapse by buckling occurs when the stress exceeds

$$\tilde{\sigma}_c \approx 0.05 \left(\frac{\tilde{\rho}}{\rho_s}\right)^2 E_s \quad \text{(buckling of foams)} \tag{8}$$

We identify the compressive strength $\tilde{\sigma}_c$ with the lesser of Eqs (7) and (8). We further set the yield strength $\tilde{\sigma}_y$ the flexural strength $\tilde{\sigma}_{flex}$ equal to $\tilde{\sigma}_c$.

(E) Strength of Stretch-Dominated Lattices: Collapse occurs when the cell edges yield, giving the collapse stress

$$\tilde{\sigma}_c \approx \frac{1}{3} \left(\frac{\tilde{\rho}}{\rho_s}\right) \sigma_{y,s} \quad \text{(yield of lattices)} \tag{9}$$

This is an upper bound because it assumes that the struts yield in tension or compression when the structure is loaded. If the struts are slender, they may buckle before they yield. They do so at the stress

$$\tilde{\sigma}_C \approx 0.2 \left(\frac{\tilde{\rho}}{\rho_s}\right)^2 E_s \tag{10}$$

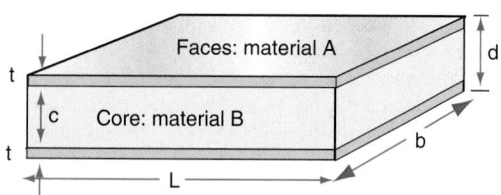

Fig. 8. The sandwich. The face thickness is t, the core thickness c, and the panel thickness d.

We identify the compressive strength $\tilde{\sigma}_c$ with the lesser of Eqs (9) and (10) .We further identify the yield strength $\tilde{\sigma}_y$, the flexural strength $\tilde{\sigma}_{flex}$ with $\tilde{\sigma}_c$.

(F) Fracture Toughness of Bending-Dominated Foams: Foams that contain crack-like flaws that are long compared with the cell size ℓ fail by fast fracture, meaning that the crack propagates unstably if the stress intensity factor exceeds the critical value $\tilde{K}_{Ic}$, which is the lesser of

$$\tilde{K}_{Ic} = 0.5 \left(\frac{\ell}{a}\right)^{1/2} \left(\frac{\tilde{\rho}}{\rho_s}\right)^{3/2} K_{Ic,s} \quad \text{and} \quad K_{Ic,s} \tag{11}$$

where a is the intrinsic flaw size of the material of the cell edges, $K_{Ic,s}$ is its fracture toughness.[9]

(G) Fracture Toughness of Stretch-Dominated Lattices: Lattices that contain crack-like flaws long compared with the cell size ℓ fail by fast fracture if the stress intensity factor exceeds the critical value $\tilde{K}_{Ic}$, which is the lesser of

$$\tilde{K}_{Ic} = 0.5 \left(\frac{\ell}{a}\right)^{1/2} \left(\frac{\tilde{\rho}}{\rho_s}\right) K_{Ic,s} \quad \text{and} \quad K_{Ic,s} \tag{12}$$

ℓ/a is the ratio of the cell size of the foam to the flaw size in the material.

(2) Sandwich Structures

A sandwich panel epitomizes the concept of a hybrid. It combines two materials in a specified geometry and scale, configured such that one forms the faces, the other the core, to give a structure of high bending stiffness and strength at low weight (Fig. 8). The separation of the faces by the core increases the moment of inertia of the section, I, and its section modulus, Z, producing a structure that resists bending and buckling loads well. The faces, each of thickness t, carry most of the load, so they must be stiff and strong; and they form the exterior surfaces of the panel so they must also tolerate the environment in which it operates. The core, of thickness c, occupies most of the volume, it must be light, and stiff and strong enough to carry the

shear stresses necessary to make the whole panel behave as a load-bearing unit.[10–15]

(A) A Sandwich as a "Material": So far we have spoken of the sandwich as a *structure*: faces of material A supported on a core of material B, each with its own density, modulus and strength. But we can also think of it as a *material* with its own set of properties, and this is useful because it allows comparison with more conventional materials. To do this we calculate equivalent material properties for the sandwich, identifying them, as with foams, by a tilde (e.g. $\tilde{\rho}, \tilde{E}$). The quantities $\tilde{\rho}$ and $\tilde{E}$ can be plotted on the modulus-density chart, allowing a direct comparison all the other materials on the chart. All the constructions using material indices apply unchanged. We base the analysis on the symmetric sandwich with the dimensions defined in Fig. 8. The symbols that appear in this section are defined in Table I.

(B) Density: The equivalent density of the sandwich (its mass divided by its volume) is

$$\tilde{\rho} = f\rho_f + (1 - f)\rho_c \tag{13}$$

here f is the volume fraction occupied by the faces: $f = 2t/d$.

In-plane and through-thickness moduli are simply the arithmetic and harmonic means of the components.

$$\tilde{E} = fE_f + (1 - f)E_c \quad \text{(in-plane)} \tag{14}$$

$$E_{tt} = \frac{1}{f/E_f + (1 - f)/E_c} \quad \text{(through thickness)} \tag{15}$$

(C) Equivalent Flexural Modulus: The flexural properties are quite different. The flexural compliance (the reciprocal of the stiffness) has two contributions, one from the bending of the panel as a whole, the other from the shear of the core (Fig. 9). They add. The bending stiffness is

$$EI = \frac{b}{12}(d^3 - c^3)E_f + \frac{bc^3}{12}E_c$$

The shear stiffness is

$$AG = \frac{bd^2}{c}G_c$$

Summing deflections gives

$$\delta = \frac{12PL^3}{B_1 b\{(d^3 - c^3)E_f + c^3 E_c\}} + \frac{PLc}{B_2 d^2 bG_c} \tag{16}$$

Table I. The Symbols Used in Describing Sandwich Structures

Symbol	Meaning and usual units
t, c, d	Face thickness, core thickness, and overall panel thickness (m)
L, b	Panel length and width (m)
m_a	Mass per unit area of the panel (kg/m²)
$f = 2t/d$	Relative volumes occupied by the faces
$(1-f) = c/d$	Relative volume occupied by the core
I	Second moment of area (m⁴)
ρ_f, ρ_c	Densities of face and core material (kg/m³)
$\tilde{\rho}$	Equivalent density of panel (kg/m³)
E_f	Young's modulus of the faces (GN/m²)
E_c, G_c	Young's modulus and shear modulus of the core (GN/m²)
$\tilde{E}, \tilde{E}_{tt}, \tilde{E}_{flex}$	Equivalent in-plane, through-thickness and flexural modulus of panel (GN/m²)
$\sigma_{y,f}$	Yield strength of faces (MN/m²)
$\sigma_{y,c}, \sigma_{ts,c}, \sigma_{c,c}, \sigma_{flex,c}$	Yield strength, tensile strength, compressive strength, and flexural strength of core (MN/m²)
$\tilde{\sigma}_{in-plane}$	Equivalent in-plane strength of panel (MN/m²)
$\tilde{\sigma}_{flex1}, \tilde{\sigma}_{flex2}, \tilde{\sigma}_{flex3}$	Equivalent flexural strength of panel, depending on mechanism of failure (MN/m²)

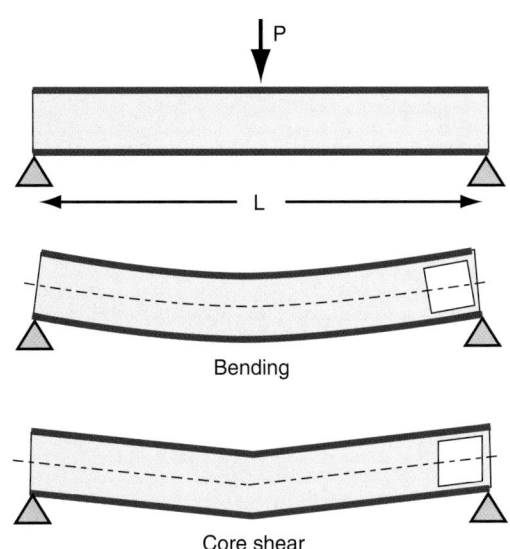

Fig. 9. Sandwich panel flexural stiffness. There are contributions from bending and from core shear.

where the dimensions, d, c, t, and L are identified in Fig. 8, E_f is Young's modulus of the face sheets and G_C is the shear modulus of the core. The load configuration determines the constant values (B_1, B_2, B_3, and B_4) as summarized in Table II. Comparison with $\delta = \frac{12PL^3}{B_1 \tilde{E}_{flex} \tilde{I}}$ (where $\tilde{I}$ is the second moment of area, $\tilde{I} = bd^3/12$) for the "equivalent" homogeneous material gives the equivalent flexural modulus[6,10–14]:

$$\frac{1}{\tilde{E}_{flex}} = \frac{1}{E_f \left\{ \left(1 - (1-f)^3\right) + \frac{E_c}{E_f}(1-f)^3 \right\}} + \frac{B_1}{B_2}\left(\frac{d}{L}\right)^2 \frac{(1-f)}{G_c} \tag{17}$$

The bending stiffness for the sandwich is recovered by forming the quantity $\tilde{E}_{flex}\tilde{I}$.

(D) In-plane and Through-Thickness Yield Strength: We equate the in-plane strength to the sum of the contributions of face and core, assuming they both must yield for the panel as a whole to yield:

$$\tilde{\sigma}_{y,ip} = f\sigma_{y,f} + (1-f)\sigma_{y,c} \tag{18a}$$

The through-thickness yield strength is the lesser of those of the faces and core:

$$\tilde{\sigma}_{y,tt} = Lesser\, of\, (\sigma_{y,f},\, \sigma_{y,c}) \tag{18b}$$

(E) Flexural Strength: Sandwich panels can fail in flexure in many different ways (Fig. 10). The failure mechanisms compete, meaning that the one that happens at the lowest load dominates. We calculate an *equivalent strength* for each mode, then seek the lowest.

(a) Face Yield: The fully plastic moment of the sandwich is

$$M_f = \frac{b}{4}\left\{ \left(d^2 - c^2\right)\sigma_{y,f} + c^2\sigma_{flex,c} \right\}$$

The equivalent flexural strength when face yielding is the dominant failure mode is then

$$\tilde{\sigma}_{flex1} = \frac{4M_f}{bd^2} = \left(1 - (1-f)^2\right)\sigma_{y,f} + (1-f)^2\sigma_{flex,c} \tag{19}$$

This is the flexural strength the equivalent "material" of a homogenous beam with the same width b, depth d and bending strength as the sandwich panel.

(b) Face Buckling: In flexure, one face of the sandwich is in compression. If it buckles, the sandwich fails. The face-stress at which this happens[10] is

$$\sigma_b = 0.57\left(E_f E_c^2\right)^{1/3} \tag{20}$$

The failure moment M_f is then well approximated by

$$M_f = \sigma_b \frac{b}{6d}\left(d^3 - (d - 2t)^3\right)$$

Table II. Constants to Describe Modes of Loading

Mode of loading	Description	B_1	B_2	B_3	B_4
	Cantilever, end load	3	1	1	1
	Cantilever, uniformly distributed load	8	2	2	1
	Three point bend, central load	48	4	4	2
	Three point bend, uniformly distributed load	384/5	8	8	2
	Ends built in, central load	192	4	8	2
	Ends built in, uniformly distributed load	384	8	12	2

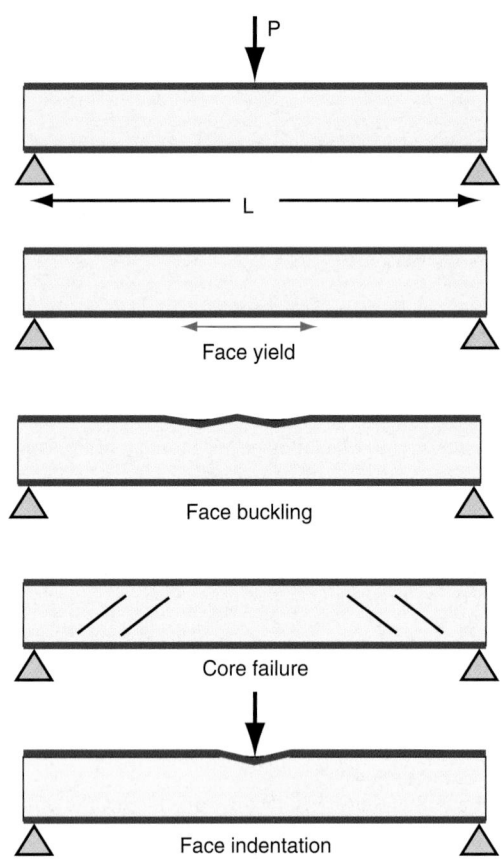

Fig. 10. Failure modes of sandwich panels in flexure.

Equating this to

$$M_f = \tilde{\sigma}_{flex2} \frac{bd^2}{6}$$

gives

$$\sigma_{flex2} = 0.57(1 - (1-f)^3)(E_f E_c^2)^{1/3} \qquad (21)$$

(c) **Core Shear:** Failure by core shear occurs at the load

$$P_f = B_4 bc \left(\tau_{y,c} + \frac{t^2}{cL}\sigma_{y,f} \right)$$

Here the first term results from shear in the core (shear strength $\tau_{y,c}$), the second from the formation of plastic hinges in the faces. Equating to

$$P_f = \frac{B_3 bd^2}{4L} \tilde{\sigma}_3$$

gives the equivalent strength when failure is by shear:

$$\tilde{\sigma}_{flex3} = \frac{B_4}{B_3}\left\{ 4\frac{L}{d}(1-f)\tau_{y,c} + f^2\sigma_{y,f} \right\} \qquad (22)$$

(The load configuration determines the constant values B_3 and B_4, as summarized in Table II) When the core material is roughly isotropic (as foams are) $\tau_{y,c}$ can be replaced by $\sigma_{c,c}/2$. When it is not (an example is that of a honeycomb core), $\tau_{y,c}$ must be retained.

(d) **Core Bending:** In the extreme case that face sheets have very low strength, or that the face-sheet thickness is set to zero, the "sandwich" still exhibits the bending strength of the core itself:

$$\tilde{\sigma}_{flex4} = \sigma_{flex,c} \qquad (23)$$

(e) **Indentation:** A load P distributed over a length a and width b of the panel exerts a local indentation pressure of

$$\frac{P}{ab} = p_{ind} = \frac{2t}{a}(\sigma_{y,f}\sigma_{y,f})^{1/2} + \sigma_{y,c}$$

(Ashby et al.,[6] eq. [10.13]) from which we find the minimum face thickness t^* to avoid indentation

$$t^* \geq \frac{a}{2}\left(\frac{P}{ab} - \sigma_{y,c}\right) \cdot \left(\frac{1}{(\sigma_{y,c} \cdot \sigma_{y,c})^{1/2}}\right)$$

The flexural strength is set equal to
Greater of σ_{flex4} and (Least of σ_{flex1}, σ_{flex2}, and σ_{flex3}). Indentation has to be treated separately as a constraint on the face-sheet thickness.

IV. Can hybrids expand the Filled Part of Material-Property Space?

The figures that follow compare the properties of hybrids with those of monolithic materials using the models of Section III. Many have the ability to expand the occupied region of property space.

(1) Foams and Lattices Based on Aluminum–SiC Composites (Figs. 11 and 12)

This is an example of the modeling of modulus and density of cellular structures. The starting point is aluminum 20% SiC(p),[‡] a mix that is the basis of one of the Cymat[15] range of metal foams. The equations describing the modulus and density of foams and lattices of Section III(1) have been evaluated for a range of relative densities. They appear on the Modulus—Density chart in Fig. 11 as lines of yellow and green ellipses. The starting material, Al 20% SiC(p), is identified at the upper right. The modeled foams, plotted yellow, are labeled in red with their relative densities in brackets. Measured data for real aluminum–SiC(p) foams, also with relative densities in brackets, are labeled in black, allowing a comparison. Note that the lattices, plotted in green, lie at higher values of modulus for the same density. At a relative density of 3.5% (0.035) the lattice is predicted to be 10 times stiffer than a foam of that density.

Figure 12 extends the comparison to a chart on which 3000 materials from all classes are plotted, using $E^{1/2}/\rho$ as a criterion of excellence. The lattice structures out-perform all of them, expanding the populated area of the chart.

(2) Stiff Sandwich Structures: Aluminum-Faced PVC Foam-Cored Sandwiches (Fig. 13)

In designing sandwich panels and shells it makes sense to choose stiff, strong materials for the thin faces and light weight materials for the core, the role of which is to separate the faces and carry shear stresses. The model equations will, however, work for other choices.

Figure 13 shows the flexural moduli and densities of a set of sandwich panels with faces of 6061 T4 aluminum alloy and cores of a PVC foam with a density of 100 kg/m³ (0.1 Mg/m³). The face-sheet thickness is varied from 0.1 to 3 mm in three (logarithmic) steps. For each of these there are five thicknesses of core, ranging from 10 to 50 mm, distinguished by color. The choice of the parameters B_1/B_2 and of B_3/B_4 are determined by the way the panel is loaded, as explained in Section III(2).

[‡](p) means "particulate."

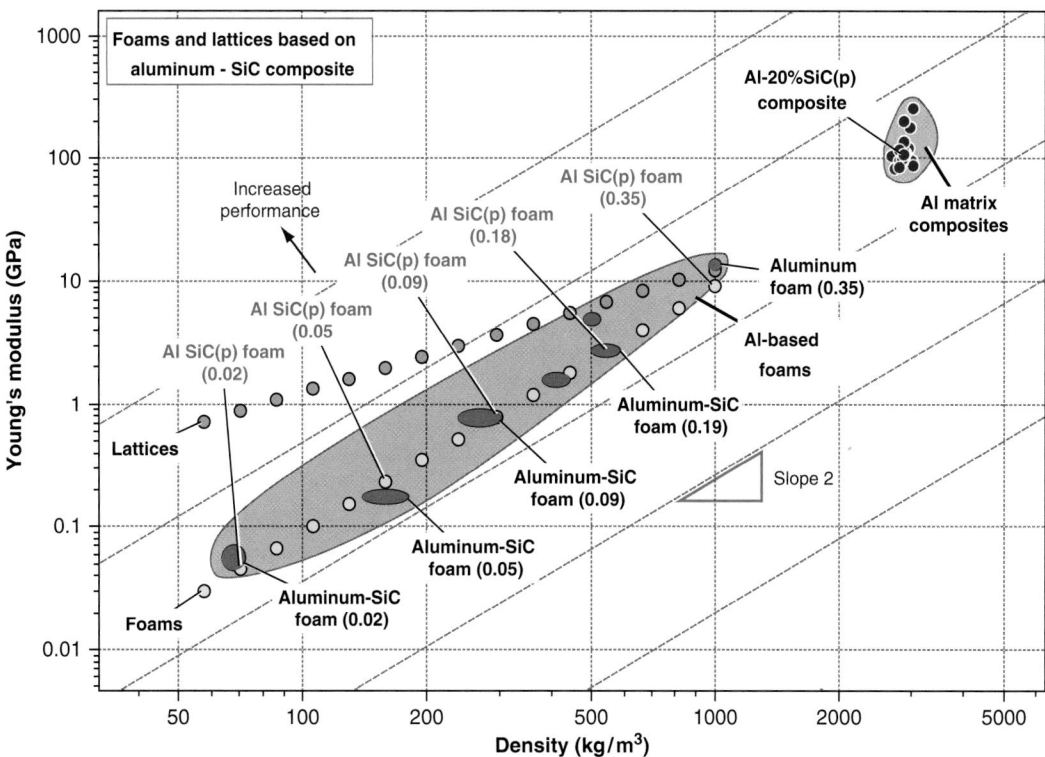

Fig. 11. The starting material,, Al 20% SiC(p) appears at the upper right of this chart. The line of yellow ellipses with red labels show the modulus and density of foams made from Al–SiC(p) They should be compared with the measured values for real aluminum SiC(p) foams, shown in red with black labels. The relative densities are listed in brackets. The modulus and density of lattices made of the same material are shown in green for comparison.

Here they are set at those for a panel centrally loaded in 3-point bending. The face-sheet material is identified at the top right, the core material at the lower left, both labeled in black. The sandwich panels, labeled in red, lie on an arc linking the two.

How good are they? Here we need a criterion of excellence. The index M for selecting materials for a light, stiff panel[4] is

$$M = \frac{E_{\text{flex}}^{1/3}}{\rho}$$

that plots as a set of lines of slope 3 in Fig. 13 (here E_{flex} is the flexural strength and ρ is the density). A contour that is tangent

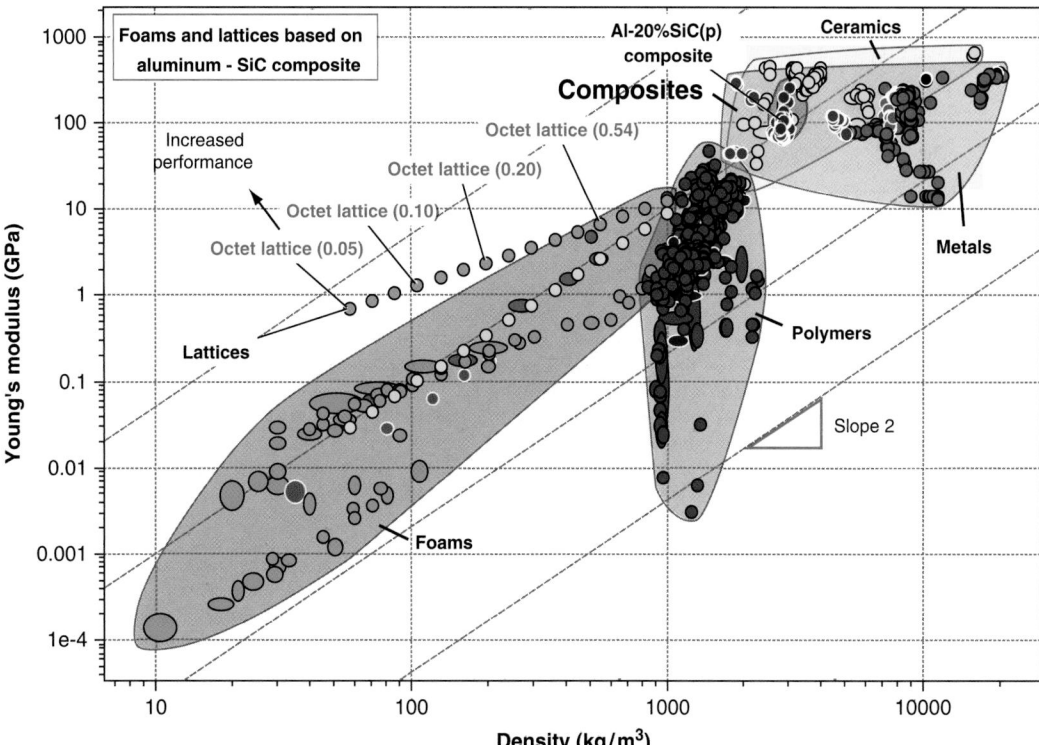

Fig. 12. The moduli of Al–SiC lattices compared with those of the other materials of engineering. The lattices lie outside the currently populated areas of the chart. Numbers in brackets are relative densities.

Fig. 13. A set of aluminum-faced sandwiches with PVC foam cores, with 2800 other materials of engineering for comparison. The sandwiches expand the filled part of the space.

to the arc of sandwiches identifies the optimum choice of face and core thicknesses.

The best of the sandwiches ranks more highly than any other material by this criterion. The chart illustrates how effectively sandwich structures can populate holes in material-property space.

(3) Strong Sandwich Structures: Aluminum-Faced PVC Foam-Cored Sandwiches (Fig. 14)

Figure 14 is a chart in which strength, rather than stiffness, is explored. The core is PVC foam with a density of 100 kg/m^3, as in the previous figure. The face-sheet material, in this case, is a high strength aluminum alloy: 7075 in the T6 condition.

Fig. 14. The arc of orange ellipses shows the flexural strengths of sandwich panels with faces of aluminum 7075 T6 and cores of PVC foam with a density of 100 kg/m^3.

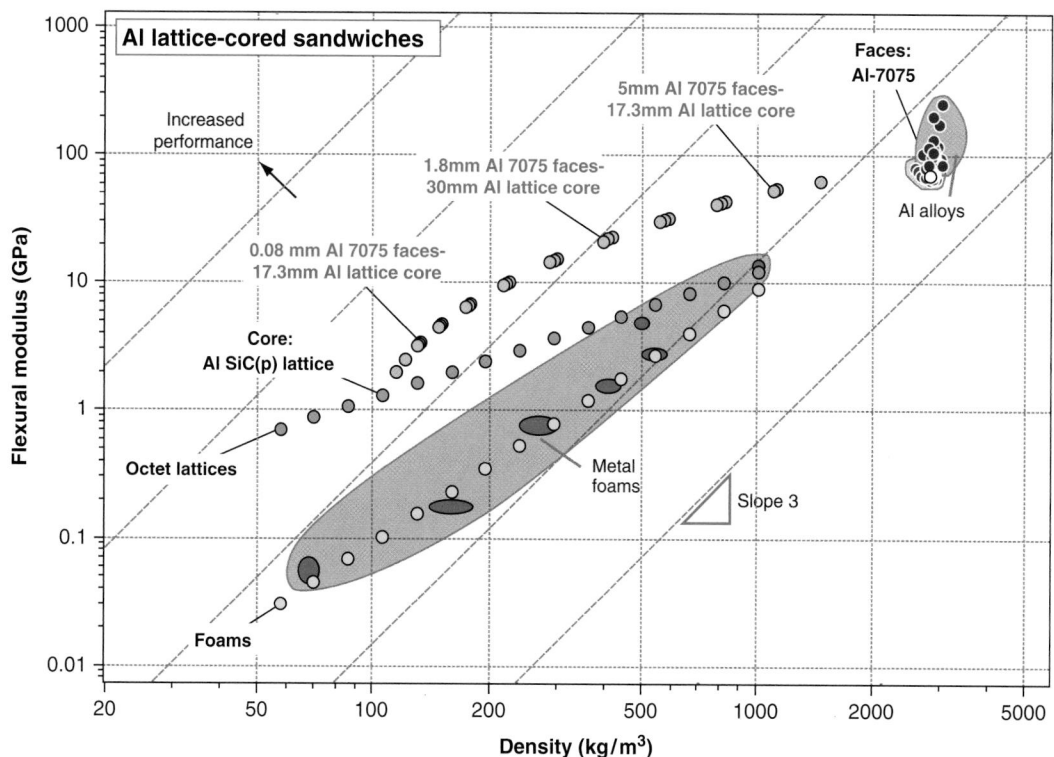

Fig. 15. The flexural moduli and densities lattice-cored aluminum-faced sandwich panels.

The other parameters are set at the same values as in Example 4.2.

The chart shows the flexural strength σ_{flex} and density ρ of this set. The face-sheet material is identified at the top right, the core material at the lower left, both labeled in black. The modeled panels, labeled in red, lie on an arc linking the two.

The criterion for selecting materials for a light, strong panel is the index

$$M = \frac{\sigma_{flex}^{1/2}}{\rho}$$

Fig. 16. The same data and axes as Fig. 15, with 2800 other engineering materials added for comparison. The lattice-cored sandwiches offer significant gain in performance.

which plots as a set of lines of slope 2 in Fig. 14. In this example, too, sandwiches perform exceptionally well expanding the filled part of the property space.

(4) Advanced sandwich structures: Aluminum-Faced Lattice-Cored Sandwiches (Figs. 15 and 16)

Sandwich panels are often made with foam cores. Figs. 11 and 12 demonstrated, however, that lattice structures offered considerably better performance than foams at the same density. Here we first calculate properties of lattice structures, then use the one of these as input for the core for a sandwich.

Figure 15 shows the result. The figure has axes of flexural modulus and density. It shows the predicted properties of a string of Al–SiC octet lattices as green circles. One of these, identified on the figure, was selected as the core material and combined with face sheets of aluminum 7075 T6 to create sandwiches. The result is the arc of sandwiches that appear as orange ellipses.

The index M for selecting materials for a light, stiff panel is

$$M = \frac{E_{\text{flex}}^{1/3}}{\rho}$$

which plots as a set of lines of slope 3 in Fig. 15 (here E_{flex} is the flexural strength and ρ is the density). A contour that is tangent to the arc of sandwiches identifies the optimum choice of face and core thicknesses.

In Fig. 16 some 2800 representative engineering materials have been added in Fig. 15. They map out the part of the space that is occupied by conventional monolithic engineering materials. The lattices already expand the occupied area. The lattice-cored sandwiches expand it further.

V. Summary and Conclusions

Hybrid materials have the capacity to expand the ranges of stiffness and strengths of structural materials, spreading the occupied areas of material-property space. This is demonstrated here by comparing the properties of lattice and sandwich structures with those of some 2800 conventional materials. The comparison is made possible by evaluating the "effective" properties of the hybrids: the properties that would be attributed to a monolithic material with the same density, axial and flexural stiffness and strength as the hybrid. The performance of the hybrids, when measured by an appropriate criterion of excellence, surpasses those of any conventional material.

Acknowledgments

I would like to recognize the insights and suggestions that have been provided by discussions with Professors John Hutchinson, Lorna Gibson, Norman Fleck, Vikram Deshpande and David Cebon. In particular I wish to acknowledge the profound influence of a close collaboration with Tony Evans.

References

[1]F. X. Kromm, J. M. Quenisset, R. Harry, and T. Lorriot, "An Example of Multimaterial Design"; *Proceedings Euromat'01*, Rimini, Italy, 2001.

[2]M. F. Ashby and Y. Brechet, "Designing Hybrid Materials," *Acta Mater.*, **51** [19] 5801–2 (2003).

[3]M. F. Ashby, D. Cebon, C. Bream, C. Cesaretto, and N. Ball, "The CES Hybrids Synthesizer—a White Paper. Granta Design, Cambridge, U.K., 2010.

[4]M. F. Ashby, *Materials Selection in Mechanical Design. ISBN 978-1-85617-663-7*, 4th edition, Butterworth Heinemann, Oxford, U.K., 2011.

[5]L. J. Gibson and M. F. Ashby, *Cellular Solids, Structure and Properties. ISBN 0-521-49560-1*, 2nd edition, Cambridge University Press, Cambridge, U.K., 1997.

[6]M. F. Ashby, A. G. Evans, N. A. Fleck, L. J. Gibson, J. W. Hutchinson, and H. N. G. Wadley, *Metal Foams: a Design Guide. ISBN 0-7506-7219-6*. Butterworth Heinemann, Oxford, 2000.

[7]H. N. G. Wadley, N. A. Fleck, and A. G. Evans, "Fabrication and structural performance of periodic metal sandwich structures," *Compos. Sci. Technol.*, **63**, 2331–43 (2003).

[8]V. S. Deshpande, M. F. Ashby, and N. A. Fleck, "Foam Topology: Bending Versus Stretching Dominated Architectures," *Acta Mater.*, **49**, 1035–40 (2001).

[9]R. G. Hutchinson and N. A. Fleck, "The structural performance of the periodic truss," *J. Mech. Phys. Solids.*, 756–82 (2006).

[10]H. G. Allen, *Analysis and Design of Structural Sandwich Panels*. Pergamon Press, Oxford, U.K., 1969.

[11]D. Zenkert, *An Introduction to Sandwich Construction. ISBN 0 947817778*. Engineering Advisory Services Ltd., Solihull, U.K., 1995 published by Chameleon Press Ltd., London, U.K.

[12]J. Pflug, B. Vangrimde, and I. Verpoest, "Material Efficiency and Cost Effectiveness of Sandwich Materials"; *Sampe US Proceedings*, Orlando, FL, 2003.

[13]J. Pflug, I. Verpoest, and D. Vandepitte. (2004) SAND.CORE Workshop, Brussels, December, 2004.

[14]J. Pflug, "Sandwich Materials Selection Charts," *J. Sandwich Struct. Mater.*, **8** [5] 407–21 (2006).

[15]Cymat Technologies Ltd. Available at http://www.cymat.com, 2010 (accessed December 2, 2010). □

J. Am. Ceram. Soc., **94** [S1] S15–S34 (2011)
DOI: 10.1111/j.1551-2916.2011.04599.x
© 2011 The American Ceramic Society

journal

Protocols for the Optimal Design of Multi-Functional Cellular Structures: From Hypersonics to Micro-Architected Materials

Lorenzo Valdevit,[‡,†] Alan J. Jacobsen,[§] Julia R. Greer,[¶] and William B. Carter[§]

[‡] Mechanical and Aerospace Engineering Department and Chemical Engineering and Materials Science Department, University of California, Irvine, California, 92697

[§] HRL Laboratories, Malibu, California, 90265

[¶] Materials Science Department, California Institute of Technology, Pasadena, California, 91125

Cellular materials with periodic architectures have been extensively investigated over the past decade for their potential to provide multifunctional solutions for a variety of applications, including lightweight thermo-structural panels, blast resistant structures, and high-authority morphing components. Stiffer and stronger than stochastic foams, *periodic* cellular materials lend themselves well to geometry optimization, enabling a high degree of tailorability and superior performance benefits. This article reviews a commonly established optimal design protocol, extensively adopted at the macro-scale for both single and multifunctional structures. Two prototypical examples are discussed: the design of strong and lightweight sandwich beams subject to mechanical loads and the combined material/geometry optimization of actively cooled combustors for hypersonic vehicles. With this body of literature in mind, we present a motivation for the development of *micro-architected materials*, namely periodic multiscale cellular materials with overall macroscopic dimensions yet with features (such as the unit cell or subunit cell constituents) at the micro- or nano-scale. We review a suite of viable manufacturing approaches and discuss the need for advanced experimental tools, numerical models, and optimization strategies. In analyzing challenges and opportunities, we conclude that the technology is approaching maturity for the development of micro-architected materials with unprecedented combinations of properties (e.g., specific stiffness and strength), with tremendous potential impact on a number of fields.

I. Introduction

STOCHASTIC cellular materials (i.e., foamed materials that contain significant amounts of porosity) have long been used for their low weight, high sound absorption, crashworthiness, and thermal properties.[1] Approximately 15 years ago, advances in manufacturing technologies spearheaded a large academic and industrial interest in metallic foams,[2] which combine all the properties listed above with increased specific strength and stiffness and high-temperature capabilities. An important feature of open-cell foams is the interconnected open space, which can be employed to enable additional capabilities, such as active cooling[2,3] or energy storage,[2,4–6] thus enabling multifunctionality. More recently, detailed mechanical experiments on metallic foam-based sandwich panels under bending and compressive loads revealed that all foams are bending-dominated, i.e., they deform by compliant and weak bending modes of the cell walls and ligaments, inefficiently using the base constituent material in the foam by leaving much of it out of the load path.[7–9] In addition, their stochastic nature inevitably introduces imperfections that further depress their mechanical properties.[10] Vastly increased specific stiffness and strength (i.e., stiffness and strength per unit weight) can be obtained in *periodic cellular architectures* (such as those depicted in Fig. 1); if designed properly, under global bending and compressive loadings these architectures will deform by stretching of the ligaments, a stiff and strong local deformation mode that makes maximal use of the base constituent and maximizes load carrying capacity.[4,11] As an additional benefit over "semi-engineered" open-cell foams, periodic cellular topologies have many more geometrical features that can be engineered and optimized. A large body of research has been published in the past decade on optimally designed metallic periodic cellular systems, with emphasis on specific strength,[11–17] active cooling,[18] and combinations thereof,[19–21] combined strength and thermal conductivity (through a heat pipe design),[22] high-velocity impact absorption,[23–26] and high-authority shape morphing potential.[27–31]

In spite of the variety of applications (each imposing different objective functions and constraints), the same protocol for optimal design has been consistently (and successfully) adopted in nearly all cases. This protocol consists of a combination of analytical, numerical, and experimental techniques, and is reviewed in Section II of this article, with emphasis on mechanical and thermo-mechanical optimization. In Section III of this article, we pose three questions: (i) Are there any mechanical benefits in designing *micro-architected materials* (namely, macro-scale periodic cellular materials with unit cells at the micro-scale)? (ii) Are there suitable and cost-effective manufacturing processes for micro-architected materials? (iii) Is the optimal design protocol (including analytical, numerical and experimental techniques) which has been successfully adopted for large-scale structures appropriate to harness the full potential of micro-architected materials? By answering these questions, we conclude that the technology is approaching maturity for the development, characterization, and optimal design of a novel class of multifunctional materials with the potential to achieve unprecedented combination of properties.

II. Micro-Architected Materials

This section briefly reviews the well-established optimal design protocol for cellular periodic structures. Manufacturing approaches are described first, to offer a flavor of the topologies and materials combination that are readily available. The optimal design protocol (consisting of a combination of

T. M. Pollock—contributing editor

Manuscript No. 28794. Received October 16, 2010; approved April 02, 2011.
[†]Author to whom correspondence should be addressed. e-mail: valdevit@uci.edu

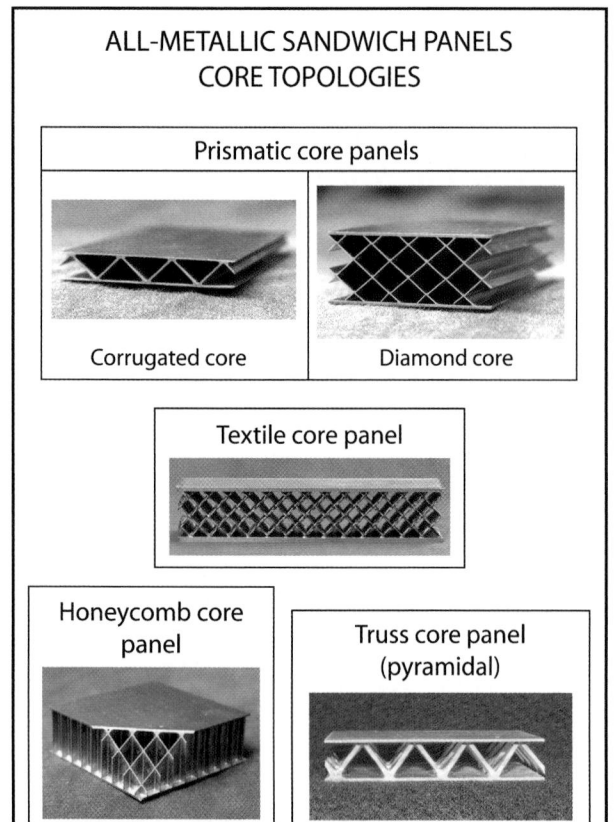

ALL-METALLIC SANDWICH PANELS CORE TOPOLOGIES

Prismatic core panels

Corrugated core | Diamond core

Textile core panel

Honeycomb core panel

Truss core panel (pyramidal)

Fig. 1. Examples of all-metallic sandwich panels manufactured by Transient Liquid Phase (TLP) bonding. Modified from Lu *et al.*[3]

analytical, numerical, and experimental techniques) is presented for two archetypal structures: a simple single-function structure (a lightweight periodic sandwich beam with a prismatic corrugated core, designed for resistance to mechanical loads—bending and transverse shear), and a more complex multifunctional structure (a lightweight actively cooled sandwich plate, designed for resistance to mechanical and thermal loads—and additional design constraints). The latter is presented in the context of materials development for hypersonic vehicles. In both cases, materials selection is addressed.

(1) Manufacturing Approaches

Reliable manufacturing techniques have been developed for metallic sandwich structures with a number of core topologies. Cores are generally assembled by folding a plate (as in case of corrugated cores or truss cores) or slotting and assembling a large number of beam and/or plate elements (honeycomb cores, diamond prismatic cores, textile cores); in the latter case, the constituents need to be metallurgically bonded to impart strength and stiffness to the structure.[6] In both approaches, face sheets are bonded to the core. As bonded nodes are inevitably subjected to substantial in-service loads (in tension, compression and shear), a manufacturing process that results in strong nodes is essential. For a number of materials such as steels, copper, and aluminum alloys, Transient Liquid Phase Bonding (TLP), a high-temperature brazing process involving significant inter-diffusion at the joints, is the ideal technique.[32] TLP enables much stronger structures than conventional, lower-temperature brazing because the resulting nodes have nearly the same chemical composition as the base metal. TLP is also better than welding, both for simplicity and scalability (all the nodes are formed at once, without need for line-of-sight access) and because solidification at the brazing temperature ensures much lower residual stresses in the bonded region

compared to welding. Examples of sandwich panels with various core topologies obtained with TLP are provided in Fig. 1. Most of the experimental work published to date on all metallic sandwich panels pertains to steel panels, for which TLP brazing agents are readily available. TLP recipes for aluminum and copper alloys also exist.[33]

Recently, alternative manufacturing approaches were developed for aerospace-relevant high-temperature alloys. Titanium (Ti6Al4V) panels were manufactured with diffusion bonding, resulting in good nodal microstructure and strengths.[34] The extension to even higher temperatures requires nickel superalloys. Unfortunately, high-strength, γ'-rich nickel superalloys are not formable at room temperature. A clever solution for thin-gage panels was recently proposed, whereby a formable, single-phase γ-Ni superalloy is assembled in the right shape, all the components are assembled via TLP bonding, and subsequently the finished structure is aluminized and precipitation hardened, resulting in a high-strength, γ'-rich alloy.[35]

For lower temperature applications, polymer-matrix composites (e.g., Carbon-epoxy) are available with significantly higher weight efficiency than other metals.[36] The primary manufacturing issue is ensuring sufficient nodal strength. Carbon-epoxy honeycomb core panels obtained by a slotting procedure were recently demonstrated and optimized for compressive loads.[37] For corrugated core panels, 3D weaving is a natural option, albeit at an increase in cost and manufacturing complexity.[38] A number of simpler, prepreg-based approaches are currently under consideration.[36]

(2) Design Protocol for Maximum Specific Strength

Periodic cellular materials have the prominent feature of being naturally suitable to optimization. In addition to selecting the ideal base material (or combinations thereof in the case of a hybrid), the architecture can be optimized for a specific objective (or multiple objectives) subject to a number of constraints. The general multi-step procedure can be summarized as follows:

1. Fundamental properties of the base material(s) are obtained, either from data sheets or through appropriate experimental characterization, e.g., dog-bone tensile testing resulting in a stress-strain curve—possibly including temperature and time effects: visco-elasticity, visco-plasticity, fatigue, etc.
2. The evolution of the variables of interest (stress, strain, temperature, electric potential, etc.) is modeled analytically as a function of the structure geometry and the applied loads (mechanical, thermal, electrical, etc.). Constraints are formulated for the specific application under consideration (e.g., no yielding or buckling anywhere in the structure, no melting of the material).
3. Numerical analyses are performed, typically employing commercial Finite Elements packages, to verify the validity of the simplifying assumptions underlying step (2).
4. A combination of steps (2) and (3) is coupled with an optimization routine (quadratic optimizers for convex problems, discrete algorithms for problems featuring many local minima) and the structure geometry and/or material are optimized subject to all the prescribed constraints. The objective function strongly depends on the specific problem.
5. A prototype (possibly to scale) of the *entire* optimal or near-optimal structure (or at least substructure) is manufactured and its performance is verified *experimentally* to verify all the modeling assumptions (underlying both (2) and (3)).

As an example, we examine optimization of a metallic corrugated-core sandwich panel loaded by any combination of bending and transverse shear loads (Fig. 2(a)).[12,13] The objective is maximum strength at a prescribed weight. Nondimensional load intensity and weight are defined as:

$$\Pi = \frac{V^2}{EM}$$
$$\psi = \frac{W}{\rho \ell^2} = 2\frac{d_f}{\ell} + \frac{1}{\cos \theta}\frac{d_c}{\ell} \quad (1)$$

where V and M are the maximum shear force and bending moment per unit width of the panel, respectively, $\ell = M/V$ is the governing length-scale in the problem, E and ρ are the Young's modulus and density of the base material, respectively, and the geometric variables d_f, d_c, θ are defined in Fig. 2(a). The length-scale ℓ defines the actual loading condition (e.g., $\ell = L/2$ for three-point bending, L being the span of the panel between the loading points); normalizing all the dimensions with ℓ renders generality. Four possible failure mechanisms are identified: face (FY) and core (CY) yielding, and face (FB) and core (CB) buckling. For transverse loadings (bending about an axis parallel to the corrugation—Fig. 2(a)), analytical expressions are readily derived:

$$\left(\frac{V^2}{EM}\right)_{FY} = \frac{\varepsilon_Y d_f}{\ell}\left(\frac{H_c}{\ell} + \frac{d_f}{\ell}\right)$$
$$\left(\frac{V^2}{EM}\right)_{CY} = \frac{\varepsilon_Y d_c \sin \theta}{\ell}$$
$$\left(\frac{V^2}{EM}\right)_{FB} = \frac{k_f \pi^2 \tan^2 \theta}{48}\left(\frac{H_c}{\ell} + \frac{d_f}{\ell}\right)^{-1}\left(\frac{d_f}{\ell}\right)^3 \quad (2)$$
$$\left(\frac{V^2}{EM}\right)_{CB} = \frac{k_c \pi^2 \sin^3 \theta}{12}\left(\frac{H_c}{\ell} + \frac{d_f}{\ell}\right)^{-2}\left(\frac{d_c}{\ell}\right)^3$$

where ε_Y is the yield strain of the constituent material. Similar equations can be derived for longitudinal loadings. Note that ε_Y is the *only* material property governing the problem. The implication is that the optimal material for a corrugated-core sandwich panel subject to any combination of bending and transverse shear loads is simply the material with the largest yield strain. The same conclusion applies to any other core topology and a number of mechanical loading conditions. This important result allows separation of materials selection and optimal topological design. Multifunctional problems involving more complex physics often lack this feature, requiring material and topology to be concurrently optimized (see Section II(3)).

The interplay of failure mechanisms is best illustrated with failure mechanisms maps (Fig. 2(b)). If the corrugation angle, θ, and the panel weight, Ψ, are fixed, panel geometry is entirely defined by the thickness of the core, H_c and face sheet, d_f. Hence, each point on the map represents a possible panel design, with all designs having the same weight. The various regions denote design spaces where panel strength is governed by each failure mechanism (core yielding is never active under these loading conditions and weight). Strength contours (expressed in nondimensional form) clearly identify that the best design occurs at the intersection of three failure mechanisms (Incidentally, the confluence of three failure mechanisms at the optimal design point is a recurring feature for many core topologies,[11,12,15–17,39] but this condition is not universal (C. A. Steeves, Personal Communication).

Before it can be used with confidence, this model needs to be verified with a combination of numerical (FE) analyses and validated with a selected set of experiments. Figure 2(c) shows excellent agreement between analytical (white dot) and numerical predictions with experimental results for one particular design loaded in three-point bending (black dot in Fig. 2(b)). The inset in Fig. 2(c) compares the deformed shape of the panel at the end of the experiment. Note that both face and core buckling are evident (face yielding was also

verified with a strain gage during the experiment), consistent with the analytical predictions of Fig. 2(b).

Numerical and experimental validation of the analytical model allows computationally efficient design optimization for a wide range of applied load intensities. With reference to Eqs. (1–2), the problem can be stated as follows: for any given load intensity, Π, minimize the panel weight, Ψ, subject to four constraints ($\Pi < \Pi_{FY}$, $\Pi < \Pi_{FB}$, $\Pi < \Pi_{CY}$, $\Pi < \Pi_{CB}$). As all functions are convex, a simple quadratic optimizer was successfully used. Results for aluminum panels are presented in Fig. 2(d). This master figure compares the weight efficiency of a number of optimally designed core topologies; the corrugated core panel loaded transversely (discussed herein) is much lighter than solid plates, but more efficient topologies can be devised (hexagonal honeycombs are optimally efficient in this loading condition, and have often been used as benchmarks).

(3) Design Protocol for Multifunctional Structures: An Example From Hypersonics

(A) Preliminaries: For multifunctional applications, the challenge is choosing a cellular material with the best combination of constituent material and architecture to optimize all the desired objective functions under a series of design constraints. The multi-step protocol of Section II(2) can be adapted to this more challenging scenario, although the computational intensity quickly grows as the physics of the problem becomes more complex. Herein, we discuss the optimization of a simple architected material (a prismatic sandwich panel with hollow rectangular channels) for minimum weight under the simultaneous application of mechanical and thermal loads (subject to a number of design constraints). The motivation is a feasibility study for metallic actively cooled combustors for hypersonic vehicles (Fig. 3(a)). Details beyond this concise treatment are provided in a number of references.[19–21,40]

(B) Thermo-Mechanical Loads on Combustor Wall of a Hypersonic Vehicle: Whereas acreage (and to some extent, leading edges) of thermally balanced hypersonic vehicles can be engineered to passively dissipate heat by radiation, combustor walls inevitably require active cooling strategies to contend with the large heat fluxes arising from the combustion process. The prototypical structure, a sandwich panel with prismatic channels that provide active cooling by the fuel before injection (Fig. 3(b)), is subjected to significant thermo-mechanical loads. The thermal loads are represented by a heat transfer coefficient and a hot gas temperature on one side of the panel. The mechanical loads are (i) pressure in the combustion chamber (which, depending on the boundary conditions, can induce panel-level bending), and (ii) pressure in the cooling channels (dictated by fuel injection requirements). As a result, significant thermo-mechanical stresses arise. Withstanding these stresses at operating temperatures necessitates careful design. A viable solution must resist several failure modes: yielding or rupture due to (a) thermal stresses, (b) pressure or inertial stresses, (c) combined thermo-mechanical stresses, as well as (d) softening of the material, (e) coking of the coolant, and (f) excessive pressure drop in the cooling ducts. The objective is twofold: (i) identify the optimal material and (ii) optimize the structure for minimum weight. The challenge is to assure that none of the failure modes is active over the pertinent ranges of coolant flow rate, V_{eff} (often nondimensionally expressed in terms of the air/fuel mixture richness, ϕ, relative to stoichiometric combustion) and thermal loads (expressed by the heat-transfer coefficient between the combustion gas and the solid surface, h_G). The intensity of the mechanical loads is assumed constant for simplicity. Geometry and loads are depicted in Fig. 3(b).

(C) Optimal Design Protocol: The multi-step approach of Section II(2) is applicable, albeit with the complication that even a simplistic analytical model precludes the extraction of a

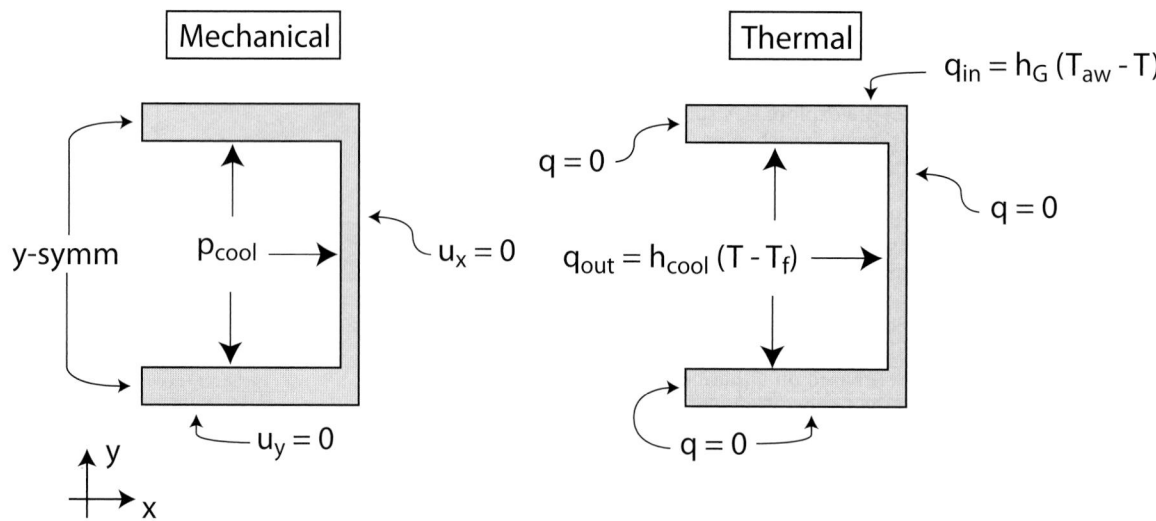

Fig. 5. Comparison of analytical and numerical (FE) von Mises stress distributions for an optimal actively cooled Inconel X-750 panel. The insets show the results for thermal, mechanical, and combined thermomechanical stresses along the four paths depicted. With the exception of Points 2 and 3, clearly affected by stress intensification, the agreement is very satisfactory, validating the optimization results presented in Fig. 6. Modified from Valdevit *et al.*[19]

Fig. 7. Weight efficiency of different material candidates for actively cooled walls of hypersonic combustors. The equivalence ratio, ϕ, is defined as for Fig. 6. From Valdevit *et al.*[19]

Fig. 6. (a) Performance maps for a number of candidate high-temperature materials for actively cooled walls of hypersonic combustors. The normalizing parameter for the coolant flow equivalence ratio ($\phi = f/f_{st}$) is that expected for steady-state Mach 7 flight conditions. Each map is the result of two independent optimizations. One (yellow) is performed absent a coating. When solutions exist only with a coating (light gray), the optimization is conducted using the TBC thickness as a variable. Areas without a feasible solution are in white. (b) Reclaimed feasibility space provided by two notional materials based on INCONEL X-750. The first notional material (designated S), examines the effect of elevating the yield strength at intermediate temperatures by 25%, whereas the second (designated T*) probes the effect of extending the maximum use temperature by 25%; analyses are shown both with (light) and without (dark) a TBC. For comparison, the notional results are superimposed on the original performance maps for INCONEL X-750 (Modified from Vermaak *et al.*[21])

is not readily accessible in any other forms (i.e., it does not transparently appear from the equations for temperature and stress distributions). This is a perfect example of the need for efficient optimal design tools to tackle these inherently multi-functional problems.

In Fig. 7, we show weight efficiency for a specific thermal load (realistic for a Mach 7 hydrocarbon-powered vehicle). Each curve tracks the minimum weight of an optimized structure for a given material. Notice that different materials have vastly different weight efficiencies, even if their robustness (the area of the curve in Fig. 6(a)) is somewhat similar. The benchmark material, C-SiC, offer by far the lowest weight. High-temperature titanium alloys (Ti-β 215) offer the lightest metallic systems, but their feasibility is limited to low thermal loads. Among the metals that offer robust solutions over a range of thermal loads, Niobium alloys (e.g., Nb-cb752, C-103) and Nickel superalloys (e.g., Inconel X-750, MAR-M246) are the most promising materials. Allowing the metals to shakedown upon thermo-mechanical cycling results in lighter systems (by as much as 30%–40%, depending on the boundary conditions), further increasing the competitiveness of metallic solutions.[40]

In closing, notice that all these conclusions are not transparently available from the equations, as thermal and mechanical properties are deeply intertwined in this heavily constrained thermo-mechanical problem.

III. Optimal Design of Micro-Architected Cellular Materials

In this section, we present the case for the development of a new class of multifunctional materials, characterized by a periodic cellular architecture with unit cell at the micro-scale and a characteristic dimension for the constituent material in the sub-micrometer region. If manufactured and designed correctly, *these micro-architected materials enable exploitation of potentially useful nano-scale mechanical effects (e.g., size effects in plasticity and fracture) that enhance mechanical properties relative to bulk macro-scale structures.* After reviewing viable manufacturing schemes to exploit this vision, we present the technical rationale for the expected performance and assess both the applicability of the multi-step optimal design approach described in Section II and the availability of suitable experimental and computational tools. In the interest of brevity we focus on mechanical design, although similar concepts can be extended to other functionalities.

(1) Manufacturing Approaches

A viable manufacturing approach for lattice-based micro-architected materials must possess the following key features: (i) dimensional control down to the 0.1–10 μm range; (ii) scalability to macroscopic part dimensions; and (iii) acceptable throughput to enable cost-effective manufacturing. Architectural flexibility (i.e., the capability to generate different unit cell topologies) and a wide suite of base materials are additional desirable attributes. To the best of the authors' knowledge, three families of manufacturing approaches exist today for the fabrication of lattice-based micro-architected materials: scaled-down versions of wire layup[41] and other modular assembly methods[42] (discussed in Section II(1)), stereolithography[43] (including the most advanced 2-photon approach[44]), and a new self-propagating photopolymer waveguide (SPPW) process,[45] recently developed at HRL Laboratories. Key attributes of each method are assessed in Table 1. Modular assembly methods can be useful for a wide range of end-materials, but are currently limited by the achievable resolution (minimum unit cell sizes ~100 μm and minimum feature sizes ~10 μm) and scalability (as the number of unit cells becomes very large, the assembly procedure becomes more and more cumbersome). Stereolithography allows incredible resolution (sub-micron feature sizes and unit cells of the order of a few microns for the most recent two-photon process) and nearly infinite architectural freedom (virtually anything that can be CAD drawn can be made). As a serial process, stereolithography is extremely slow: for a given sample size, the total processing time roughly scales with the inverse of the minimum feature size, implying that macroscopic quantities of micro-architected materials could take days to make. For proof-of-concept and basic research, stereolithography is a very powerful technique, but its difficult scalability makes it currently inadequate for industrial processing.

For a wide range of desirable end-geometries, the best compromise among resolution, architectural freedom, and scalability may be provided by the SPPW process.[45] Polymeric lattices are formed by UV exposure of a two-dimensional photolithographic mask with a pattern of circular apertures that is covering a reservoir containing an appropriate photomonomer (Fig. 8). Within the photomonomer, self-propagating polymer waveguides originate at each aperture in the direction of each collimated UV beam, forming a three-dimensional array of polymer fibers that polymerize together at all points of intersection. After removing the uncured monomer, three-dimensional lattice-based open-cell polymeric materials can be rapidly fabricated. Although this method does not allow for *arbitrary* shapes to be formed within the starting resin bath, it has the potential to form a wide range of free-standing 3D polymer structures based on linear mechanically efficient truss-type elements. In striking contrast with stereolithography, the optical waveguide process can form all truss-type elements in the structure in parallel with a single exposure step, typically lasting less than 1 min. With current UV exposure capabilities, cellular materials with truss member diameters ranging from ~10 μm to >1 mm and a relative density <5% up to 30% have been demonstrated.[46] The overall material thickness, H, can range from 100 μm to over 25 mm (although generally $H < 100 \cdot d$, where d is the truss diameter). Examples of ordered unit cell architectures with different symmetries are shown in Fig. 9; however, this process is not limited to such architectures. Nonsymmetric architectures, functionally graded materials, and hierarchical micro-lattice structures are all easily obtained.

For manufacturing techniques that result in a polymer template, such as stereolithography and the SPPW process, a number of postprocessing techniques are available to replicate the micro-architectural features with a metal or a ceramic[47,48] (Fig. 10). Continuous metallic film, such as nickel, can be deposited by electroplating or electroless process on the surface of the polymer micro-lattice structure and the polymer template can be subsequently removed with a chemical etch.[49] Controlled thickness coatings are obtained by varying the plating time. Ceramic films can be deposited with chemical vapor deposition (CVD) techniques. To withstand the high temperatures required for CVD of refractory metals or ceramics, the polymeric template must be pyrolyzed with minimal geometric distortion as recently demonstrated.[50] After the CVD process, the carbon micro-lattice template can be removed by oxidation (>600°C in air), leaving a hollow tube ceramic micro-lattice structure, such as the SiC sample shown in Fig. 10.

One key advantage of the polymer→metal or polymer→ceramic conversion process is to capture the strengthening effects associated with a constituent material in thin-film form factor in a bulk form (Section III(2)). These "film form" properties generally require film thicknesses in the micro- (or even nano-) scale, dictating truss diameters ~10–100 μm. This makes the fabrication approach described above ideally suited for fabricating optimal open-cell periodic architectures with exceptionally strong metallic or ceramic constituent materials.

(2) Challenges and Opportunities

The lattice materials manufactured with the SPPW process described in Section III(1) possess two distinct features, not readily available with competing concepts: (i) *hollow truss configurations* and (ii) *multi-scale architectures*, with global sample size on the order of several inches and sub-millimeter unit cell dimensions. These two features provide unique opportunities to expand the current bounds of material properties spaces and achieve combinations of properties currently unavailable in any existing material (including the macro-scale architected materials described in Section II). The target regions for specific strength and stiffness are depicted in Fig. 11. Importantly, micro-architected materials fabricated as described herein maintain an open core architecture, enabling multifunctionality: with reference to Section II(3), strong and stiff structures amenable to efficient active cooling are obviously an attractive possibility. Herein, we briefly review the rationale for these opportunities (limiting our attention to mechanical properties), and summarize the outstanding challenges that must be overcome in order to exploit the full potential of micro-architected materials.

(A) Advantages of a Hollow Truss Configuration:
Under any mechanical loadings, the strength of metallic lattice materials designed to operate in the elastic regime is limited by the onset of either yielding or elastic buckling. A simple analysis reveals the benefits of a hollow truss configuration; uniform compressive loading is assumed for simplicity, but the same conclusions qualitatively apply to other loading conditions. Consider a lattice material with a solid truss pyramidal unit cell, defined by truss member length, l, truss diameter, $2a$, and truss angle, ω. The relative density can be expressed as[51]:

$$\bar{\rho} = \frac{2\pi}{\cos^2 \omega \sin \omega} \left(\frac{a}{l}\right)^2 \tag{3}$$

and the compressive strength is:

$$\frac{\sigma_{\text{comp}}}{\sigma_{\text{bar}}} = \bar{\rho} \sin^2 \omega \tag{4}$$

where $\sigma_{bar} = \min\{\sigma_Y, \sigma_b\}$ represents the strength of the individual truss member, with σ_Y the yield strength of the base material and $\sigma_b = k^2 \pi^2 E a^2 / 4 l^2$ the elastic buckling strength. For conservativeness, it is customary to idealize each bar as pin-jointed, resulting in $k = 1$. Solid truss lattice materials are buckling-limited at low relative density, and transition to the yielding-limited regime at $\bar{\rho}_{trans} = 8\varepsilon_Y / (\pi \sin \omega \cos^2 \omega)$ (Fig. 12). For most metals, assuming a truss angle of 45–70°, the yield strain $\varepsilon_Y \sim 10^{-3}$, indicating a transition at $\bar{\rho} \sim 1 - 2\%$. As trusses with relative densities <<1% can be manufactured with the approach described in Section III(1), the implication is that the lightest lattice materials based on solid trusses will be inevitably buckling-dominated. The situation improves when hollow trusses are employed. Invoking a thin-wall approximation, the relative density of hollow truss structures scales as:

$$\bar{\rho} = \frac{4\pi}{\cos^2 \omega \sin \omega} \left(\frac{a}{l}\right) \left(\frac{t}{l}\right) \tag{5}$$

with t the truss wall thickness. The compressive strength scales as before, but the strength of the bar, σ_{bar}, is now:

$$\sigma_{\text{bar}} = \min \begin{cases} \sigma_Y & \text{yielding} \\ \sigma_{\text{gb}} = \dfrac{k^2 \pi^2 E a^2}{2 l^2} & \text{global (Euler) buckling} \\ \sigma_{\text{lb}} = \dfrac{E}{\sqrt{3(1 - v^2)}} \dfrac{t}{a} & \text{local buckling} \end{cases} \tag{6}$$

The local buckling load corresponds to the chessboard mode.[52] Again, we assume $k = 1$ in the global buckling

Table I. Comparison of Known Fabrication Methods for Open-Cellular Lattice-Type Materials. Reported Values Have Been Extrapolated Based on Published Techniques. Modified From Jacobsen et al.[46]

Approach to lattice-type cellular structures	Wire or textile layup; modular assembly[41,42]	Stereolithography[43]	2-photon stereo-lithography[44]	Self-propagating polymer waveguide (SPPW) process[45]
Unit cell type	3D Periodic	3D Periodic or aperiodic	3D Periodic or aperiodic	3D, Periodic or aperiodic
Multiple sizes of trusses in unit cell	Difficult	Yes (maximum flexibility)	Yes	Yes (Intrinsic to process)
Control of solid member diameter	Limited	Independent control	Independent control	Independent control
Member angle control*	0–90°	0–90°	0–90°	~50–90°
Fabrication area	0.1 m^2	0.25 m^2	0.005 m^2	~0.4 m^2
Thickness range	1–20 cm	1 mm–0.1 m	10–100 μm	100 μm–5 cm
Nominal max. volume	0.02 m^3	1 m^3	10^{-6} m^3	0.02 m^3
Max no. unit cells through thickness.	~10	10's	~10	~10
Min. unit cell size	~100 μm	<10 μm	<2 μm	<50 μm
Min. feature size	~10 μm	<1 μm	<<0.1 μm	<5 μm
Max. no. truss elements in a cubic volume	~4000	>10 000	~1000	~4000
Potential to grade properties	Limited	Yes (highest flexibility)	Yes	Yes
Possible base materials	Metals, polymers	Polymers	Polymers	Polymers
Post-processing material options	Annealing	Template for CVD/CVI, electroplating, casting, slurry coating, carbonization	Limited for very small structures: electroplating or CVD are possible	Template for CVD/CVI, electroplating, investment casting, slurry coating, carbonization
Rate of manufacture	Hours	Hours to days	Minutes to hours	Minutes
Potential for scalable manufacturing	Medium (single parts only)	Low (single parts only)	Low (single parts only)	High (>1 m^2/min for continuous process)

*(relative to horizontal).

Fig. 8. Schematic representation of the process used to form micro-truss structures from self-propagating polymer waveguides (SPPW) and a prototypical structure formed by this process.[45]

load for conservativeness. For optimal structures in the buckling-dominated regime, $\sigma_{lb} = \sigma_{gb}$, resulting in

$$(t/l) = \frac{\pi^2 \sqrt{3(1 - v^2)}}{2}(a/l)^3.$$

The transition between buckling- and yielding-dominated regimes now occurs at:

$$\bar{\rho}_{trans} = \frac{8\sqrt{3(1 - v^2)}}{\pi \sin \omega \cos^2 \omega} \varepsilon_Y^2.$$

For a metal, $\bar{\rho}_{trans} \sim 0.002\%$. Hence, metallic hollow trusses are yielding-dominated in the entire range of feasible relative densities, with substantial benefits on the strength (Fig. 12). For applications exploiting the local plastic buckling modes of hollow trusses, such as energy absorption, the advantage is even more significant. The amount of energy dissipated in crushing a bar by global (Euler) buckling is insignificant compared to the amount of energy absorbed in local modes. The implication is that hollow truss lattice materials will exhibit unique properties as cores of impact resistant sandwich structures.[49]

The situation is qualitatively identical for ceramic materials, whereby the yield strength is replaced by a defect-sensitive fracture strength. See Section III(2)(B) for more details.

(B) Advantages of Small-Scale Architecture: The strength-density relation derived in Section III(2)(A) for both solid and hollow trusses is *length-scale independent*: proportional scaling of all dimensional geometric variables (truss bar length, l, radius, a, and wall thickness, t) has no effect on either relative density or specific strength. The fundamental assumption is that constituent material properties are themselves scale-independent. Although reasonable for wall thicknesses as small as a few microns, powerful strengthening effects will

Fig. 10. Suitable fabrication routes for hollow tube metallic and ceramic micro-lattice structures (Images from Jacobsen *et al.*[45,50]).

emerge as sub-micron dimensions are approached. These recently documented effects arise from three phenomena: (a) yield strength elevation in metals due to strain gradient effects and/or (b) dislocation/surface interactions, and (c) fracture strength elevation in ceramics due to reduced average flaw size. Herein, we briefly review all the three mechanisms.

Scale Effects in Plasticity in the Presence of Strain Gradients: A large body of experimental investigations reveals the presence of size effects in plastic response that become more pronounced as the size of the sample (or the relevant length scale) approaches μm or sub-μm dimensions. Notable examples are the increase of indentation strength at shallower indentation depths,[53] increase in flow stress and

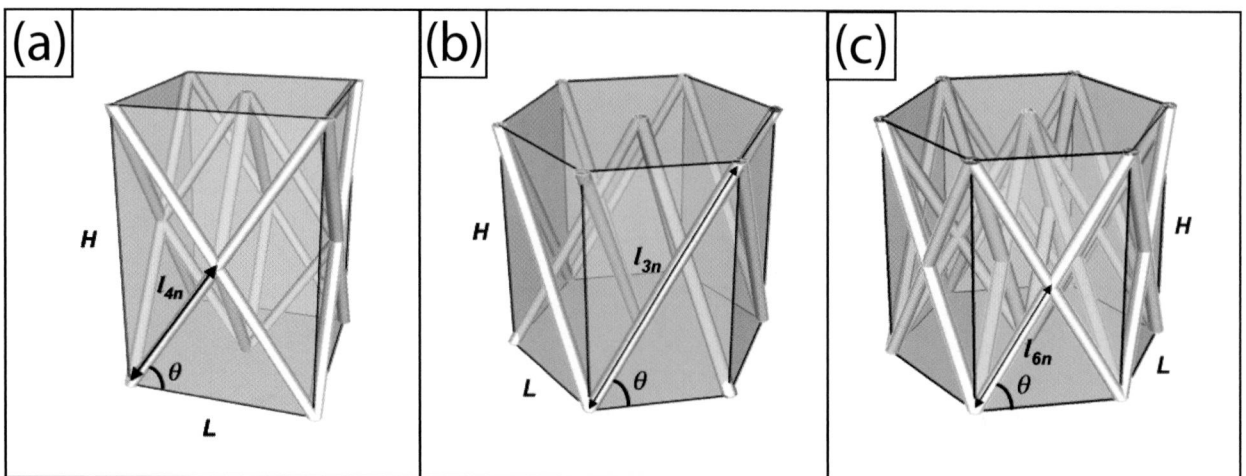

Fig. 9. Archetypal unit cell architectures with (a) 4-fold symmetry, (b) 3-fold symmetry, and (c) six-fold symmetry, as examples of structures that can be manufactured with the SPPW process depicted in Fig. 8 (from Jacobsen *et al.*[56]).

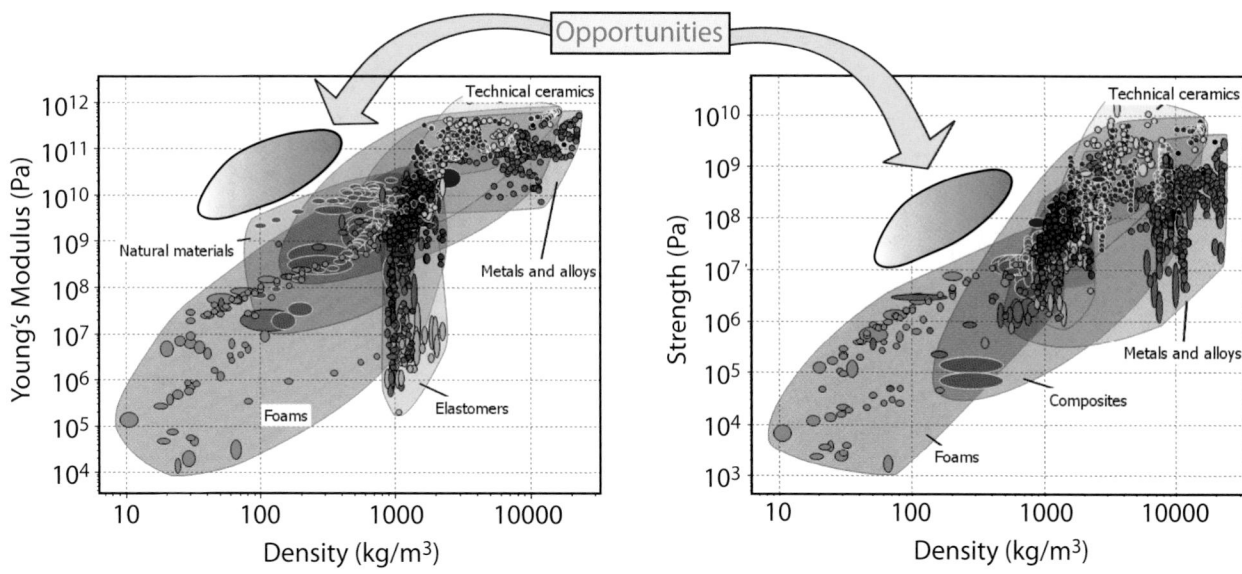

Fig. 11. Ashby charts for stiffness versus density, and compressive strength versus density, depicting all existing materials and emphasizing two target areas for micro-architected materials.

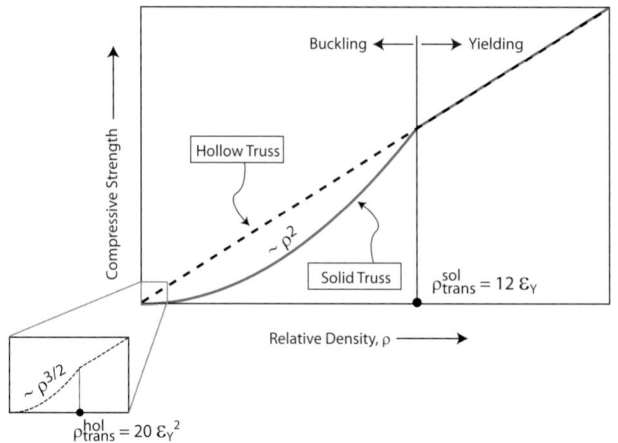

Fig. 12. Maximum compressive strength of solid and hollow metallic trusses. Notice that hollow trusses are yielding-dominated for any reasonable relative density.

hardening rate for thinner wires in torsion[54] and bending,[55] the classic Hall-Petch effect on the grain size dependence of flow stress,[56] and the increase in particle strengthening as the reinforcement size is reduced.[57] While these problems are vastly different, they all require a natural length scale for interpretation. A fundamental commonality in all the aforementioned situations is the presence of substantial plastic strain gradients during deformation. The development of constitutive laws that capture stress dependence on both strain and strain gradients is a natural modeling strategy. A number of strain gradient plasticity theories that reduce to the classic J_2 theory as the strain gradients are progressively reduced have been proposed over the past 25 years, most prominently by Fleck and Hutchinson,[58,59] and Nix and Gao.[60–62] The fundamental differences between the two theories in predicting experimental results were recently reviewed by Evans and Hutchinson.[63] Regardless of the differences, central to both theories is the concept of geometrically necessary dislocations (GND), initially introduced by Ashby.[64] Geometric considerations demonstrate that plastic strain gradients often require the storage of GNDs to maintain displacement compatibility. The GND density, ρ_G (total GND line length per unit volume) can be calculated once the active

slip systems are identified. In an averaged sense, ρ_G can be related to the strain gradient, ε_p^*, through a natural length scale: $\rho_G \sim \varepsilon_p^*/\ell$, where ℓ is generally extracted from experimental results. Both theories predict hardening effects (and in the case of Fleck/Hutchinson, initial yield strength elevation) increasing with ρ_G.

Although a comprehensive strain gradient plasticity theory capable of capturing all the experimental phenomena while reducing to J_2 theory at large scale is still incomplete, there exists a general agreement on the marked effect of strain gradients on flow stress. These effects appear even at relative large sample sizes ($\sim$10 μm) and have the potential to substantially elevate the performance of micro-architected materials relative to their macro-scale counterparts. Although the lattice structures manufactured as described in Section III(1) (Fig. 10) will initially experience nearly zero strain gradients when loaded in compression and/or bending (all the truss members will uniformly compress or stretch), as the deformation progresses beyond first yield and the hollow truss members plastically buckle, substantial plastic strain gradients will arise. Although no initial yield strength elevation due to strain gradients is anticipated, both the collapse strength and the crushing energy (plastic dissipation) of micro-architected materials may be significantly higher than for conventional macro-scale materials. Recently, significant size effects have also been observed even in the absence of strain gradients, with further potential benefits to micro-architected materials. These effects are reviewed in the following subsection.

Size Effects in Plasticity in the Absence of Strain Gradients: Over the past 5 years, a multitude of room-temperature uniaxial compression and tension experiments have been performed on a wide range of single-crystalline metallic nano-pillars and nano-dogbones, including fcc metals (Ni and Ni-based superalloys,[65–67] Au,[68–71] Cu,[72–80] and Al[81,82]), bcc metals (W, Nb, Ta, and Mo[83–89]), hcp metals (Mg[90,91] and Ti[92]), tetragonal low-temperature metals (In[93]), Gum metal[94,95], nanocrystalline metals (Ni[96,97]), shape memory alloys (NiTi[98–102] and Cu-Al-Ni[103,104]), and a variety of metallic glasses[105–107]. For samples with nonzero initial dislocation densities (i.e., excluding whiskers and nano-fibers), a strong size effect on the flow strength was ubiquitously demonstrated as the sample size approached μm and sub-μm dimensions.[108] The compressive strength data for all single crystalline face-centered cubic (fcc) metals (Au, Al, Ni, and Cu) show a unique trend, suggesting the existence of a universal law of the form:

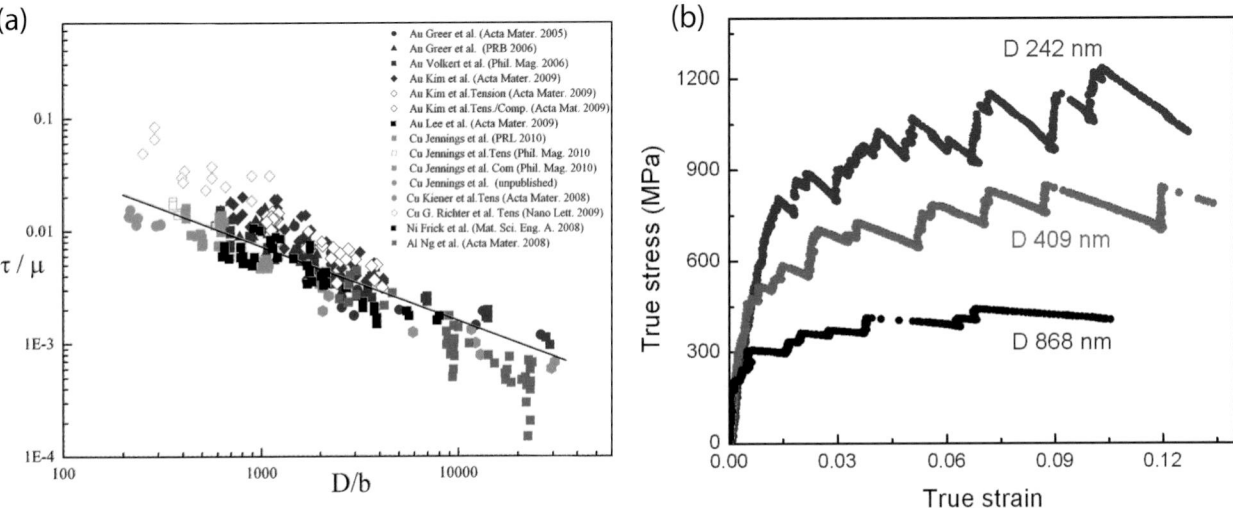

Fig. 13. Sample size effects on the flow stress in nanoscale experiments. (a) Shear flow stress (normalized with the shear modulus) VS sample diameter for a number of small-scale experiments on FCC metals reported in the literature. Resolving the stress on the relevant slip system and normalizing the sample diameter with the Burger's vector allow comparison of different metals on the same chart.[157] Reprinted from Greer and De Hosson (2011), with permission. (b) Compressive stress versus strain for uniaxial compression of single crystal Nb nano-pillars of different diameters (above each curve).[86]

$$\sigma_{res}/\mu = A(d/b)^{-m} \qquad (7)$$

where μ is the shear modulus, σ_{res} is the resolved shear stress onto the $\{111\}/<110>$ slip system, d is the pillar diameter, b is the Burgers vector, and A and m are constants (Fig. 13(a)). A similar observation was reported by Dou and Derby.[109] Based on the existing data for Au, Al, and Ni, $A{\sim}0.71$, and $m{\sim}0.66$. This exponent is nearly identical to those reported for nearly all other fcc micro- and nano-pillars where the samples contain initial dislocations.[108]

Figure 13(b) depicts some representative stress-strain curves for single crystalline Nb nano-pillars subjected to uni-axial compression.[86] The size-dependent strengthening effect cannot be explained through well-known thin-film mechanisms, such as grain size hardening,[110] the confinement of dislocations within a thin film by the substrate,[111] or the presence of strong strain gradients.[112] Intriguingly, unlike Taylor hardening, the flow strength does not appear to scale with the evolving density of mobile dislocations. Several models attempting to explain the causality between the declining dislocation density and attained strengths have been put forth. For example, the *dislocation starvation* model, first proposed by Greer and Nix[113] hypothesizes that the mobile dislocations inside a small nano-pillar have a greater probability of annihilating at a free surface than of interacting with one another, thereby shifting plasticity into nucleation-controlled regime.[66,68,69,114] Other models include source exhaustion hardening,[115,116] source truncation,[117,118] and weakest link theory.[116,118] The general commonality in all these theories is the representation of dislocation source operations in a discrete fashion, enabling an evaluation of the effect of sample size on the source lengths, and therefore on their operation strengths. Some of these models also capture the ubiquitously observed stochastic signature of the experimental results, showing either marginal dislocation storage[114,116,119] or no storage at all.[66,69,80]

The vast majority of samples for the nano-mechanical characterization described above have been produced by Focused Ion Beam (FIB). Figure 14 shows a number of examples. Unfortunately, the effect of the ion implantation introduced as a result of the FIB processing is not well characterized, and hinders an accurate interpretation of the experimentally observed size effects in plasticity. Although several investigators have reported Ga+ ion bombardment damage resulting in altered microstructural features (e.g., dislocation sources,

Fig. 14. SEM images of FIB-fabricated samples. (a) 400 nm nano-crystalline Ni-W nano-pillar. (b) Compressed 600 nm Nb pillar with significant slip offsets. (c) Typical dog-bone shaped tensile Au sample,[139,140] and (d) Mo anti-pillar with a hollow center.[158]

lattice rotations), attempts to attribute the observed strain hardening to these features have been inconclusive. Recent evidence (both experimental and computational, within and outside of the authors' groups) convincingly demonstrates that the size effect is a strong function of the initial dislocation density rather than the fabrication technique.[80,120] The size effects observed in the FIB-prepared specimens, for example, are identical to those obtained for the samples fabricated by a completely FIB-less technique, which contain similar initial dislocation densities.[121] Further, it has been reported that introducing dislocations into an initially pristine structure actually weakens rather that strengthens the sample.[122,123] In a recent computational study, it was reported that FIB-induced damage could contribute up to ~10% of the observed flow stress increase only for a particular size range between 500 nm and 1 μm, whereas for the larger and smaller specimens, the effects of the FIB on strength are marginal.[124]

Based on these arguments and the now ubiquitously reported presence of power-law size effects for all non-pristine pillars produced with or without the use of FIB, the authors are confident that the size effects are real and are *not* a function of the fabrication technique.

The ability to manufacture a macro-scale micro-architected material with a hollow truss topology characterized by truss wall thickness in the μm (and sub-μm) scale has the potential to exploit these beneficial size effects to achieve exceptional constituent materials properties. When combined with optimal design of the truss architecture, this approach should result in a macro-scale material with unprecedented specific strength. Admittedly, the metallic films deposited on the trusses will be polycrystalline, likely with a nanoscale grain size. The strengthening effects described above for fcc single crystals are much less understood in multi-grain surface-dominated small-scale systems. In fact, both homogeneous (grain boundaries, twin boundaries, etc.) and heterogeneous (i.e., phase boundaries, precipitate-matrix boundaries, free surfaces, and passivated surfaces) interfaces in size-limited features are crucial elements in the structural reliability of most modern materials. Yet very little work has been done on characterizing the combined effects of interfaces and surfaces—extrinsic (sample size in a surface-dominated structure) and intrinsic (microstructural features like grain boundaries, twin boundaries, phase boundaries, etc.)—on the mechanical response of materials. Furthermore, a vast majority of the above-mentioned experiments was conducted at room temperature, limiting our understanding of athermal versus thermal contribution to size-dependent strength. Significant efforts must be focused on investigating mechanical properties and identifying particular deformation mechanisms operating in boundary-containing metallic material systems with reduced dimensions (for example, nano-pillars containing two or three grains, twin boundaries, and homo- and heterogeneous nano-laminates). The knowledge of the specific deformation mechanisms as a function of feature size and initial microstructure will be essential for the design, manufacturing, and property control of new, revolutionary lightweight metallic micro-architected materials with unprecedented combinations of properties.

Fracture Strength Elevation in Ceramics at Small Scales: Ceramic thin films (e.g., carbon, silicon carbide, silicon nitride) possess yield strengths >10 GPa. Unless the constraining environment is such that crack growth is impeded (as would be the case for a uniform film adhered to a substrate and loaded in compression normal to the plane of the wafer), failure will generally occur by fracture. Linear elastic fracture mechanics predicts a fracture strength $\sigma_\mathrm{f} \sim K_\mathrm{c}^{\mathrm{solid}}/\sqrt{a}$, with $K_\mathrm{c}^{\mathrm{solid}}$ the fracture toughness of the constituent material and a the size of the largest crack. Assuming a statistical distribution of crack directions, mode I conditions will generally dominate the strength, whereby $K_\mathrm{c}^{\mathrm{solid}} = K_\mathrm{Ic}^{\mathrm{solid}}$ and a is the size of the largest crack oriented favorably to mode I propagation. The smaller the sample dimension, the smaller its largest crack. The implication is that the strength of a ceramic material will substantially increase as the sample length scale is reduced. As an example, a 5–10 μm thick polycrystalline diamond film deposited on a hollow truss might have[125] $a \sim 1\mu m$, $K_\mathrm{c} \sim 4.6\,\mathrm{MPa}\sqrt{\mathrm{m}}$, resulting in $\sigma_f \sim 4\,\mathrm{GPa}$. The relationship between the fracture toughness of a lattice and that of its constituent material has been recently elucidated by Fleck *et al.* for the case of planar lattices[126]: $K_\mathrm{Ic}^{\mathrm{lattice}}/K_\mathrm{Ic}^{\mathrm{solid}} \sim \rho^d\sqrt{\ell/a}$, with ρ the relative density of the lattice, ℓ the unit cell size, a a typical crack size in the constituent material, and the exponent d is a strong function of the lattice topology ($0.5 < d < 2$). Similar relationships can be derived for 3D lattices. If the architecture is properly chosen to minimize d, and assuming that the constituent crack size can be reduced together with the unit cell size, micro-architected lattice materials can have substantial fracture toughness, at a fraction of

the weight of solid materials. At the same time, ceramics are exceptionally stiff (E~1TPa, for polycrystalline diamond[125]). Such strengths and stiffnesses are unattainable with any metallic system, offering ceramic micro-architected materials the potential to leap into currently unclaimed areas in a number of materials property charts. Importantly, the manufacturing technology described in Section III(1) is a key enabler for this vision: only an approach capable of manufacturing *large-scale* materials with *micron-level control* of the lattice architecture allows the base material to be deposited in the form of a sub-μm thin film. This has two enormous benefits: (i) it allows use of materials not available in the bulk (e.g., polycrystalline diamond), and (ii) it allows accurate control of the maximum flaw size, with enormous increases in the fracture strength relative to bulk values.

(3) Optimal Design Protocol for Micro-Architected Materials

The multi-step optimal design protocol presented in Section II in the context of large-scale periodic cellular structures is generally applicable to micro-architected materials. Although the general methodology is unchanged, a fundamental complexity emerges. *The bulk properties of micro-architected materials are a strong function of phenomena occurring across three length scales* (Fig. 15): a macroscopic level (bulk), a mesoscopic level (unit-cell) and a micro/nanoscopic level (the characteristic length scale of the constituent materials). Unique critical phenomena occur at each length scale, requiring experimental investigation. The size effects on plastic flow stress, strain hardening, and fracture strength (discussed above) are clearly micro/nano-scale phenomena, as are microstructure (and properties) anisotropy possibly arising from the film deposition process. Film thickness/microstructure variation along the truss members and details of the node topology—and their effects on the mechanical properties of the material—occur at the unit-cell level. The same length-scale can also affect the fracture strength, as variation in flaw distributions along members and around nodes can play a substantial role. Finally, the vast number of unit cells composing the bulk material may introduce large-scale effects previously unnoticed in macro-scale lattice materials: geometric imperfections (i.e., deviation from a perfect lattice) and the possible occurrence of buckling modes with characteristic length scale much larger than the unit cell level (and hence not captured with the type of analysis presented in Section II(2)) might play a significant role on the overall stiffness and strength of the bulk material. Novel characterization techniques and numerical strategies must be implemented to enable multi-scale studies. These are discussed below.

(4) Experimental Characterization of Micro-Architected Materials

(A) Macro-Scale Mechanical Characterization: At the macro-scale, conventional techniques traditionally employed for the characterization of large-scale lattice materials (and in fact, any other material) are perfectly adequate to analyze micro-architected materials. Traditional universal test frames (e.g., INSTRON, MTS) equipped with tensile, bending and shear fixtures can be used to measure stress-strain response in different loading scenarios and validate/calibrate analytical and numerical models for failure prediction.

(B) Micro-Scale Mechanical Characterization: As the unit cells of micro-architected materials can take many shapes and sizes, and several base materials can be used, the ideal device for mechanical characterization at this scale should have the following features: (i) be adaptable to samples of vastly different sizes and shapes; (ii) allow controlled displacement actuation and independent load measurement;

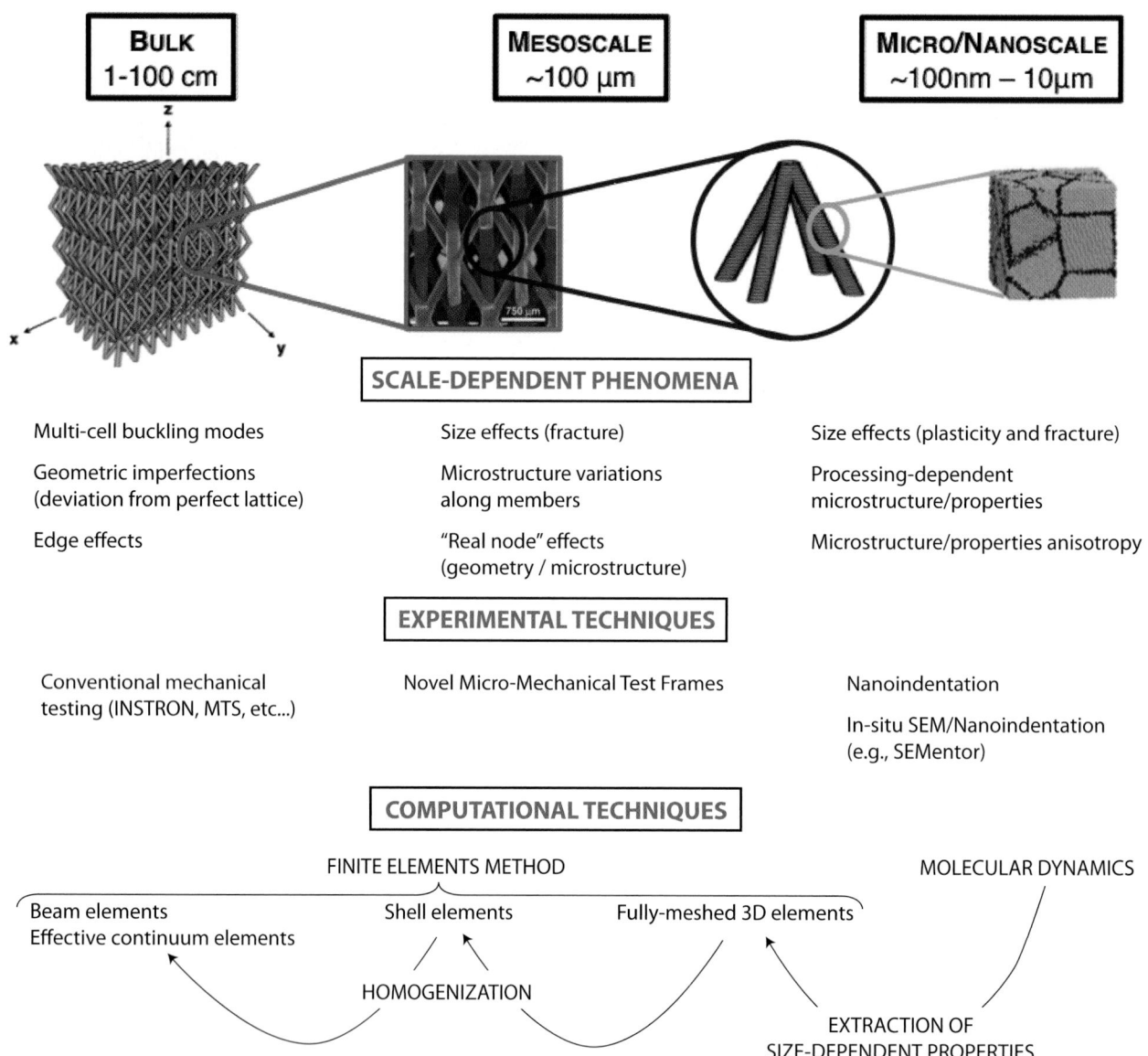

Fig. 15. Expected phenomena, mechanical characterization tools and computational techniques for the three length-scales of interest in micro-architected materials. (SEM images from Jacobsen *et al.*[45,159])

(iii) be capable of extreme force resolution (~1–100 nN) and range (~1 N), displacement resolution (~10–100 nm) and range (1–10 mm); and (iv) allow optical (or SEM) access to the test coupon with potential for strain mapping (via Digital Image Correlation). Hybrid micro-test frames, encompassing a MEMS force sensor and an off-chip displacement actuator, are necessary to meet the requirements listed above. A number of such devices have been developed in the past two decades.[127,128] An economical and versatile device, capable of covering the entire range of force and displacement described above, has been recently introduced by one of the authors (Fig. 16).[129] In the proposed design, the sensor is a micro-fabricated Silicon double-ended tuning fork (DETF), whose working principle is the change in natural frequency of vibration in a pair of parallel and connected beams upon application of an external axial force. DETF sensors deflect axially rather than laterally, hence exhibiting essentially infinite stiffness relative to the sample being tested and exceptional force range (in the Newton range). This is in stark contrast with the more commonly employed capacitive or visual force detection schemes, in which load cell and sample have comparable compliance. At the same time, the strong dependence of the natural frequency of a beam on the axial load and the extreme

precision available in frequency measurement (a change of a fraction of a Hertz is easily detected in a 100 kHz signal) grant the device nN resolution. Samples of different geometry can be handled with micro-fabricated custom fixtures, enabling a variety of testing conditions (bending, compression, tension, etc....).

(C) Nanoscale Mechanical Characterization: Uniaxial Mechanical Testing at the Nanoscale: The two best-established techniques for nanoscale mechanical characterization are Atomic Force Microscopy (AFM) and Nanoindentation. The former controls the displacement through a piezo-actuator and senses the force through the deflection of a micro-cantilever, typically measured optically. Although extreme force and deflection resolutions are possible,[1,130–132] the force range is small, on the order of a few pico-Newtons, with a vertical distance resolution smaller than ~0.1 nm. Such a small load range limits the applicability of AFM to the characterization of stiff materials (e.g., metals and ceramics). The mechanical properties of stiff materials can be well characterized by nanoindentation, whose premise involves forcing a generally sharp diamond indenter tip into the surface of a material, while measuring the imposed force, the corresponding displacement of the indenter, and in some cases, the contact stiffness.[62,133–138] Nanoindenters are inherently

Fig. 16. (a) Micro-mechanical test frame (μ-MTF) for unit-cell level characterization. The displacement actuator is a commercially available nano-stage, while the load cell (inset) is microfabricated. (b) Detail of the microfabricated load cell (From Torrents *et al.*[129] and Azgin *et al.*[160])

load-controlled instruments, where the load is applied through an electromagnetic coil assembly, and the displacement is measured by capacitive gages. Typical modern-day nanoindenters have load resolution of several nano-Newtons and sub-nanometer displacement resolution. From a measurement of the indentation depth, the local hardness, H, of the material is readily accessed, and the yield strength is generally estimated as $\sigma_Y \approx H/3$. Obtaining more detailed mechanical information is unfortunately very difficult: the stress and strain fields induced by the indenter are complex and tri-axial, rendering the interpretation of experimental results challenging. Carving nanoscale samples in the shape of pillars or dog-bones addresses this difficulty, by enabling the introduction of nearly uniaxial stress and strian fields (see Section III(2)(B)). To allow *in-situ* observation of the sample deformation, nanoindenters have been coupled with electron microscopes. A unique such instrument (called the "SEMentor") was recently developed by one of the authors.[139,140] The SEMentor is comprised of a nanomechanical module, similar to the DCM assembly of a commercial Agilent nanoindenter, inside of a SEM. The former offers a precise control and high resolution of load (~1 nN) and displacement (<1 nm) and their rates, as well as contact stiffness during the experiment, while the latter allows for visualization of the process. Custom-made grips were fabricated to conduct nanoscale *in-situ* experiments in uniaxial compression and tension. Uniaxial tensile investigations will be essential in determining nanoscale yield and ultimate tensile strength and fracture toughness of nanoscale materials, as well as in elucidating the origins of tension/compression asymmetry likely to be observed in nanoscale polycrystalline samples. All these features are critically important to the development of micro-architected materials. Furthermore, this *in-situ* testing technique will allow correlation of the macroscopic stress-strain behavior with some microstructural activity by direct observation of, for example, the dislocation glide "avalanches" manifested by multiple slip lines, shear offsets, and phase delamination (if any).

Microstructural Characterization: A key analytical technique allowing direct observation of dislocations and their interactions with various boundaries and surfaces is High Resolution Transmission Electron Microscopy (HR-TEM). While, of course, the post-mortem TEM analysis is not capable of providing any information about the mobile defect activity, it is powerful in revealing the post-deformation micro-

structure, i.e., the evolved dislocation networks, the final grain configurations, and most importantly it sheds light on the particular interactions of dislocations with the individual interfaces and surfaces in the deformed samples. This information is useful in uncovering some of the fundamental mechanisms that might have operated during deformation of these nano-volumes with specified interfaces.

(5) Numerical Modeling of Micro-Architected Materials

(A) Continuum-Based Approaches (Finite Elements Analysis): Once the properties of the thin film base materials are known (including their size effects), traditional contin-

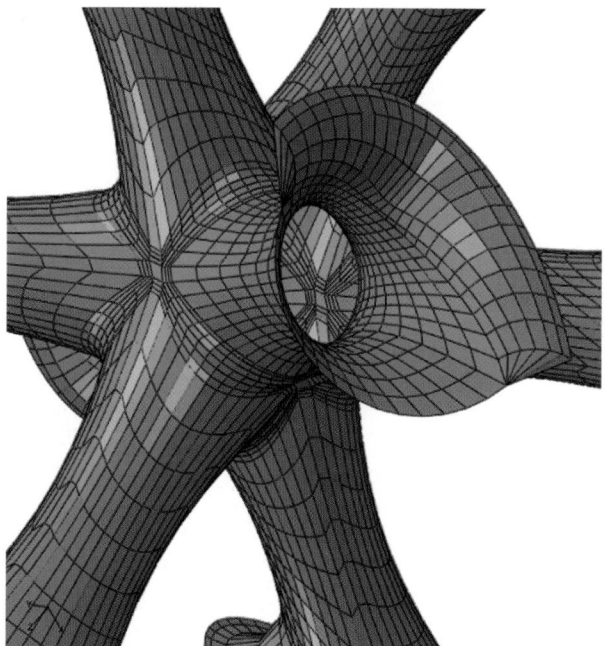

Fig. 17. Example of Finite Elements mesh for micro-architected hollow truss structures. Node fillet radii, nonuniform wall thicknesses, noncircular cross-sections and bar tapers are parametrically defined with geometric modeling tools, for efficient integration with optimization algorithms.

uum finite elements approaches can be used to extract both the unit cell properties as well as the averaged bulk properties of micro-architected materials. Although the numerical tools are identical to those used for large-scale lattice materials (see Section II), some additional steps must be taken to merge the unit cell and the bulk length scales (Fig. 15).

Finite Elements Modeling of Individual Trusses and Unit Cells: Individual hollow truss members and unit cells can be modeled with shell elements and/or solid elements. Besides validating the analytical predictions for an ideal structure, FE modeling must quantify the effect of three potentially critical factors on stiffness, strength, ductility, and collapse mechanisms. *(i) Geometric non-uniformities* (e.g., curviness in nominally straight truss members, non-ideal node size and shape, and wall thickness variations along members and within nodes). *(ii) Heterogeneity and anisotropy in materials properties* (arising from variations in grain size and texture along the trusses and around the nodes). *(iii) Interface strength and toughness in multi-materials systems* (e.g., a ceramic film deposited on a metallic wall). These phenomena are in principle present in large-scale lattice structures as well, but their effect on the overall properties is typically negligible. Geometric algorithms for automatic meshing are essential to capture point (i) above (Fig. 17).

Large-Scale (Multi-Cells) Finite Elements Modeling: The vast difference between the unit cell and the bulk scales requires numerical strategies for efficient modeling. Generally, the plethora of Finite Elements results extracted from the fully meshed unit-cell models described above must be condensed into a lower-order model (homogenization). Two sequential approaches can be envisioned, in increasing order of complexity and computational efficiency: (a) beam-elements unit-cell model; (b) effective 3D solid elements model. For approach (a), each unit cell is modeled with a number of beam elements and the effects of nodal size/shape and local buckling that are lost in transitioning from shell to beam elements can be incorporated by introducing fictitious (non-uniform) stress-strain response for the material. Approach (b) follows homogenization procedures similar to those implemented for large-scale period core sandwich panels.[26,141] The challenge is to define the constitutive behavior of the material in a way that a single solid element (possibly with size spanning several unit cells) elastically and plastically deforms consistently with buckling phenomena that are dominated by length scales of the order of the unit cell size, truss radius and wall thickness.

(B) Atomistic Approaches (Molecular Statics and Dynamics): Standard molecular dynamics (MD) and molecular statics (MS) techniques can be used to develop quantitative predictions of the elastic behavior of brittle ceramics and the elasto-plastic behavior of ductile metals and nanostructured composites of ductile and brittle materials. The Embedded Atom Method (EAM) potential[142] and the Modified EAM (MEAM) potential[143] are appropriate for metals, whereas materials with significant angular bonding (such as some ceramics) require dedicated MEAM potentials.[144] Molecular statics (MS) approaches are adequate for stiffness calculations, both in single crystal and nanocrystalline samples, and enable investigation of the role of all strain components on grain size and orientation at T~0K (this is acceptable as stiffness is not expected to be a strong function of temperature). If necessary, temperature effect can be quantified with MD simulations at non-zero temperature. Prediction of the plastic behavior of micro-architected materials at the nanoscale is significantly more challenging, as significant size and orientation effects might dominate strength and ductility. Controlled defects can be introduced to examine trade-offs between homogeneous and heterogeneous dislocation and void nucleation. As plastic flow is highly temperature dependent, MS is inadequate for elasto-plastic investigations, and MD simulations must be performed at nonzero temperature (typically at T~273 K). A substantial challenge is bridging the simulation and experimental time scales:

computational resources demand strain rates $>10^6 \, s^{-1}$ for all MD simulations, whereas experiments typically take place at rates of $1 \, s^{-1}$ or less. To avoid recoding fictitious strain rate effects in the data, high strain rate MD simulations must be extrapolated to lower strain rates, possibly using mechanism-based modeling. Validation of these calculations with nano-scale experiments (Section III(4)(C)) is critical to ensure that deformation mechanisms unveiled by the MD simulations persist even at much lower strain rates. MD capabilities to accurately predict the initial yield point have been well established. Quantitatively capturing phenomena subsequent to failure initiation (e.g., hardening and ultimate failure) is less straightforward. Once again, nanomechanical experiments will be essential to validate and supplement MD models. MD and FEM analyses can be interfaced *off-line* with phenomenological constitutive laws: such laws are informed by MD (e.g., yield strength VS sample size) and can be easily imported in commercial FE packages (either directly or via user-defined subroutines). Naturally, combining MD numerical schemes at the nanoscale with FE calculations at the micro and macro scale in *real-time* in a truly multi-scale algorithm is the Holy Grail of computational mechanics. After decades of research in multi-scale mechanics, this is still a daunting task. Nonetheless, *off-line* combinations of atomistic simulations at the nanoscale and Finite Elements and analytical continuum mechanics at the micro/macro-scale have been capable of macro-scale properties predictions.[145] Experimental results as those described in Section III(4) are often necessary for model calibration.

In addition to MD and MS simulations, Discrete Dislocations Simulations (DDS) have been introduced to compute the effect of dislocation motion and interaction on plastic flow in crystals.[146–148] These models generally utilize a FE continuum framework to solve for stresses, strains and displacements, and treat dislocations as singularities, which both affect and are affected by the global strain and stress fields. Dislocations motion is governed by the Peach-Koehler equation and standard dislocation-dislocation interaction laws. Dislocation sources like Frank-Read and single-arm sources are introduced at statistically random locations. These models (both in 2D and 3D) are computationally more efficient than full-scale MD simulations, but have not yet fully succeeded in duplicating key aspects of the experimentally measured size effects in plasticity.[149]

(C) Optimization Algorithms: The large number of variables in the optimal design of micro-architected materials, coupled with the possible difficulty in obtaining closed-

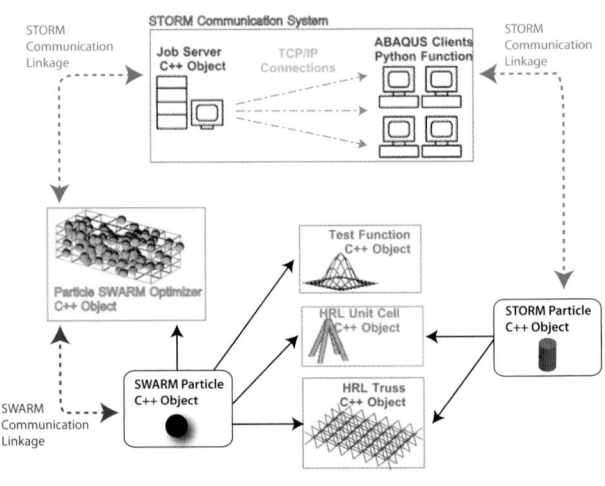

Fig. 18. Schematic of a possible efficient optimization code for the optimal design of complex micro-architected materials.[36] The optimizer can be coupled with commercial Finite Elements packages for on-line fully numerical optimization.

form solutions for some of the objectives and constraints, require an optimization procedure more sophisticated than that typically used for the optimization of macro-scale structures (Section II). Importantly, most of the structural optimization work performed to date assumes that the *constituent material(s)* and the *architecture* of the cellular structure is decided *a priori,* and the optimizer must simply select the best values for all the *geometric parameters* that define that architecture (e.g., unit-cell size, truss angle, wall thickness). A large number of architectures were independently studied at the macroscopic scales.[11,12,15,16,150] Standard quadratic optimizers were typically used, and relied on close-form expressions for the objective function and all the constraints. Although easy and convenient, this approach might be unsuitable for the optimization of micro-architected multifunctional materials, for two reasons: (i) complex geometric features (e.g., details of the node architecture) might strongly affect stress and strain distributions, and hence impede close-form expressions of objective and constraints, and (ii) a large number of local optima might appear, which could confuse quadratic optimizers. Discrete optimizers (e.g., genetic or particle swarm optimizers) do not require close-form expressions for objectives and constraints and are rather insensitive to local minima. Unfortunately, their robustness and efficiency are not as good as for gradient-based algorithms. A discrete optimization protocol that allows on-line interaction with a commercial Finite Elements program for objective function evaluation is schematically described in Fig. 18.[36] The on-line integration of FE analyses within the optimization loops is extremely resource-intensive. Although novel software technology (e.g., Microsoft HPC) will help, this remains a key bottleneck in the optimization process.

Ultimately, to take full advantage of the nearly infinite freedom in designing with micro-architected materials, the topological architecture of the material should be itself a variable in the optimization process. Topology optimization algorithms were developed over the past couple of decades,[151–154] with the scope of defining the ideal arrangement of two or more phases of matter (one might be air to define a cellular solid) to achieve extremal values in one or more macroscopic properties. These algorithms are naturally more complex and resource-intensive than the simple procedure described in Section II. In addition, although exceptionally interesting and unexpected results might occur (e.g., when the Kagome structure was identified by Torquato's group as an ideal stiff and statically determinate lattice,[155] spearheading dozens of studies on its mechanical properties and technological importance,[28,30,150]) in many cases the resulting topologies are nearly impossible to manufacture. Nonetheless, incorporation of topology optimization concepts within the computational framework described in Section II is a promising way to explore a massive design space.

IV. Conclusions

Periodic cellular materials and structures have been extensively investigated over the past decade for a large number of multifunctional applications. A common optimal design protocol was successfully adopted in several studies to select the geometry, architecture, and base material that maximizes the relevant objective function(s). The protocol employs a combination of analytical, numerical, and experimental techniques, and was reviewed herein in the context of mechanical and thermo-mechanical structures. We extended this concept to the multi-scale design of a micro-architected material. Enabling this vision are newly available manufacturing approaches for suitable periodic cellular architectures with unit cell sizes in the sub-millimeter scale and constituent material characteristic length in the micro/nano scale. The resulting micro-architected material is characterized by three different length scales: a bulk scale (~1–100 cm), a unit-cell scale (~100 μm–1 mm) and a constituent material scale

(~100 nm–10 μm). We surmise that micro-architected materials will possess unique superior mechanical properties (primarily specific stiffness and strength), by exploiting recently observed size effects at the micro and nano scales. Extending optimal design protocols for large-scale periodic cellular structures to the case of micro-architected materials entails the added difficulty that bulk mechanical properties are now a strong function of physical phenomena occurring at three different length scales. Existing modeling tools must be adapted to handle such difficulty. We briefly reviewed the state-of-the-art in small-scale mechanical characterization and numerical modeling and concluded that the necessary techniques are sufficiently mature to underpin the development of micro-architected materials. As the length scale ℓ of a periodic cellular structure shrinks to sub-millimeter dimensions, and the number of unit cells in the bulk material increases as ℓ^{-3}, the difference between a *structure* and a *material* gets progressively blurred: in this sense, micro-architected materials represent a new class of materials, characterized by a unique variable amenable to optimization: the *topological architecture*. We hope that this overview will stimulate intense research efforts aimed at developing new micro-architected materials with properties outside of the boundaries of current materials properties spaces and into new, previously unclaimed regions.

Acknowledgments

The authors are grateful to DARPA for financial support through grant no. W91CRB-10-C-0305 on Materials with Controlled Microstructural Architecture (Judah Goldwasser, program manager). LV acknowledges partial funding from the California-Catalonia Engineering Program. JRG acknowledges the financial support from NSF CAREER Award (DMR-0748267) and ONR Grant no. N000140910883. WBC and AJJ also acknowledge prior DARPA support through contract no. W911NF-08-C-0038 and internal support funds from HRL. The authors dedicate this article to the memory of Anthony G. Evans, who inspired and directed much of the work on periodic cellular materials. His leadership, enthusiasm, and mentorship are dearly missed.

References

[1]L. J. Gibson and M. F. Ashby, *Cellular Solids: Structure and Properties.* Cambridge University Press, Cambridge, UK, 1999.
[2]M. F. Ashby, A. G. Evans, N. A. Fleck, L. J. Gibson, J. W. Hutchinson, and H. N. G. Wadley, Metal Foams: A Design Guide. Butterworth-Heinemann, Oxford, UK, 2000.
[3]T. J. Lu, L. Valdevit, and A. G. Evans, "Active Cooling by Metallic Sandwich Structures With Periodic Cores," *Prog. Mater. Sci.,* **50** [7] 789–815 (2005).
[4]A. G. Evans, J. W. Hutchinson, N. A. Fleck, M. F. Ashby, and H. N. G. Wadley, "The Topological Design of Multifunctional Cellular Metals," *Prog. Mater. Sci.,* **46** [3-4] 309–27 (2001).
[5]H. N. G. Wadley, "Cellular Metals Manufacturing," *Adv. Eng. Mater.,* **4** [10] 726–33 (2002).
[6]H. N. G. Wadley, N. A. Fleck, and A. G. Evans, "Fabrication and Structural Performance of Periodic Cellular Metal Sandwich Structures," *Compos. Sci. Technol.,* **63** [16] 2331–43 (2003).
[7]H. Bart-Smith, A. F. Bastawros, D. R. Mumm, A. G. Evans, D. J. Sypeck, and H. N. G. Wadley, "Compressive Deformation and Yielding Mechanisms in Cellular Al Alloys Determined Using X-ray Tomography and Surface Strain Mapping," *Acta Mater.,* **46** [10] 3583–92 (1998).
[8]H. Bart-Smith, J. W. Hutchinson, and A. G. Evans, "Measurement and Analysis of the Structural Performance of Cellular Metal Sandwich Construction," *Int. J. Mech. Sci.,* **43** [8] 1945–63 (2001).
[9]A. F. Bastawros, H. Bart-Smith, and A. G. Evans, "Experimental Analysis of Deformation Mechanisms in a Closed-Cell Aluminum Alloy Foam," *J. Mech. Phys. Solids,* **48** [2] 301–22 (2000).
[10]H. Bart-Smith, J. W. Hutchinson, N. A. Fleck, and A. G. Evans, "Influence of Imperfections on the Performance of Metal Foam Core Sandwich Panels," *Int. J. Solids Struct.,* **39** [19] 4999–5012 (2002).
[11]N. Wicks and J. W. Hutchinson, "Optimal Truss Plates," *Int. J. Solids Struct.,* **38** [30–31] 5165–83 (2001).
[12]L. Valdevit, J. W. Hutchinson, and A. G. Evans, "Structurally Optimized Sandwich Panels With Prismatic Cores," *Int. J. Solids Struct.,* **41** [18–19] 5105–24 (2004).

[13]L. Valdevit, Z. Wei, C. Mercer, F. W. Zok, and A. G. Evans, "Structural Performance of Near-Optimal Sandwich Panels With Corrugated Cores," *Int. J. Solids Struct.*, **43** [16] 4888–905 (2006).

[14]N. Wicks and J. W. Hutchinson, "Performance of Sandwich Plates With Truss Cores," *Mech. Mater.*, **36** [8] 739–51 (2004).

[15]F. W. Zok, H. Rathbun, M. He, E. Ferri, C. Mercer, R. M. McMeeking, and A. G. Evans, "Structural Performance of Metallic Sandwich Panels With Square Honeycomb Cores," *Philos. Mag.*, **85** [26–27] 3207–34 (2005).

[16]F. W. Zok, H. J. Rathbun, Z. Wei, and A. G. Evans, "Design of Metallic Textile Core Sandwich Panels," *Int. J. Solids Struct.*, **40** [21] 5707–22 (2003).

[17]F. W. Zok, S. A. Waltner, Z. Wei, H. J. Rathbun, R. M. McMeeking, and A. G. Evans, "A protocol for Characterizing the Structural Performance of Metallic Sandwich Panels: Application to Pyramidal Truss Cores," *Int. J. Solids Struct.*, **41** [22–23] 6249–71 (2004).

[18]L. Valdevit, A. Pantano, H. A. Stone, and A. G. Evans, "Optimal Active Cooling Performance of Metallic Sandwich Panels With Prismatic Cores," *Int. J. Heat Mass Transfer*, **49** [21–22] 3819–30 (2006).

[19]L. Valdevit, N. Vermaak, F. W. Zok, and A. G. Evans, "A Materials Selection Protocol for Lightweight Actively Cooled Panels," *J. Appl. Mech.*, **75**, 061022 (2008).

[20]N. Vermaak, L. Valdevit, and A. G. Evans, "Influence of Configuration on Materials Selection for Actively Cooled Combustors," *J. Propul. Power*, **26** [2] 295–302 (2010).

[21]N. Vermaak, L. Valdevit, and A. G. Evans, "Materials Property Profiles for Actively Cooled Panels: An Illustration for Scramjet Applications," *Metall. Mater. Trans. A.*, **40A** [4] 877–90 (2009).

[22]C. S. Roper, "Multiobjective Optimization for Design of Multifunctional Sandwich Panel Heat Pipes With Micro-Architected Truss Cores," *Int. J. Heat Fluid Flow*, **32**, 239–48 (2011).

[23]H. J. Rathbun, D. D. Radford, Z. Xue, M. Y. He, J. Yang, V. Deshpande, N. A. Fleck, J. W. Hutchinson, F. W. Zok, and A. G. Evans, "Performance of Metallic Honeycomb-Core Sandwich Beams Under Shock Loading," *Int. J. Solids Struct.*, **43** [6] 1746–63 (2006).

[24]M. T. Tilbrook, V. S. Deshpande, and N. A. Fleck, "The Impulsive Response of Sandwich Beams: Analytical and Numerical Investigation of Regimes of Behaviour," *J. Mech. Phys. Solids*, **54** [11] 2242–80 (2006).

[25]H. N. G. Wadley, K. P. Dharmasena, M. Y. He, R. M. McMeeking, A. G. Evans, T. Bui-Thanh, and R. Radovitzky, "An active Concept for Limiting Injuries Caused by Air Blasts," *Int. J. Impact Eng.*, **37** [3] 317–23 (2010).

[26]Z. Y. Xue and J. W. Hutchinson, "Crush Dynamics of Square Honeycomb Sandwich Cores," *Int. J. Numer. Meth. Eng.*, **65** [13] 2221–45 (2006).

[27]L. H. Han, T. J. Lu, and A. G. Evans, "Optimal Design of a Novel High Authority SMA Actuator," *Mech. Adv. Mater. Struct.*, **12** [3] 217–27 (2005).

[28]R. G. Hutchinson, N. Wicks, A. G. Evans, N. A. Fleck, and J. W. Hutchinson, "Kagome Plate Structures for Actuation," *Int. J. Solids Struct.*, **40** [25] 6969–80 (2003).

[29]T. J. Lu, J. W. Hutchinson, and A. G. Evans, "Optimal Design of a Flexural Actuator," *J. Mech. Phys. Solids*, **49** [9] 2071–93 (2001).

[30]S. Lucato, J. Wang, P. Maxwell, R. M. McMeeking, and A. G. Evans, "Design and Demonstration of a High Authority Shape Morphing Structure," *Int. J. Solids Struct.*, **41** [13] 3521–40 (2004).

[31]N. Wicks and J. W. Hutchinson, "Sandwich Plates Actuated by a Kagome Planar Truss," *J. Appl. Mech.-Trans. Asme*, **71** [5] 652–62 (2004).

[32]W. D. Macdonald and T. W. Eagar, "Transient Liquid-Phase Bonding," *Annu. Rev. Mater. Sci*, **22**, 23–46 (1992).

[33]Wall-Colmonoy Corporation. Available at http://www.wallcolmonoy.com

[34]D. T. Queheillalt and H. N. G. Wadley, "Titanium alloy Lattice Truss Structures," *Mater. Des.*, **30** [6] 1966–75 (2009).

[35]S. J. Johnson, B. Tryon, and T. M. Pollock, "Post-Fabrication Vapor Phase Strengthening of Nickel-Based Sheet Alloys for Thermostructural Panels," *Acta Mater.*, **56** [17] 4577–84 (2008).

[36]S. W. Godfrey, "Optimal Design of Orthotropic Fiber-Composite Corrugated-Core Sandwich Panels Under Axial Compression"; MS Thesis, Mechanical and Aerospace Engineering, University of California, Irvine, (2010).

[37]B. P. Russell, V. S. Deshpande, and H. N. G. Wadley, "Quasi-Static Deformation and Failure Modes of Composite Square Honeycombs," *J. Mech. Mater. Struct.*, **3** [7] 1315–40 (2008).

[38]A. P. Mouritz and B. N. Cox, "A Mechanistic Interpretation of the Comparative In-Plane Mechanical Properties of 3D Woven, Stitched and Pinned Composites," *Compos. Part A - Appl. S.*, **41** [6] 709–28 (2001).

[39]V. S. Deshpande, N. A. Fleck, and M. F. Ashby, "Effective Properties of the Octet-Truss Lattice Material," *J. Mech. Phys. Solids*, **49** [8] 1747–69 (2001).

[40]N. Vermaak, "Thermostructural Design Tools for Hypersonic Vehicles," Ph.D. Thesis, Materials Department, University of California, Santa Barbara, 2010.

[41]D. Queheillalt and H. Wadley, "Cellular Metal Lattices With Hollow Trusses," *Acta Mater.*, **53** [2] 303–13 (2005).

[42]Q. Li, E. Y. Chen, D. R. Bice, and D. C. Dunand, "Mechanical Properties of Cast Ti-6Al-2Sn-4Zr-2Mo Lattice Block Structures," *Adv. Eng. Mater.*, **10** [10] 939–42 (2008).

[43]F. P. W. Melchels, J. Feijen, and D. W. Grijpma, "A Review on Stereolithography and its Applications in Biomedical Engineering," *Biomaterials*, **31** [24] 6121–30 (2010).

[44]H.-B. Sun and S. Kawata, "Two-Photon Photopolymerization and 3D Lithographic Microfabrication," pp. 169–273 in *NMR, 3D Analysis, Photopolymerization*, Vol. 190. Edited by N. Fatkullin. Springer, Berlin, Germany, 2004.

[45]A. J. Jacobsen, W. Barvosa-Carter, and S. Nutt, "Micro-Scale Truss Structures Formed From Self-Propagating Photopolymer Waveguides," *Adv. Mater.*, **19** [22] 3892–6 (2007).

[46]A. J. Jacobsen, J. A. Kolodziejska, R. Doty, K. D. Fink, C. Zhou, C. S. Roper, and W. B. Carter, "Interconnected Self-Propagating Photopolymer Waveguides: An Alternative to Stereolithography for Rapid Formation of Lattice-Based Open-Cellular Materials" in *Twenty First Annual International Solid Freeform Fabrication Symposium - An Additive Manufacturing Conference*, University of Texas, Austin (2010).

[47]J. Banhart, "Manufacture, Characterisation and Application of Cellular Metals and Metal Foams," *Prog. Mater. Sci.*, **46**, 559–632 (2001).

[48]M. Scheffler and P. Colombo, *"Cellular Ceramics: Structure, Manufacturing, Properties and Applications.* WILEY, Weinheim, 2005.

[49]A. G. Evans, M. Y. He, V. S. Deshpande, J. W. Hutchinson, A. J. Jacobsen, and W. B. Carter, "Concepts for Enhanced Energy Absorption Using Hollow Micro-Lattices," *Int. J. Impact Eng.*, **37** [9] 947–59 (2010).

[50]A. J. Jacobsen, S. Mahoney, W. B. Carter, and S. Nutt, "Vitreous Carbon Micro-Lattice Structures," *Carbon*, **49** [3] 1025–32 (2011).

[51]V. S. Deshpande and N. A. Fleck, "Collapse of Truss Core Sandwich Beams in 3-Point Bending," *Int. J. Solids Struct.*, **38** [36–37] 6275–305 (2001).

[52]H. G. Allen and P. S. Bulson, *"Background to Buckling."* McGraw-Hill, London, 1980.

[53]M. R. Begley and J. W. Hutchinson, "The Mechanics of Size-Dependent Indentation," *J. Mech. Phys. Solids*, **46** [10] 2049–68 (1998).

[54]N. A. Fleck, G. M. Muller, M. F. Ashby, and J. W. Hutchinson, "Strain Gradient Plasticity – Theory and Experiment," *Acta Metall. Mater.*, **42** [2] 475–87 (1994).

[55]J. S. Stolken and A. G. Evans, "A microbend Test Method for Measuring the Plasticity Length Scale," *Acta Mater.*, **46** [14] 5109–15 (1998).

[56]N. J. Petch, "The Cleavage Strength of Polycrystals," *J. Iron Steel I.*, **174** [1] 25–8 (1953).

[57]D. J. Lloyd, "Particle Reinforced Aluminum and Magnesium Matrix Composites," *Int. Mater. Rev*, **39**, 1–23 (1994).

[58]N. A. Fleck and J. W. Hutchinson, "Strain Gradient Plasticity," *Adv. Appl. Mech.*, **33**, 295–361 (1997).

[59]N. A. Fleck and J. W. Hutchinson, "A Reformulation of Strain Gradient Plasticity," *J. Mech. Phys. Solids*, **49** [10] 2245–71 (2001).

[60]H. Gao, Y. Huang, W. D. Nix, and J. W. Hutchinson, "Mechanism-Based Strain Gradient Plasticity – I. Theory," *J. Mech. Phys. Solids*, **47** [6] 1239–63 (1999).

[61]Y. Huang, H. Gao, W. D. Nix, and J. W. Hutchinson, "Mechanism-Based Strain Gradient Plasticity – II. Analysis," *J. Mech. Phys. Solids*, **48** [1] 99–128 (2000).

[62]W. D. Nix and H. J. Gao, "Indentation Size Effects in Crystalline Materials: A Law for Strain Gradient Plasticity," *J. Mech. Phys. Solids*, **46** [3] 411–25 (1998).

[63]A. G. Evans and J. W. Hutchinson, "A Critical Assessment of Theories of Strain Gradient Plasticity," *Acta Mater.*, **57** [5] 1675–88 (2009).

[64]M. F. Ashby, "Deformation of Plastically Non-Homogeneous Materials," *Philos. Mag.*, **21** [170] 399 (1970).

[65]D. M. Dimiduk, M. D. Uchic, and T. A. Parthasarathy, "Size-Affected Single-Slip Behavior of Pure Nickel Microcrystals," *Acta Mater.*, **53**, 4065–77 (2005).

[66]Z. W. Shan, R. Mishra, S. A. Syed, O. L. Warren, and A. M. Minor, "Mechanical Annealing and Source-limited Deformation in Submicron-diameter Ni Crystals," *Nat. Mater.*, **7** [116] 115–9 (2008).

[67]M. D. Uchic, D. M. Dimiduk, J. N. Florando, and W. D. Nix, "Sample Dimensions Influence Strength and Crystal Plasticity," *Science*, **305** [5686] 986–9 (2004).

[68]J. R. Greer, W. C. Oliver, and W. D. Nix, "Size Dependence of Mechanical Properties of Gold at the Micron Scale in the Absence of Strain Gradients," *Acta Mater.*, **53**, 1821–30 (2005).

[69]J. R. Greer and W. D. Nix, "Nanoscale Gold Pillars Strengthened through Dislocation Starvation," *Phys. Rev. B*, **73**, 245410–6 (2006).

[70]C. A. Volkert and E. T. Lilleodden, "Size Effects in the Deformation of Sub-Micron Au Columns," *Philos. Mag.*, **86**, 5567–79 (2006).

[71]A. Budiman, S. Han, J. R. Greer, N. Tamura, J. Patel, and W. D. Nix, "A Search for Evidence of Strain Gradient Hardening in Au Submicron Pillars Under Uniaxial Compression Using Synchrotron X-ray Microdiffraction," *Acta Mater.*, **56** [3] 602–8 (2007).

[72]D. Kiener, C. Motz, M. Rester, M. Jenko, and G. Dehm, "FIB Damage of Cu and Possible Consequences for Miniaturized Mechanical Tests," *Mater. Sci. Eng. A*, **459** [1–2] 262–72 (2006).

[73]G. Dehm, "Miniaturized Single-Crystalline fcc Metals Deformed in Tension: New Insights in Size-Dependent Plasticity," *Prog. Mater. Sci.*, **54** [6] 664–88 (2009).

[74]D. Kiener, C. Motz, T. Schoberl, M. Jenko, and G. Dehm, "Determination of Mechanical Properties of Copper at the Micron Scale," *Adv. Eng. Mater.*, **8** 1119–25 (2006).

[75]D. Kiener, W. Grosinger, G. Dehm, and R. Pippan, "A Further Step Towards an Understanding of Size-Dependent Crystal Plasticity: In Situ Tension Experiments of Miniaturized Single-Crystal Copper Samples," *Acta Mater.*, **56**, 580–92 (2008).

[76]D. Kiener, W. Grosinger, and G. Dehm, "On the Importance of Sample Compliance in Uniaxial Microtesting," *Scr. Mater.*, **60** [3] 148–51 (2009).

[77]D. Kiener, C. Motz, and G. Dehm, "Micro-Compression Testing: A Critical Discussion of Experimental Constraints," *Mater. Sci. Eng. A*, **505**, 79–87 (2009).

[78]R. Maass, S. Van Petegem, D. Grolimund, H. Van Swygenhoven, D. Kiener, and G. Dehm, "Crystal Rotation in Cu Single Crystal Micropillars:

In Situ Laue and Electron Backscatter Diffraction," *Appl. Phys. Lett.*, 017905 (2008).

[79]A. T. Jennings and J. R. Greer, "Tensile Deformation of Electroplated Copper Nanopillars," *Phil. Mag. A.*, **91** [7–9] 1108–20 (2010).

[80]A. T. Jennings, M. J. Burek, and J. R. Greer, "Size Effects in Single Crystalline cu Nano-Pillars Fabricated Without the Use of Fib," *Phys. Rev. Lett.*, **104**, 135503 (2010).

[81]K. S. Ng and A. H. W. Ngan, "Breakdown of Schmid's Law in Micropillars," *Scr. Mater.*, **59** [7] 796–9 (2008).

[82]K. S. Ng and A. H. W. Ngan, "Effects of Trapping Dislocations Within Small Crystals on Their Deformation Behavior," *Acta Mater.*, **57**, 4902–10 (2009).

[83]H. Bei, E. P. George, and G. M. Pharr, "Small-Scale Mechanical Behavior of Intermetallics and Their Composites," *Mater. Sci. Eng. A*, **483–484**, 218–22 (2008).

[84]H. Bei, S. Shim, G. M. Pharr, and E. P. George, "Effects of Pre-Strain on the Compressive Stress–Strain Response of Mo-Alloy Single-Crystal Micropillars," *Acta Mater.*, **56** [17] 4762–70 (2008).

[85]H. Bei and S. Shim, "Effects of Focused Ion Beam Milling on the Mechanical Behavior of a Molybdenum-Alloy Single Crystal," *Appl. Phys. Lett.*, **91** [11] 11915 (2007).

[86]J.-Y. Kim, D. Jang, and J. R. Greer, "Insights into Deformation Behavior and Microstructure Evolution in Nb Single Crystalline Nano-Pillars Under Uniaxial Tension and Compression," *Scr. Mater.*, **61** [3] 300–3 (2009).

[87]J. Y. Kim and J. R. Greer, "Tensile and Compressive Behavior of Gold and Molybdenum Single Crystals at the Nano-Scale," *Acta Mater.*, **57** [17] 5245–53 (2009).

[88]J. Y. Kim, D. C. Jong, and J. R. Greer, "Tensile and Compressive Behavior of Tungsten, Molybdenum, Tantalum and Niobium at the Nanoscale," *Acta Mater.*, **58** [7] 2355–63 (2010).

[89]A. S. Schneider, D. Kaufmann, B. G. Clark, C. P. Frick, P. A. Gruber, R. Monig, O. Kraft, and E. Arzt, "Correlation Between Critical Temperature and Strength of Small-Scale bcc Pillars," *Phys. Rev. Lett.*, **103**, 105501 (2009).

[90]E. Lilleodden, "Microcompression Study of Mg (0001) Single Crystal," *Scr. Mater.*, **62** [8] 532–5 (2010).

[91]C. M. Byer, B. Li, B. Cao, and K. T. Ramesh, "Microcompression of Single-Crystal Magnesium," *Scr. Mater.*, **62** [8] 536–9 (2010).

[92]Q. Yu, Z.-W. Shan, J. Li, X. Huang, L. Xiao, J. Sun, and E. Ma, "Strong Crystal Size Effect on Deformation Twinning," *Nature*, **463**, 335–8 (2010).

[93]G. Lee, J.-Y. Kim, A. S. Budiman, N. Tamura, M. Kunz, K. Chen, M. J. Burek, J. R. Greer, and T. Y. Tsui, "Fabrication, Structure, and Mechanical Properties of Indium Nanopillars," *Acta Mater.*, **58** [4] 1361–8 (2010).

[94]E. A. Withey, A. M. Minor, Jr, D. C. Morris, J. W. Morris, and S. Kuramot , "The Deformation of Gum Metal Through In Situ Compression of Nanopillars," *Acta Mater.*, **58** [7] 2652–65 (2010).

[95]E. A. Withey, J. Ye, A. M. Minor, S. Kuramoto, D. C. Chrzan, and J. W. Morris Jr, "Nanomechanical Testing of Gum Metal," *Exp. Mech.*, **50** 37–45 (2010).

[96]A. Rinaldi, P. Peralta, C. Friesen, and K. Sieradzki, "Sample-Size Effects in the Yield Behavior of Nanocrystalline Nickel," *Acta Mater.*, **56** 511–7 (2008).

[97]D. Dang and J. R. Greer, "Size-Induced Weakening and Grain Boundary-Assisted Deformation in 60nm Grained Ni Nanopillars," *Scripta Mater.*, **64**, 77–80 (2011).

[98]B. G. Clark, D. S. Gianola, O. Kraft, and C. P. Frick, "Size Independent Shape Memory Behavior of Nickel-Titanium," *Adv. Eng. Mater.*, **12** [8] 805–15 (2010).

[99]C. P. Frick, S. Orso, and E. Arzt, "Loss of Pseudoelasticity in Nickel-Titanium Submicron Compression Pillars," *Acta Mater.*, **55**, 3845–55 (2007).

[100]C. P. Frick, B. G. Clark, S. Orso, A. S. Schneider, and E. Arzt, "Size Effect on Strength an Strain Hardening of Small-Scale (111), Nickel Compression Pillars," *Mater. Sci. Eng. A*, **489** [1–2] 319–29 (2008).

[101]C. P. Frick, B. G. Clark, S. Orso, P. Sonnweber-Ribic, and E. Arzt, "Orientation-Independent Pseudoelasticity in Small-Scale NiTi Compression Pillars," *Scr. Mater.*, **59**, 7–10 (2008).

[102]C. P. Frick, B. G. Clark, A. S. Schneider, R. Maaß, S. Van Petegem, and H. Van Swygenhoven, "On the Plasticity of Small-Scale Nickel-Titanium Shape Memory Alloys," *Scr. Mater.*, **62**, 492–5 (2010).

[103]J. M. San Juan, M. L. Nó, and C. A. Schuh, "Nanoscale Shape-Memory Alloys for Ultrahigh Mechanical Damping," *Nat. Nanotechnol.*, **4**, 415–9 (2009).

[104]J. M. San Juan, M. L. Nó, and C. A. Schuh, "Superelasticity and Shape Memory in Micro- and Nanometer-Scale Pillars," *Adv. Mater.*, **20**, 272–8 (2008).

[105]D. Jang and J. R. Greer, "Transition From a Strong-Yet-Brittle to a Stronger-and-Ductile State by Size Reduction of Metallic Glasses," *Nat. Mater.*, **9**, 215–9 (2010).

[106]C. A. Volkert, A. Donohue, and F. Spaepen, "Effect of Sample Size on Deformation in Amorphous Metals," *J. Appl. Phys.*, **103** [8] 083539 (2008).

[107]C. Q. Chen, Y. T. Pei, and J. T. M. De Hosson, "Strength of Submicrometer Diameter Pillars of Metallic Glasses Investigated with in Situ Transmission Electron Microscopy," *Philos. Mag. Lett.*, **89** [10] 633–40 (2009).

[108]M. D. Uchic, P. A. Shade, and D. M. Dimiduk, "Plasticity of Micrometer-Scale Single Crystals in Compression," *Annu. Rev. Mater. Res.*, **39** [1] 361–86 (2009).

[109]R. Dou and B. Derby, "A Universal Scaling Law for the Strength of Metal Micropillars and Nanowires," *Scr. Mater.*, **61** [5] 524–7 (2009).

[110]C. V. Thompson, "The Yield Stress of Polycrystalline Thin Films," *J. Mater. Res.*, **8**, 237 (1993).

[111]B. von Blanckenhagen, P. Gumbsch, and E. Arzt, "Dislocation Sources and the Flow Stress of Polycrystalline Thin Metal Films," *Philos. Mag. Lett.*, **83** [1] 1–8 (2003).

[112]M. S. De Guzman, G. Neubauer, P. Flinn, and W. D. Nix, "The Role of Indentation Depth on the Measured Hardness of Materials," *Mater. Res. Soc. Symp. Proc.*, **308** 613 (1993).

[113]J. R. Greer, W. C. Oliver, and W. D. Nix, "Size Dependence of Mechanical Properties of Gold at the Micron Scale in the Absence of Strain Gradients," *Acta Mater.*, **53** [6] 1821–30 (2005).

[114]K. S. Ng and A. H. W. Ngan, "Stochastic Nature of Plasticity of Aluminum Micro-Pillars," *Acta Mater.*, **56** [8] 1712–20 (2008).

[115]S. I. Rao, D. M. Dimiduk, T. A. Parthasarathy, M. D. Uchic, M. Tang, and C. Woodward, "Athermal Mechanisms of Size-Dependent Crystal Flow Gleaned From Three-Dimensional Discrete Dislocation Simulations," *Acta Mater.*, **56** [13] 3245–59 (2008).

[116]D. M. Norfleet, D. M. Dimiduk, S. J. Polasik, M. D. Uchic, and M. J. Mills, "Dislocation Structures and Their Relationship to Strength in Deformed Nickel Microcrystals," *Acta Mater.*, **56** [13] 2988–3001 (2008).

[117]T. A. Parthasarathy, S. I. Rao, D. M. Dimiduk, M. D. Uchic, and D. R. Trinkle, "Contribution to size effect of yield strength from the stochastics of dislocation source lengths in finite samples," *Scr. Mater.*, **56** [4] 313–6 (2007).

[118]S. I. Rao, D. M. Dimiduk, M. Tang, T. A. Parthasarathy, M. D. Uchic, and C. Woodward, "Estimating the Strength of Single-Ended Dislocation Sources in Micron-Sized Single Crystals," *Philos. Mag. A*, **87** [30], 4777–94 (2007).

[119]D. M. Dimiduk, C. Woodward, R. LeSar, and M. D. Uchic, "Scale-Free Intermittent Flow in Crystal Plasticity," *Science*, **312**, 1188–90 (2006).

[120]O. Kraft, P. A. Gruber, R. Monig, and D. Weygand, "Plasticity in Confined Dimensions," *Annu. Rev. Mater. Res.*, **40**, 293–317 (2010).

[121]A. T. Jennings, M. J. Burek, and J. R. Greer, "Size Effects in Single Crystalline cu Nano-Pillars Fabricated Without the Use of Fib," *Phys. Rev. Lett.*, **104**, 135503 (2010).

[122]S. Lee, S. Han, and W. D. Nix, "Uniaxial Compression of fcc Au Nanopillars on an MgO Substrate: The Effects of Prestraining and Annealing," *Acta Mater.*, **57** [15] 4404–15 (2009).

[123]S. Shim, H. Bei, M. K. Miller, G. M. Pharr, and E. P. George, "Effects of Focused Ion Beam Milling on the Compressive Behavior of Directionally Solidified Micropillars and the Nanoindentation Response of an Electropolished Surface," *Acta Mater.*, **57**, 503–10 (2009).

[124]J. A. El-Awady, C. Woodward, D. M. Dimiduk, and N. M. Ghoniem, "Effects of Focused Ion Beam Induced Damage on the Plasticity of Micropillars," *Phys. Rev. B*, **80**, 104104 (2009).

[125]R. Ikeda, H. Ogi, T. Ogawa, and M. Takemoto, "Mechanical Properties of CVD Synthesized Polycrystalline Diamond," *Ind. Diamond Quat.*, **69**, 35–8 (2009).

[126]N. A. Fleck, V. Deshpande, and M. F. Ashby, "Micro-Architected Materials: Past, Present and Future," *Proc. Royal Soc. A*, **466**, 2495–516 (2010).

[127]D. J. Bell, T. J. Lu, N. A. Fleck, and S. M. Spearing, "MEMS Actuators and Sensors: Observations on Their Performance and Selection for Purpose," *J. Micromech. Microeng.*, **15** [7] S153–64 (2005).

[128]M. A. Haque and T. Saif, "Mechanical Testing at the Micro/Nano Scale." in *Springer Handbook of Experimental Solid Mechanics*. Edited by W. N. Sharpe. Springer, New York, 2008.

[129]A. Torrents, K. Azgin, S. W. Godfrey, E. S. Topalli, T. Akin, and L. Valdevit, "MEMS Resonant Load Cells for Micro-Mechanical Test Frames: Feasibility Study and Optimal Design," *J. Micromech. Microeng.*, **20** 125004 (17pp) (2010).

[130]F. Giessibl, "Advances in Atomic Force Microscopy," *Rev. Mod. Phys.*, **75** [3] 949–83 (2003).

[131]P. Hinterdorfer and Y. Dufrene, "Detection and Localization of Single Molecular Recognition Events Using Atomic Force Microscopy," *Nat. Methods*, **3** [5] 347–55 (2006).

[132]G. Schitter and M. J. Rost, "Scanning Probe Microscopy at Video-Rate," *Mater. Today*, **11**, 40–8 (2008).

[133]M. D. Kriese, W. W. Gerberich, and N. R. Moody, "Quantitative Adhesion Measures of Multilayer Films: Part I. Indentation Mechanics," *J. Mater. Res.*, **14** [7] 3007–18 (1999).

[134]M. D. Kriese, W. W. Gerberich, and N. R. Moody, "Quantitative Adhesion Measures of Multilayer Films: Part II. Indentation of W/Cu, W/W, Cr/W," *J. Mater. Res.*, **14** [7] 3019–26 (1999).

[135]E. T. Lilleodden, W. Bonin, J. Nelson, J. T. Wyrobek, and W. W. Gerberich, "In-Situ Imaging of MU-N Load Indents into GAAS," *J. Mater. Res.*, **10** [9] 2162–5 (1995).

[136]W. D. Nix, "Elastic and Plastic Properties of Thin Films on Substrates: Nanoindentation Techniques," *Mat. Sci. Eng. A - Struct.*, **234**, 37–44 (1997).

[137]W. C. Oliver and G. M. Pharr, "An Improved Technique for Determining Hardness and Elastic Modulus Using Load and Displacement Sensing Indentation Experiments," *J. Mater. Res.*, **7** [6] 1564–83 (1992).

[138]W. C. Oliver and G. M. Pharr, "Measurement of Hardness and Elastic Modulus by Instrumented Indentation: Advances in Understanding and Refinements to Methodology," *J. Mater. Res.*, **19** [1] 3–20 (2004).

[139]J. R. Greer, J.-Y. Kim, and M. J. Burek, "The In-Situ Mechanical Testing of Nanoscale Single-Crystalline Nanopillars," *J. Mater.*, **61** [12] 19–25 (2009).

[140]J.-Y. Kim and J. R. Greer, "Tensile and Compressive Behavior of Gold and Molybdenum Single Crystals at the Nano-Scale," *Acta Mater.*, **57**, 5245–53 (2009).

[141]Z. Y. Xue and J. W. Hutchinson, "Constitutive Model for Quasi-Static Deformation of Metallic Sandwich Cores," *Int. J. Numer. Meth. Eng.*, **61** [13] 2205–38 (2004).

[142]J. E. Angelo, N. R. Moody, and M. I. Baskes, "Trapping of Hydrogen to Lattice-Defects in Nickel," *Modelling Simul. Mater. Sci. Eng.*, **3** [3] 289–307 (1995).

[143]M. I. Baskes, "Modified Embedded-Atom Potentials for Cubic Materials," *Phys. Rev. B*, **46** [5] 2727–42 (1992).

[144]W. Xiao, M. I. Baskes, and K. Cho, "MEAM Study of Carbon Atom Interaction With Ni Nano Particle," *Surf. Sci.*, **603** [13] 1985–98 (2009).

[145]Y. Jiang, Y. G. Wei, J. R. Smith, J. W. Hutchinson, and A. G. Evans, "First Principles Based Predictions of the Toughness of a Metal/Oxide Interface," *Int. J. Mater. Res.*, **101** [1] 8–15 (2010).

[146]V. S. Deshpande, A. Needleman, and E. Van der Giessen, "Plasticity Size Effects in Tension and Compression of Single Crystals," *J. Mech. Phys. Solids*, **53** [12] 2661–91 (2005).

[147]H. Tang, K. W. Schwarz, and H. D. Espinosa, "Dislocation-Source Shutdown and the Plastic Behavior of Single-Crystal Micropillars," *Phys. Rev. Lett.*, **100**, 185503 (2008).

[148]E. Van der Giessen and A. Needleman, "Discrete Dislocation Plasticity – A Simple Planar Model," *Modelling Simul. Mater. Sci. Eng.*, **3** [5] 689–735 (1995).

[149]M. D. Uchic, P. A. Shade, and D. M. Dimiduk, "Plasticity of Micrometer-Scale Single Crystals in Compression," *Annu. Rev. Mater. Res.*, **39**, 361–86 (2009).

[150]J. Wang, A. G. Evans, K. Dharmasena, and H. N. G. Wadley, "On the Performance of Truss Panels With Kagome Cores," *Int. J. Solids Struct.*, **40** [25] 6981–8 (2003).

[151]M. P. Bendsoe and O. Sigmund, "*Topology Optimization*," pp. 370. Springer, Berlin, Germany, (2002).

[152]P. W. Christensen and A. Klabring, "*An Introduction to Structural Optimization*," pp. 214. Springer, Berlin, Germany, (2008).

[153]O. Sigmund and S. Torquato, "Design of Materials With Extreme Thermal Expansion Using a Three-Phase Topology Optimization Method," *J. Mech. Phys. Solids*, **45** [6] 1037–67 (1997).

[154]O. Sigmund and S. Torquato, "Design of Smart Composite Materials Using Topology Optimization," *Smart Mater. Struct.*, **8** [3] 365–79 (1999).

[155]S. Hyun and S. Torquato, "Optimal and Manufacturable Two-Dimensional, Kagome-Like Cellular Solids," *J. Mater. Res.*, **17** [1] 137–44 (2002).

[156]A. J. Jacobsen, W. Barvosa-Carter, and S. Nutt, "Micro-Scale Truss Structures With Three-Fold And Six-Fold Symmetry Formed From Self-Propagating Polymer Waveguides," *Acta Mater.*, **56** [11] 2540–8 (2008).

[157]J. R. Greer and J. T. M. De Hosson, "Plasticity in Small-Sized Metallic Systems: Intrinsic Versus Extrinsic Size Effects," *Prog. Mater. Sci.*, **56**, 654–724 (2011).

[158]J.-Y. Kim and J. R. Greer, "Size-Dependent Mechanical Properties of Mo Nanopillars," *Appl. Phys. Lett.*, **93**, 101916 (2008).

[159]A. J. Jacobsen, W. Barvosa-Carter, and S. Nutt, "Compression Behavior of Micro-Scale Truss Structures Formed From Self-Propagating Polymer Waveguides," *Acta Mater.*, **55** [20] 6724–33 (2007).

[160]K. Azgin, C. Ro, A. Torrents, T. Akin, and L. Valdevit, "A Resonant Tuning Fork Sensor with Unprecedented Combination of Resolution and Range." in *IEEE MEMS Conference*, Paper # 0700. Cancun, Mexico, 2011.

J. Am. Ceram. Soc., **94** [S1] S35–S41 (2011)

DOI: 10.1111/j.1551-2916.2011.04561.x

© 2011 The American Ceramic Society

journal

A Shape-Morphing Ceramic Composite for Variable Geometry Scramjet Inlets

Richard Miles,[†] Phillip Howard, Christopher Limbach, and Syed Zaidi

Department of Mechanical and Aerospace Engineering, Princeton University, Princeton, New Jersey 08544

Sergio Lucato, Brian Cox, and David Marshall

Teledyne Scientific Company, Thousand Oaks, California 91360

Angel M. Espinosa, and Dan Driemeyer

Boeing Phantom Works, Saint Louis, Missouri 63166

The development of ceramic composites with three-dimensional fiber reinforcement architectures formed by textile methods has led to the potential for active shape-morphing surfaces that can operate in high temperature and variable pressure environments. This technology is of particular interest for hypersonic applications, where SCRAM jet engines require variable inlet geometry to achieve efficient flight over realistic flight profiles and variable flight conditions. The experiments reported here show that significant shape morphing can be achieved and good control of the shape sustained even in the presence of large temperature and pressure gradients. Experiments were carried out using a subscale morphing hypersonic inlet with rectangular cross-section in a Mach 8 wind tunnel facility with a total temperature of 800 K. The upper surface of the inlet consisted of a C–SiC composite plate (0.7 mm thick, 37.5 cm long, and 11 cm wide) connected to five actuators through a triangular truss support structure. The lower surface was a flat plate instrumented with an array of pressure taps along the flow centerline. As the shape varied, the surface contour was reliably controlled for high efficiency, low loss compression. A factor of six inlet area ratio variation was achieved and good agreement with model predictions was observed.

I. Introduction

HIGH-TEMPERATURE shape-morphing structural materials could potentially revolutionize the design of air breathing hypersonic vehicles. The performance of the RAM/SCRAM engine is strongly affected by shapes and contours of the compression ramp and inner surfaces, including the inlet, the isolator, the combustor and the nozzle.[1,2] A contour that is designed for operation at one Mach number can be very inefficient at off-design Mach numbers. Even weak shock waves can substantially reduce engine performance, while instabilities in the flow can lead to upstream shock motion and engine unstart. High-temperature shape-morphing materials could (i) enable continuous variation of the inlet area ratio to accommodate changing Mach number and air density associated with climb and descent; (ii) improve engine and isolator performance by accommodating area and shape changes associated with engine mode transition from subsonic to supersonic combustion (RAM to SCRAM); (iii) optimize the compression ramp and inlet contour to minimize shock formation and reduce pressure losses; (iv) permit dynamic adaptation to compensate for flight maneuvers and correct for distortion due to thermal expansion, as well as suppress upstream shock motion to reduce the probability of engine unstart, and if unstart occurs, reduce the recovery time; and (v) improve nozzle performance by controlling the nozzle area ratio and contour for efficient thrust generation and minimum pressure loss. On both internal and external surfaces of the vehicle, high-temperature morphing materials could replace hinged control surfaces with smoothly varying surfaces, thereby eliminating flow disturbances associated with discontinuous changes in slope.[3]

High-temperature shape-morphing materials are also of interest for hypersonic and supersonic ground test facilities.[4] In this case, the shape-morphing capability would allow for dynamic optimization of the expansion contour of the facility, improving the flow quality, reducing shock and expansion waves and thus minimizing losses and improving performance. Shape morphing would also provide the opportunity for real time control of the facility Mach number by changing the expansion area ratio continuously. Low-temperature variable Mach capability has been previously demonstrated for a Mach 2.5–3.2 in-draft wind tunnel nozzle.[5] This leads to the possibility of simulating portions of the flight mission including climb and descent. Asymmetric shape morphing can be used to simulate asymmetric inlet flows for direct connect engine test applications.[6] Optimizing the performance of flow deflectors and other aero appliances and the diffuser is also important for ground test facilities because they affect the flow quality and the pressure recovery and thus the facility operating envelope and run time. High-temperature shape-morphing aero appliance surfaces and diffuser geometries could accommodate configuration variation of test articles and allow for dynamic changes to occur during the ground test, such as changes in angle of attack.

To operate in these applications, a high-temperature shape-morphing material must be both flexible and structurally stiff, with strength sufficient to withstand the stresses associated with high thermal gradients and pressure gradients. The material must be capable of withstanding temperatures in excess of 1500 K for even modest Mach numbers (Mach 8), resistant to oxidation, and resistant to thermomechanical fatigue. The purpose of this paper is to describe the design and fabrication of a shape-morphing ceramic composite (CMC) structure that satisfies these requirements and to present results of preliminary testing of its performance under hypersonic conditions, as the wall of a subscale inlet in a Mach 8 wind tunnel.

B. McMeeking—contributing editor

Manuscript No. 29043. Received December 13, 2010; approved March 9, 2011.

Funding at Princeton was provided by the Office of Naval Research Multi University Research Initiative through the University of California, Santa Barbara. Funding at TSC was provided by Boeing Phantom Works for development of the morphing CMC structures and by the Air Force Office of Scientific Research and NASA through the National Hypersonic Science Center for Materials and Structures for collaboration during testing at Princeton.

[†]Author to whom correspondence should be addressed. e-mail: miles@princeton.edu

II. CMC Structure

(1) General Design Considerations for Morphing Ceramic Composite

The challenge of designing morphing surfaces that are sufficiently flexible to change shape by bending or twisting yet capable of supporting large pressure loads has been addressed by combining thin face sheets with stiff, statically determinate truss structures.[7–10] A class of those high authority shape-morphing structures has been initially described by Lu *et al.*[11] and further generalized by Hutchinson *et al.*[10] Previous studies have also suggested that these face sheet truss structures are a lighter-weight alternative to metallic foam core structures.[12] A simple example building on Lu's work is shown in Fig. 1, where a thin face sheet attached to a triangular truss structure is actuated by heating selected truss elements with an electric current. The active elements consist of a shape memory alloy, which undergoes a reversible strain on heating and cooling. This class of morphing structures can be actuated by internal active elements, as in Fig. 1, or by forces applied externally. Previous studies have been directed at optimizing morphing structures with other truss structures capable of more intricate shape changes with metallic or polymer face sheets.[7–10]

With recent advances in forming textile-based CMCs,[10] hybrid systems consisting of a CMC face sheet attached to metallic truss structure are also feasible. In this case, with the high-temperature CMC face sheet facing the hot gas flow, the temperature of the metallic truss structure may be controlled if necessary by active or passive cooling to maintain sufficient temperature gradient across the hot face sheet. Structures of this type are enabled by several characteristics of textile-based composites. One is the ability to form mechanically robust thin skins with through-thickness fiber reinforcements that eliminate delamination as a failure mode. Thin C–SiC and SiC–SiC composite skins (thickness <1 mm) with woven reinforcement architectures have been fabricated and shown to survive combustion environments with surface temperatures of 1800 K and temperature gradients >1000 K/mm across the hot skin.[13] The face sheet of the structure in Fig. 1 consists of such a composite (SiC–SiC), with two-layer, angle-interlock reinforcement architecture.[14]

The other enabling characteristic is the ability to form arrays of connections between the backside of the hot skin and the truss structure capable of transmitting both tensile and compressive loads. A simple example of a smoothly pinned joint, from the structure in Fig. 1, is shown in Figs. 2(a) and (b). The connectors consist of fiber tows drawn out from the weave structure of the composite skin to form loops held in shape by carbon rods during weaving of the fiber preform. After using chemical vapor infiltration (CVI) to deposit a thin coating of BN on all the fibers (which serves as a debond layer to enable toughening), the preform was rigid and the carbon rods were removed. The SiC matrix was then infiltrated into the preform, a

Fig. 2. Smoothly pinned joints connecting CMC skin and truss structure: (a) schematic of weave structure; (b) configuration used in the morphing structure in Fig. 1 with superalloy struts attached; (c) configuration used in the morphing structure in Fig. 3.

superalloy wire was inserted into the loops, and the truss elements were attached to the wire by laser welding.

The maximum load capability (or lifetime in oxidizing environments) of the attachment feature shown in Fig. 2(b) is limited by damage generated by local stress concentrations at the contact between the superalloy wire and the inside of the CMC loops. A more robust design is shown in Fig. 2(c). In this case, the fiber tows forming the connectors were drawn over molybdenum tubes (outer diameter 2 mm, inner diameter 1 mm) during weaving. The tubes remained in place during subsequent processing of the composite, whereupon selected sections of the tubes between the loops were cut and removed, leaving the sections of the tubes beneath the loops as an integral part of the attachment structure. The insides of these tube sections provide a smoother, better aligned, and more damage resistant bearing surface for the superalloy rods.

The achievable deflections of a structure such as Fig. 1 are dictated by the face sheet thickness and strain limit. The critical radius of curvature ρ_c for bending of an initially flat face sheet is given by

$$\rho_c = \frac{t}{2\varepsilon_c} \tag{1}$$

where t is the face sheet thickness and ε_c the critical strain. CMCs generally show relatively high fatigue limits in cyclic loading.[15,16] If we take a conservative limit $\varepsilon_c \sim 0.25\%$, less than half of the typical failure strain for SiC–SiC and C–SiC composites, the minimum radius of curvature for a face sheet of thickness of 1 mm is ~200 mm. The radius of curvature in Fig. 1(b), with center deflection of 19 mm, is ~600 mm.

(2) Fabrication of morphing duct

A morphing CMC structure that would serve as the top wall of a model inlet duct of length 37.5 cm and width 11 cm was constructed as shown in Fig. 3. The CMC skin was fabricated using a woven preform of carbon fibers (T300-3K) with a two-layer angle-interlock structure and attachment sites formed with the method shown in Fig. 2(c). The composite was processed by CVI in two steps, the first to deposit a coating of pyrolytic carbon (~0.5 μm thickness) on individual fibers, and the second to form a matrix of SiC.

The supporting metal truss structure, consisting of a triangular array of solid superalloy plates, was connected to the CMC skin with superalloy rods passing through the molybdenum tubes of the attachment sites and matching holes in lugs that had been cut at the edges of the plates in a hinge-like configuration. This structure is easily flexed in the longitudinal direction but has high stiffness in the lateral direction.

Fig. 1. Shape-morphing structure with SiC–SiC face sheet, superalloy support structure, and shape memory actuating struts.

Fig. 3. Subscale morphing inlet. (a) Schematic showing sequential target shapes of top CMC wall during morphing; (b) and (c) side view of inlet showing CMC wall in positions for maximum and minimum inlet area ratio.

Fig. 4. Diagram of the Princeton Mach 8, 800 K, 1050 psi (7.24 MPa) facility showing details of the morphing inlet insertion and the morphing inlet geometry. A copper plate is placed between the surface and the five actuators to isolate them from the high temperature. Inset shows morphing inlet seen through the schlieren side windows.

The target family of deflected shapes for the wall, illustrated in Fig. 3(a), was achieved using five actuators connected to the support structure through pivot plates visible in Fig. 3. The actuator rods were driven by stepper motors controlled through a Labview program, so that all actuators moved in synchronization and the shape transitioned smoothly between successive contours. The position control relied on simple counting of steps in 3.05 μm/step increments starting from an initial calibrated reference state. The family of shape contours satisfied the constraints that the morphing CMC section of the duct must transition smoothly at the ends to a mating fixed metallic structure (i.e., the slope at positions A and E remained fixed during all actuation movements) and that the deflections in the central throat region (C) extend over a range of 25 mm. At the extreme position Fig. 3(c), the radius of curvature of the CMC wall is approximately 250 mm in the throat region (C) and approximately 500 mm at positions B and C. Note that while the actuator displacements are in the same direction at positions B, C and D, the forces applied at the CMC skin are compressive at position C and tensile at positions B and D.

III. Wind Tunnel Testing

A Mach 8 wind tunnel at Princeton University was used to assess the potential of the composite structure in Fig. 3 for hypersonic morphing applications. Issues to be examined include the performance of the material under thermal stress, the flow quality achievable, the performance of the actuators, the impact of pressure gradients on the material, including start up and shut down transients, the stiffness and dynamic stability of the material in the presence of high frequency buffeting from a hypersonic flow, and the ability of the material to deflect sufficiently far and sustain a specified contour sufficiently well to control the flow without leakage or mechanical damage. The wind tunnel operates at a total temperature of 790 K and at a stagnation pressure of 1050 psi (7.24 MPa). The tunnel is operated with an air ejector system and can be run for several minutes at a time, limited by the capacity of the electrically driven storage heater. At Mach 8, the static temperature is ∼60 K and the static pressure is ∼5 Torr (0.28 kPa). The tunnel has a 9 in. (23 cm) diameter circular test section.

The morphing CMC skin of Fig. 3 was mounted in the fixture shown in the inset of Fig. 4, which formed a rectangular duct with fixed side walls and base plate beneath the morphing CMC top wall. The upstream end of the CMC skin was held in a fixed position against the mating inlet ramp on the tunnel wall, while at the downstream edge sliding occurred between the CMC face sheet and the mating metal duct surface during actuation. The side walls were quartz to allow optical observations during testing. The fixture was placed at the top of the tunnel, with bypass air flowing below it to avoid choking of the tunnel itself. Figure

4 is a schematic of the wind tunnel showing the location of the test fixture.

As the flow is compressed in the morphing inlet, the pressure increases significantly above the free stream static pressure. The pressure at the back of the morphing surface is determined by the downstream static pressure, which remains well below the pressure of the air in the inlet duct, leading to an outward compressive force on the actuators. This force increases nonlinearly as the morphing surface approaches the opposite wall. The pressure in the duct ranged from about 0.5 psi to ∼14 psi (2.5–100 kPa). The edges of the morphing wall were not sealed against the side walls (the clearance was <0.5 mm), so some air leakage occurred from the duct to the backside cavity. However, the leak rate was sufficiently small that there was very little perturbation to the flow inside the duct.

Because the actuator driver rods could not be conveniently vacuum sealed from the morphing surface in this particular setup, the entire structure including the actuators and driving electronics was enclosed within the wind tunnel test chamber, as shown in the Fig. 5. A 7° inlet ramp was installed from the top of the curved tunnel wall to divert the flow into the morphing section. The leading edge of the morphing plate was mounted at a 5° angle to minimize discontinuity and shock formation at the joint with the inlet ramp. The rear of the morphing section was captured into a sliding retainer. The back end of the actuator section was left open to the tunnel diffuser so that large pressure gradients would not occur across the morphing panel during

Fig. 5. Morphing inlet tunnel model including the 13 pressure taps that are routed to the bottom of the wind tunnel through a thin fin. The water-cooled copper plate can be seen near the top. The curved boundary beneath the air-foil mates flush with the wind tunnel's inner perimeter.

start up and shut down. Even though the static temperature of the air in the test section is low, the boundary layer and other stagnant air rises to a temperature close to the total temperature of the plenum (800 K). Because the stagnant hot air permeates the chamber behind the morphing panel, a water-cooled copper thermal insulation plate was installed between the CMC wall and the actuators to protect the actuators from the high temperatures.

In order to quantify the static pressure profile, a series of 13 pressure tap holes were drilled along the centerline of the flat bottom plate at 16 mm intervals. The pressure tap tubes were routed across the hypersonic flow below the flat plate to the bottom of the wind tunnel through a thin fin-shaped air-foil to minimize disturbance of the wind tunnel below the inlet flow. The fin structure, with a bottom curved plate that fits the contour of the lower section of the wind tunnel, is shown in Fig. 5. The diagnostics included the static pressure profile from the 13 pressure taps as well as schlieren images of the shock structure taken through the side windows and an overall video of the motion of the surface.

IV. Flow Modeling

Initial modeling of the inlet was undertaken using the proprietary Boeing 3D BCFD flow solver, assuming laminar adiabatic flow. This modeling was undertaken to assure that the insertion of the morphing inlet did not choke the wind tunnel and to determine the impact of the tunnel boundary layer on the inlet flow field. The flow solver computed the flow along the entire wind tunnel, following the expansion contour from the throat of the tunnel through the test region. The flow was solved for half of the tunnel using 6.3 million cells assuming a symmetry plane. The computation was performed for the flow from the nozzle throat through and below the morphing inlet and it included the boundary layer evolution. Because of the conservation of enthalpy and entropy in an ideal expansion, the plenum has to be operated at high temperature and pressure to provide the kinetic energy and dynamic pressure required for testing. For the tests conducted here, the objective was to create a Mach 8 simulation so that aerodynamic properties of the flow including shock and expansion wave structure could be determined. Even though the proper Mach number is reached, the facility is not capable of producing the true flight air speed and static temperature because that requires plenum temperatures on the order of 2500 K. For Mach number duplication what is necessary is to have high enough plenum temperature to avoid condensation of the air as it cools by a factor of almost 14 in the expansion. The pressure drops by a factor of $\sim 10\,000$ (assuming ideal gas with a ratio of heat capacities of 1.4). Modeled conditions at

Fig. 7. (a) Mach number field around the test section. (b) Surface pressures computed for Mach 8 flow. All simulations were done using the Boeing 3D BCFD code with laminar flow.

the throat of the tunnel were 800 K (1440 R) and 450 psia (3.1 MPa), corresponding to plenum conditions of 960 K (somewhat higher than actual operating conditions of the facility) and 860 psi (5.9 MPa) (somewhat lower than the actual operating conditions of the facility). The geometry used for the modeling is shown in Fig. 6. This inlet geometry was an earlier configuration with the cowl plate located 2.31 s (59 mm) from the flat morphing surface. For the run reported here, the location of the cowl plate was 1.2 s (30 mm) below the flat morphing surface, significantly closer to that surface so a more substantial compression ratio could be achieved.

Figure 7 shows the predicted centerline Mach numbers (7a) and the wall pressure profiles (7b) of the upper (morphing) and low surfaces of the tunnel and the top of the cowl plate. In this realization the surface is morphed as shown in Fig. 6. The model predicts that the insertion of the inlet into the top of the wind tunnel test section will not choke the wind tunnel, although there is a significant subsonic region at the bottom of the tunnel below the inlet where the cowl shock interacts with the lower boundary layer.

The boundary layer at the top ramp of the inlet is thick due to the fact that it has grown along the wind tunnel wall throughout the expansion. This thick boundary layer and the elliptical intersection of the inlet ramp with the round surface of the tunnel

Fig. 6. Wind tunnel geometry used for the Boeing 3D BCFD code computation.

Fig. 8. Schlieren images of the flow during compression. Images are taken at 1.6-s intervals starting at 34.1 s into the run and show a 1.85 in. (4.7 cm) long section of the flow path. The weak shock structure from the upper (morphing) surface arises from the slight bumps in the woven ceramic associated with the connector pin locations.

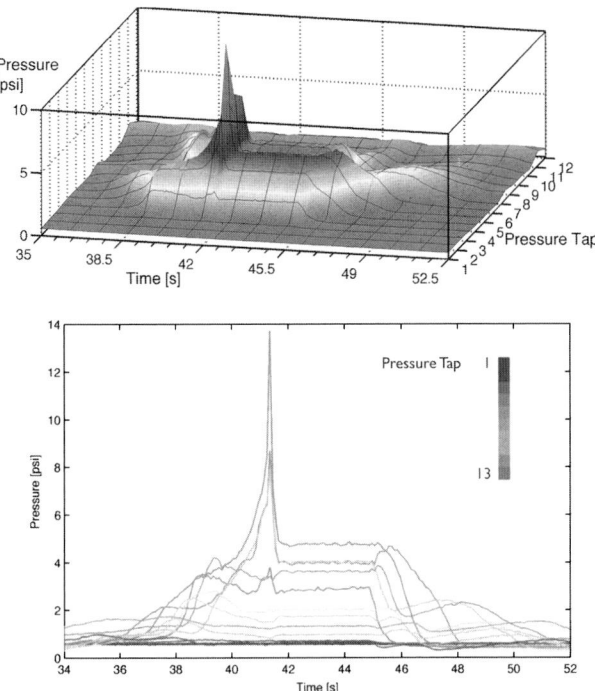

Fig. 9. Pressure versus time plots for the 13 pressure taps. Tap #1 is farthest upstream. The rapid pressure drop at 41.6 s corresponds to the slippage of an actuator due to the high surface pressure loading at maximum compression.

cause the shock from the 7° inlet ramp to be curved and unsteady, and thus not seen with schlieren imaging across the flow. The dashed ellipse in Fig. 7(b) highlights the slow pressure rise along the upper compression ramp associated with this three-dimensional (3D) interaction. Even though the shock is not apparent from the modeled schlieren images, it does lead to a reduction of the centerline Mach number and an increase in the static pressure along the morphing surface and along the upper surface of the cowl plate, as noted in Fig. 7(b). The high-pressure peak at the leading edge of the cowl plate results from the small stagnation point at that location.

A 2D method of characteristics model of the centerline shock structure and pressure contour inside the morphing inlet was developed as an on-site working tool for direct comparisons with experimental data. By implementing the rotational method of characteristics due to Ferri,[17] the expected pressure profile resulting from the morphing surface contour could be rapidly calculated. The simulations were performed using an inverse-marching scheme with 60 characteristics and 10^3 stream-wise steps. As input to the model the pressure and Mach number were specified at the inlet using experimental data and isentropic flow relationships.

V. Results

Results are shown in Figs. 8 and 9 from a series of tests in which the CMC wall was morphed as shown in Fig. 3 between positions corresponding to $h = 30$ and 5 mm, reducing the area ratio between the inlet and throat by a factor of 6. Figure 8 shows a sequence of schlieren images of the morphing inlet. These images were taken at 1.6-s interval starting at 34.1 s after the start of the run. Figures 9(a) and (b) are renderings of the time evolution of the pressure tap measurements during the shape morphing, with tap #1 the farthest upstream and tap #13 the farthest downstream. It is apparent that the morphing continued until the maximum pressure reached 13.7 psi (94.5 kPa), at which point the force on the actuators exceeded their load capacity and slippage occurred, immediately reducing the compression. The force on the central actuator at this point was estimated by multiplying the maximum pressure with the area supported by the central actuator (approximately 10 cm × 5 cm, or 4 in. × 2 in.), which results in a load of over 100 lbf (445 N). Subsequent static load tests indicated the actuators slip with a force of approximately 108 lbf (480 N).

The static pressure of 0.5 psi (3.4 kPa) in the unmorphed condition indicates that the Mach number of the flow through the inlet is 6.3. This is below the Mach 8 free stream condition due to the shock and compression waves from the front of the compression ramp and the leading edge of the cowl. Other weak shock structure in the facility upstream of the test section may also contribute to the reduction of the inlet Mach number. At the maximum displacement, the area ratio is approximately a factor of 6. The thickness of the boundary layer along the morphing surface can be estimated from the curvature of weak shocks visible in the in the schlieren images. These shocks were generated from small distortions of the woven ceramic plate beneath the actuator attachment points. The slope of these weak shocks can be used to approximate the local Mach number. These weak shocks show the Mach number increasing away from the wall and approaching the free stream angle. In the unmorphed condition, they reach the free stream angle at about 10 mm from the morphing surface. This corresponds to about 30% of the cross sectional area. In the highly deflected cases, the boundary layer shrinks due to compression, but occupies a larger fraction of the flow cross section, leading to an increase in the effective compression area ratio between the undeflected and fully deflected configurations. There is only a thin laminar

Fig. 10. Computed Mach number (a) and pressure (b) fields for inlet Mach number of 6.3 and computed and measured cowl wall pressure profiles (c).

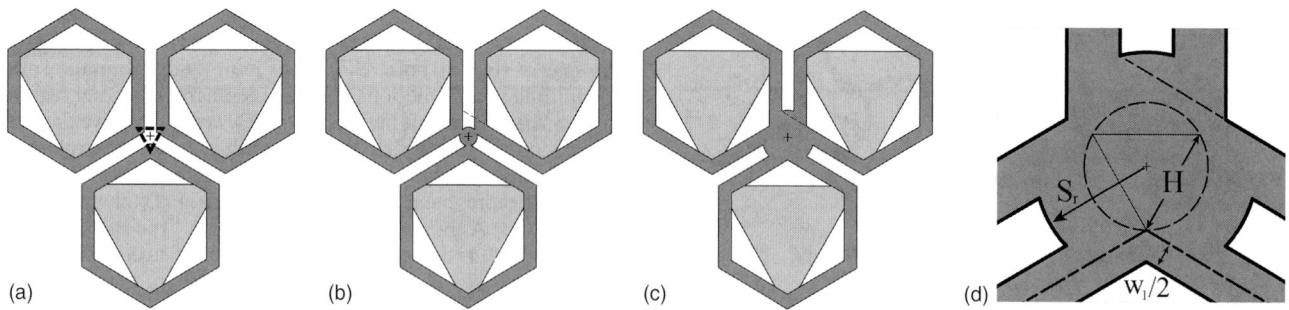

Fig. 3. Spaced lattices with $\theta = \pi/6$, with no additional material (a), undersized stiffener (b), and appropriately sized spacer (c and d) where axial member loads can be transmitted directly to adjacent unit cells. Spacer has characteristic dimension H while the stiffener has radius S_r.

where I_1, for type I members, is the second moment of area of the cross section about the neutral axis, so that E_1I_1 is the bending stiffness of type I members, specified for bending in the

$$\frac{\Delta L}{L} = \frac{\sqrt{3}\left(\varepsilon_1^t \frac{\hat{L}_1}{L_1} + \hat{\varepsilon}_1^t \frac{R_1}{L_1} = \hat{\varepsilon}_2^t \frac{R_2}{L_1}\right) - (\cos\theta + \sqrt{3}\sin\theta)\sin\theta\left(\varepsilon_2^t \frac{\hat{L}_2}{L_2} + 2\hat{\varepsilon}_2^t \frac{R_2}{L_2}\right)}{(\sqrt{3}\cos\theta - \sin\theta)\cos\theta}$$

$$+ \frac{12I_1(\cos\theta + \sqrt{3}\sin\theta)\left[\sqrt{3}\cos\theta + \left(3 + \frac{2E_1A_1\hat{L}_2}{E_2A_2\hat{L}_1}\right)\sin\theta\right]\left(\varepsilon_2^t \frac{\hat{L}_2}{L_2} + 2\hat{\varepsilon}_2^t \frac{R_2}{L_2} - \varepsilon_1^t \frac{\hat{L}_1}{L_1} - \hat{\varepsilon}_1^t \frac{R_1}{L_1} - \hat{\varepsilon}_2^t \frac{R_2}{L_1}\right)}{A_1\hat{L}_1^2(\sqrt{3}\cos\theta - \sin\theta)\cos\theta\left\{(\sqrt{3}\cos\theta - \sin\theta)^2 + \frac{12I_1}{A_1\hat{L}_1^2}\left[\frac{2E_1A_1\hat{L}_2}{E_2A_2\hat{L}_1} + (\cos\theta + \sqrt{3}\sin\theta)^2\right]\right\}} \tag{17}$$

plane of the lattice. Elimination of δ and V among Eqs. (7)–(10) provides

$$T_1 = \frac{12E_1I_1L_1(\cos\theta + \sqrt{3}\sin\theta)^2\left(\frac{\Delta L_2}{L_2} - \frac{\Delta L_1}{L_1}\right)}{(L_1 - R_1 - R_2)^3(\sqrt{3}\cos\theta - \sin\theta)^2} \tag{11}$$

and

$$T_2 = \frac{24E_1I_1L_1(\cos\theta + \sqrt{3}\sin\theta)\left(\frac{\Delta L_2}{L_2} - \frac{\Delta L_1}{L_1}\right)}{(L_1 - R_1 - R_2)^3(\sqrt{3}\cos\theta - \sin\theta)^2} \tag{12}$$

Use of Eqs. (5) and (6) allows a solution for the strut tensions as

$$T_1 =$$
$$\frac{12E_1I_1L_1(\cos\theta + \sqrt{3}\sin\theta)^2\left(\varepsilon_2^t \frac{\hat{L}_2}{L_2} + 2\hat{\varepsilon}_2^t \frac{R_2}{L_2} - \varepsilon_1^t \frac{\hat{L}_1}{L_1} - \hat{\varepsilon}_1^t \frac{R_1}{L_1} - \hat{\varepsilon}_2^t \frac{R_2}{L_1}\right)}{\hat{L}_1^3\left\{(\sqrt{3}\cos\theta - \sin\theta)^2 + \frac{12I_1}{A_1\hat{L}_1^2}\left[\frac{2E_1A_1\hat{L}_2}{E_2A_2\hat{L}_1} + (\cos\theta + \sqrt{3}\sin\theta)^2\right]\right\}} \tag{13}$$

and

$$T_1 =$$
$$\frac{24E_1I_1L_1(\cos\theta + \sqrt{3}\sin\theta)\left(\varepsilon_2^t \frac{\hat{L}_2}{L_2} + 2\hat{\varepsilon}_2^t \frac{R_2}{L_2} - \varepsilon_1^t \frac{\hat{L}_1}{L_1} - \hat{\varepsilon}_1^t \frac{R_1}{L_1} - \hat{\varepsilon}_2^t \frac{R_2}{L_1}\right)}{\hat{L}_1^3\left\{(\sqrt{3}\cos\theta - \sin\theta)^2 + \frac{12I_1}{A_1\hat{L}_1^2}\left[\frac{2E_1A_1\hat{L}_2}{E_2A_2\hat{L}_1} + (\cos\theta + \sqrt{3}\sin\theta)^2\right]\right\}} \tag{14}$$

where the notation

$$\hat{L}_1 = L_1 - R_1 - R_2 \tag{15}$$

and

$$\hat{L}_2 = L_2 - 2R_2 \tag{16}$$

has been introduced. When these results are inserted into Eqs. (5) and (6) and the outcome used in Eq. (4), the thermal expansion of the lattice becomes

When the bending stiffness is negligible ($I_1 = 0$), and the joint radii R_1 and R_2 are neglected, this result simplifies to Eq. (1).

The result in Eq. (17) highlights the two effects of bonded joints. The first involves the finite extent of the joints (R_1 and R_2 are nonzero), implied by the first term on the right hand side. The second arises because the bending of type I struts adds strain to the thermal response, signified by the second term on the right hand side.

Appropriate choices for the joint radii (Fig. 2) are

$$R_1 = \frac{3w_1}{\pi - 6\theta} \tag{18}$$

and

$$R_2 = \max\left(\frac{3(w_1 + w_2)}{2\pi - 6\theta}, \frac{3w_2}{\pi}\right) \tag{19}$$

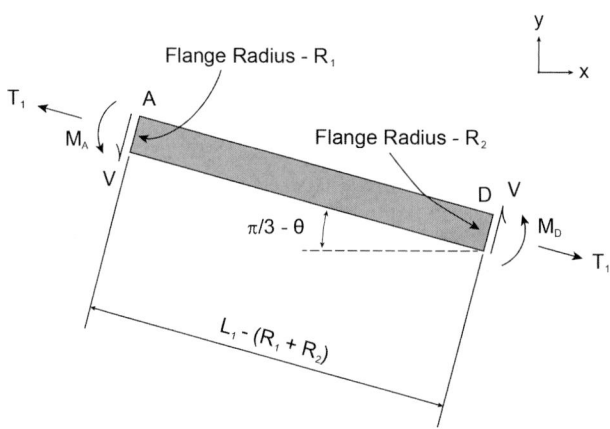

Fig. 4. Type I members sustain an axial load, T, shear, V, and bending moment, M.

The thermal expansion for nodes A, J, F is

$$\hat{\varepsilon}_1^t = \varepsilon_1^t \tag{20}$$

An appropriate choice for nodes B, C, D is a weighted average of the expansion of type I and II materials,

$$\hat{\varepsilon}_2^t = \frac{1}{3}(2\varepsilon_1^t + \varepsilon_2^t) \tag{21}$$

Results for the linear thermal expansion for various material combinations are plotted against θ in Fig. 5, with a range of values for the bending stiffness parameter $I_1/A_1L_1^2$ used for illustration.

The above results affirm that the design of joints of Type A, J, F is critical to the realization of low thermal expansion behavior, because the low angle included between type I members from adjacent unit cells leads to excessively large R_1. This limitation is obviated by using an offset design wherein the centers of neighboring unit cells are displaced, allowing R_1 to become small even when $\theta \rightarrow \pi/6$. Such offset designs are achieved by insertion of a spacer made from type I material, as shown in Fig. 3. The spacer removes the coupling of $\hat{L}_1$ and θ, for a fixed L, enabling the use of larger values of θ. Offset designs can have negligible bending, whereupon, the thermal expansion can be closely approximated by the first term on the right hand side of Eq. (17), with R_1 and R_2 neglected. The CTE for the offset lattice is thus

$$\frac{\bar{\alpha}}{\alpha_1} = \frac{\sqrt{3} - \lambda F(\theta)}{\sqrt{3} - F(\theta)}\left(\frac{L}{L+H}\right) + \frac{H}{L+H} \tag{22}$$

where H is the size of the spacer and $F(\theta) = (\cos\theta + \sqrt{3}\sin\theta)\sin\theta$. When H is zero, Eq. (1) is recovered once more. The formula in Eq. (22) can be used to guide low thermal expansion lattice designs; specifically, Eq. (22) predicts that zero thermal expansion of the lattice occurs when

$$\frac{H}{L} = \frac{\lambda F(\theta) - \sqrt{3}}{\sqrt{3} - F(\theta)} \tag{23}$$

This design has unit cells that contract upon heating, compensating for the expansion of the spacers. Such solutions are feasible when $\sqrt{3}/\lambda < F(\theta) < \sqrt{3}$ (note that, within this range, the unit cell has negative thermal expansion). The upper limit of this range coincides with $\theta = \pi/3$. For illustration, when $\lambda = 2$, the lower limit is $\theta = \pi/6$. Beyond these limits is the requirement that joints A, J, F remain physically small, requiring that $\frac{\geq H\sqrt{3}w_1\sin\theta}{F(\theta)}$, whereupon the type I elements from

neighboring unit cells have no intersection. From Eq. (23), this requirement provides

$$\frac{w_1}{L} \leq \frac{[\lambda F(\theta) - \sqrt{3}]F(\theta)}{[\sqrt{3} - F(\theta)]\sqrt{3}\sin\theta} \tag{24}$$

revealing that the slenderness of type I elements has to be respected. Consequently, when $\theta = \pi/6$, Eq. (24) gives $w_1/L \leq (\lambda - 2)$, indicating that the design can be satisfied provided that $\lambda = \alpha_2/\alpha_1 > 2$, ensuring that w_1 is positive. Moreover, any material combination with λ slightly in excess of 2 (say more than 2.1) is acceptable, because w_1/L must be small to ensure low bending stiffness.

Alternatively, Eq. (24) can be recast as a condition on θ with λ and w_1/L already selected. The outcome is not transparent because it involves a combination of trigonometric functions of θ without obvious simplification. Nevertheless, inspection indicates that any θ slightly above the lower limit $F(\theta) = \sqrt{3}/\lambda$ will satisfy Eq. (24). Specific cases should be assessed numerically to ensure a satisfactory design, as elaborated below.

A chosen design is limited in its range of temperature operation by the requirement that the triangle of type II elements has space into which material can expand when the lattice is heated. The critical condition occurs when joint C (Fig. 2) touches its counterparts from the two adjacent unit cells. It is straightforward to ascertain that the strain increment in type II bars that causes this critical condition is

$$\varepsilon_2^t = \frac{\cos\theta - \sqrt{3}\sin\theta + \frac{2H}{L}\cos\theta}{\cos\theta + \sqrt{3}\sin\theta} \tag{25}$$

Consequently, the operating temperature range must be chosen to ensure that thermal straining of type II elements is smaller. For $\theta \sim \pi/6$ and greater a careful choice of $\frac{H}{L}$ is needed, because $\cos\theta - \sqrt{3}\sin\theta \leq 0$ for $\theta \geq \pi/6$.

III. Specific, Low Expansion, Offset Designs

(1) FE Method

Lattices are modeled using a representative volume element (RVE) FE technique. Three-dimensional models are subject to periodic boundary conditions for the two in-plane dimensions in the form of uniform macroscopic strains. The relative displacements between pairs of boundary nodes are controlled by tying their displacements together consistent with the strains we wish to impose.[8,9] The third, z-direction, is left free. A single node on the interior of the model is held fixed in space to prevent rigid-body translations. The commercial FE code ABAQUS[10] is used for mesh generation and to perform analysis. MATLAB[11] code is used extensively to manipulate meshes, apply boundary conditions, and for postprocessing. Periodic boundary conditions were implemented in Cartesian coordinates. A typical RVE is pictured in Fig. 6. Typical meshes consist of $\sim 20\,000$ to $\sim 70\,000$ eight-noded linear hexahedral elements (type C3D8R). The number of elements varies greatly with the relative density of the lattice geometry being analyzed, which ranged from 8% to 98% for slender to space-filling designs, respectively. Mesh sensitivity is studied to ascertain model resolution at which solutions converge. Because the average strains are small, there is no distinction between the macroscopic Cauchy stress and the macroscopic nominal stress, so that the macroscopic stress can be computed by simply dividing force resultants by section areas for the undeformed volume of the RVE.

(2) Offset Lattices

In stretch-dominated structures, where loads are well distributed, stress contours are rather uniform, and consequently the structure is relatively efficient in its utilization of material. On the other hand, FE calculated thermomechanical stress distributions in the original lattice design reveal large nonuniform

Fig. 5. Lattice coefficient of thermal expansion as a function of skew angle θ and type I member aspect ratios.

Fig. 6. (a) Perimeter of unit cell used for computations. (b) Representative finite-element mesh that consisted of 50 000–70 000 three-dimensional elements. (c) Detail of a bimaterial joint region.

bending stresses associated with rigidity and member intersection at the joints (Fig. 7(a)). The lattices analyzed to obtain these results are modeled with the temperature-dependent properties of Ti and Al alloys listed in Table Ib and subject to a temperature excursion from 40° to 250°C. The bimaterial interface, in reality press-fit, is considered to be perfectly bonded for simplicity. Simulations that model the interface between sublattices with contact in compression, and a frictional coefficient of $\mu = 1$, give results that are negligibly different from welded models.

The maximum tensile equivalent stress (also known as the von Mises stress) is located in the bimaterial joint region and results from CTE mismatch between constituents. In the temperature range analyzed, the maximum tensile equivalent stress is lower than the tensile yield stress in both materials.

Constraints on the motion of the lattice's components drive the macroscopic CTE toward that of the constituents. By placing a spacer (Fig. 3), with characteristic dimension H, between unit cells and eliminating excess material around the bimaterial joint, the lattice geometry can be designed so that it behaves according to the original concept for low CTE, i.e. with negligible stress (Fig. 7(b)). Contours for the revised design show greatly reduced bending stresses and the overall lattice behavior is in good agreement with new analytical predictions (Fig. 8).

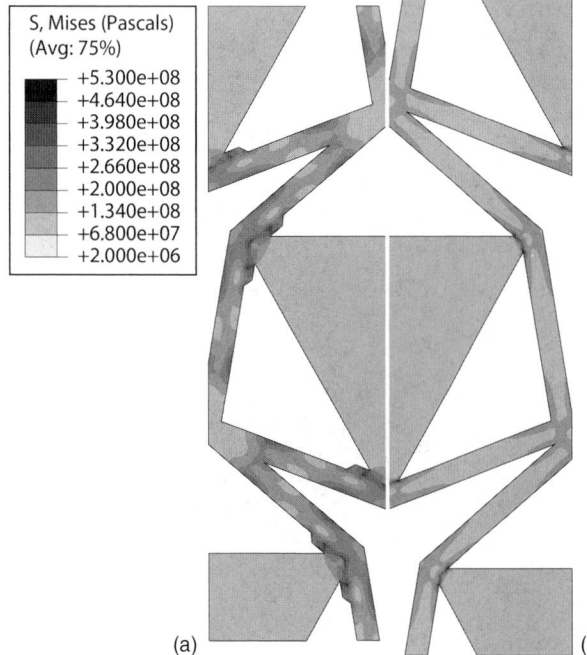

Fig. 7. Finite-element thermomechanical stress distribution in Ti and Al lattices given the material properties listed in Table I and subject to a 175°C temperature excursion—(a) original design and (b) new offset design.

Table Ia. Temperature Average Values of Material Properties—For the Range 40°–215°C

		Coefficient of thermal expansion (ppm/°C)	Young's modulus (GPa)	σ_y (MPa)
Type-1 material	Ti–6Al–4V	9.4	110	800
Type-2 material	7075-T6 aluminum	24.3	70	400

Table Ib. Temperature-Dependent Material Properties

		Coefficient of thermal expansion (ppm/°C)	Young's modulus (GPa)	σ_y (MPa)	Temperature (°C)
Type-1 material	Ti–6Al–4V	9.2	110	1100	20
		9.6		858	300
Type-2 material	7075-T6	22.9	70	434	20
		26.3		391	200
				339	300

(3) Lattice Design

For any pair of constituent materials and their associated properties, there is a range of skew angles, θ, over which a desired lattice CTE can be achieved. Properties such as stiffness, the Poisson ratio, and strength vary over this range so that any one solution may be superior to others depending on design requirements. For a unit temperature change, $\Delta T = 1$ so that $\varepsilon_i^t = \alpha_i$, Eq. (17) gives the thermal expansion coefficient of the lattice and can be inserted into Eq. (26),

$$\bar{\alpha} = \frac{\Delta L + \mathrm{H}\alpha_1}{L + \mathrm{H}} \tag{26}$$

to calculate the expansion of a spaced lattice. The macroscopic thermal strain in a spaced lattice is the weighted sum of the lattice and spacer strain, the latter having thermal expansion coefficient α_1. Equation (26) can be used to identify regions of design parameter space where lattices with the desired thermomechanical response exist. Designs in this neighborhood can then be investigated through FE to address specific geometries and to investigate their thermomechanical response in comparison with the results in Section II.

The thermomechanical strain response of the bimaterial joints in Eq. (21) is assumed to be an average of the constituents. While this approximation is sufficient for exploratory investigations, specific mechanical bimaterial interface geometries must be considered when designing real structures, as their behavior may differ substantially from this idealization. The dovetail joints used by Steeves *et al.*,[5,7] and in the current effort, are examples of bimaterial attachments. Mechanical connections capable of carrying tensile loads are necessary for transmitting all applied macroscopic loading situations except biaxial tension. The size of these joints, given by their characteristic dimension, R_2, reduce the effective length of members ($\hat{L}_1$ and $\hat{L}_2$, Eqs. (15) and (16)).

For fixed values of w_1 and w_2, the thermomechanical response of the system is a strong function of the member effective lengths, and therefore of R_1 and R_2. In many designs, a single radius R_2 does not exist by which both type I and II members are reduced equally at the Type D joints (Eqs. (15) and (16)). The actual reduction in effective length of members at these nodes is a function of member width and the relative angle at which they are incident to the joint. To identify designs with a tailored and well-predicted CTE, which is amenable to fabrication, more detailed modeling is necessary.

The addition of the spacer (Fig. 3) allows practical access to previously unachievable skew angles and corresponding higher relative densities. In previous designs without the spacer, at skew angles approaching 30°, members from adjacent unit cells intersect to an extent, and prevent the desired thermomechanical response, and the desired CTE cannot be realized. A skew angle of 30°, without a spacer present, results in type I members in adjacent cells being parallel and a hexagonal unit cell appearance. If only the area bounded by the unit cell is considered, these designs can achieve near-maximum areal density as an assembly of hexagonal cells with small gaps between them. The size of the gaps between unit cells is directly related to the size of the spacer H. Upon temperature excursion from the reference state, type I members expand, distort, and rotate, causing Type D nodes to translate away from the center of the unit cell. The size of the spacer and the corresponding gap is dictated by the maximum outward deflection of these nodes in the specified temperature range while considering the need to avoid adjacent unit cells impinging upon each other after thermal straining. For some designs with skew angles near or above 30°, an upper use temperature exists at which initially nearly parallel type I members in adjacent cells deform to contact each other causing the lattice to densify. For these designs, the upper use temperature can be increased by expanding the size of the spacer at the cost of driving the CTE of the system toward that of the type I material (Eq. (26)). Densification may also significantly influence the stiffness and strength of these lattices and may be a beneficial feature in some applications. If material continuity is beneficial, such as for aerodynamic surfaces, densified lattices can be useful.

In some spaced lattice designs where the dimension of the spacer is on the order of w_1, no additional material is needed to achieve this offset. In others, designs with skew angles in the neighborhood of 30°, unit cells may intersect minimally or not at all (Fig. 3) and the strength and stiffness joint will suffer. A disk of material centered on the spacer, a stiffener with radius S_r can be added to the lattice to maintain continuity and transmit stresses between unit cells. If the disk of material is too small, stress concentrations in these regions can be design limiting. An appropriate radius for this disk is one where axial loads in type I members can be directly transmitted to adjacent cells (Figs. 3(c) and (d)). This stiffener has a dimension independent of the spacer and may be useful in facilitating a connection between the lattice and a substructure.

As type II members are not subject to thermally induced bending stresses, members with geometries other than truss or beam-like forms can be considered without altering the mechanics of the system.[5] In Fig. 10, the lattices pictured in the middle and on the right (b and c) are variations of the lattice on the left (a) incorporating type II members that are not simple prismatic beams. Stiffening of type II members drives the thermal expansion of the system to lower values. The increased stiffness of these members can be modeled by using an effective modulus E_2^* for E_2 in Eqs. (13) and (14). The only restriction in geometry is that the type II members not impinge on, and reduce the effective length of the slender type I members when thermomechanically strained. Previous fabricated designs have used truss and solid triangular inner type II members (Figs. 9(a) and (b)).[5,7] To explore the potential space-filling properties of this lattice, a hexagonally shaped type II sublattice geometry was chosen for fabrication in this work.

If in-plane geometries of members are specified to have a finite width centered on the lines shown in Fig. 2, the width of the bimaterial joint is limited by the width of type II members, w_2 (for the geometry considered in this work). Selecting a small bimaterial joint resulted in a small value of w_2 in Eq. (17). To account for the much larger cross section of the hexagonal geometry used, an effective modulus of $E_2^* = 10E_2$ is utilized in the analytical model. The order of magnitude increase in stiffness is an estimate; further increases do not significantly influence results. By choosing a slightly skewed, but nearly hexagonal type II element, an upper use temperature densification event can be engineered between high and low CTE sublattices. When both adjacent unit cells and sublattices contact each other at the same ΔT, a nearly or completely densified and continuous structure can be formed.

Fig. 8. Comparison of results from finite-element analysis for coefficient of thermal expansion and experiment.

Fig. 9. Thermal stress distributions in Ti- and Al-fabricated lattice geometry with thin type I members (a), triangular type II member (b), and space-filling design (c).

The dovetail joint dimensions were chosen to maximize the effective length of type I members. Joint dimensions were minimized to reduce R_2 with practical consideration for the interface size and the need to maintain a robust mechanical connection between sublattices. Lattices previously investigated[5,7] used dovetail joints where the high CTE type II material composed the inner, male side of the connection. Additional material was added to the lattice at these locations to help facilitate a robust connection. This added material served to reduce the effective length of members resulting in a poorly predicted thermal response (Fig. 8). In the current design, the inner male component is now composed of the low CTE type I material, and the bimaterial joint is relocated to be enclosed in the inner type II member. Switching the low CTE material to the male side reduces thermal stresses resulting from the expansion of the high CTE material, which was previously confined by the low CTE constituent. It is hoped that plasticity can be avoided altogether with this configuration, in contrast to the previous dovetail joint design.

The test specimen designed and fabricated in this work was chosen to have a skew angle of 30° and nearly space-filling inner type II member. By specifying the geometric parameters $w_1 = 3.0$ mm, $w_2 = 0.4$ mm, $\theta = 30°$, H = 2.0 mm, $S_r = 3.80$ mm, and unit cell length $L = 50.0$ mm, the designs in Fig. 10 are achieved. FE and analytical predictions for the geometries shown are listed in Table II.

(4) Lattice Fabrication

A metallic lattice was fabricated for CTE measurement purposes. The type I lattice was electro discharge machined from 3-mm-thick sheets of Ti–6Al–4V. The type II members were machined from 7075-T6 aluminum alloy. Sublattices were press-fit together. Tolerances were such that assembly with hand pressure was possible; however, a mechanical press was used to ensure proper assembly. The structure consists of 10 unit cells arranged so that two cells in the interior are separated from the edge by another unit cell to minimize edge effects (Fig. 10).

(5) CTE Measurement Methodology

Thermal expansion measurements on the Ti and Al lattice were performed using a two-dimensional (2D) digital image correlation (DIC) system. A high contrast black and white speckle pattern was applied using spray paint. Lattices were heated on a laboratory hot plate (Wensco, model #H1818RA4000) at a rate of 60°C/h from room temperature to 220°C. A frame of

Fig. 10. Lattices of Ti alloy (struts) and Al alloy (hexagonal units) fabricated for measurement of the coefficient of thermal expansion.

Table II. Analytical and FE CTE Predictions for Thin, Triangular, and Hexagonal Type-2 Member Lattices Pictured in Fig. 9

Coefficient of thermal expansion (ppm/°C)	Thin	Triangular	Hexagonal	Fabricated
Analytical prediction	2.6	−1.4	−1.4	−1.4
Finite element	1.4	−1.6	−1.6	−1.1

Constituent properties and FE results are temperature average values from Table 1b from 20° to 70°C. Thin Type-2 members elastically buckled under compression for larger temperature excursions.

common silica insulation, approximately 1.5 in thick, was placed around the lattice with a glass plate on top. The temperature was recorded by four self-adhering K-type thermocouples (Omega, model SA1XL-K-SRTC, Omega Engineering Inc., Stamford, CT) located on the upper face of the lattice, two each on Ti and Al. Acetone was used to remove the applied paint at the location of the thermocouple attachments to increase heat transfer. The temperature of the lattice was taken to be the average of the four. Digital images were captured by CCD camera (AVT Dolphin F-201B, Allied Vision Technologies, Newburyport, MA) with a zoom lens (Tamron AF 70-300 1:4-5. 6, Tamron USA Inc., Commack, NY) positioned approximately 2 m from the specimen. The focal length was maximized subject in the confines of the laboratory space to minimize the effect of out of plane deformations on in-plane measurements. Images were taken every 5 s to record deformations. Two 300 W incandescent lights in hoods were positioned close above the glass plate for imaging purposes. Using the Vic-2D (Correlated Solutions, Correlated Solutions Inc., Columbia, SC) software, virtual extensometers placed on the reference image, and tracked through the images, measured the displacement between pairs of pixel subsets. A typical area of interest consisted of one unit cell with three Type D nodes visible. Strains were calculated from the relative displacement between pairs of subsets that were typically 23^2 pixels. Strains were measured in the Al and Ti and were calculated as the average of six virtual extensometers. Lattice strains were calculated as the average of three virtual extensometers placed between the three Type D nodes surrounding an interior unit cell. Temperature average CTE was measured by linear fitting to strain–temperature plots, where the average CTE over the temperature range is the slope of the resulting straight line. Extensometers were placed upon relatively unstressed material regions to measure the CTE of the constituents.

Image distortion from convective currents emanating from the specimen is a common problem when using DIC to record thermomechanical strains. A frame of silica fiber insulation was placed around the lattice and a glass plate on top to help thermally isolate the specimen from the camera. A fan is used to mix the air directly above the glass plate and carry hot air away from the lights.

(6) CTE Results

A typical experimental strain versus temperature plot is shown in Fig. 11 along with FE prediction. Temperature average CTE for the lattice and the constituent materials is reported in Table III. Although the DIC software is capable of measuring microstrains, scatter on the order of 1000 microstrain was present in the measurements due to image distortion. Convective currents emanating from the lighting were sufficient to cause visible inhomogeneous lensing in successive images. Measurements on the Ti were limited to the Type D joints resulting in short gauge lengths and larger scatter.

The average value of CTE for the constituent materials is measured to be ∼4% higher than reported values (24.3 ppm/C for Al, and 9.4 ppm/C for Ti) (Table Ia), but are within experimental error. The lattice has an average measured CTE

Fig. 11. Thermal strains measured in Al alloy, Ti alloy, and the lattice. Significant scatter in Ti measurements is due to the relatively short gauge lengths that had to be used.

of −0.9 ppm/C that is well predicted by FE of −1.1 ppm/C, and close to the analytical prediction of −1.4 ppm/C.

3.7 Discussion—Thermal Expansion

Measured thermal expansions of the lattice agree well with analytical and FE predictions. Differences between measured and FE prediction are believed to be due to the reduced effective length of the type I members resulting from fabrication imperfections causing premature contact between sublattices. The hexagonally shaped aluminum members had visible machining imperfections on some edges where they were cut from stock. The small wing-shaped gap between high and low CTE members results in a structure that is more imperfection sensitive than a lattice with triangular or prismatic beam-type members. Geometric imperfections in this gap can cause sublattices to impact each other before the predicted upper use temperature, reducing the effective length of type I members and driving up the CTE. Space-filling lattices with these features, such as aerodynamic surfaces, may be adversely affected by debris in these small gaps.

No plasticity is predicted in the revised dovetail geometry. The limited plasticity present in other designs[7] was a function of the bimaterial joint geometry used and is not inherent to the functional mechanics of the lattice itself.

Scatter in CTE measurements results from several sources. Convective currents in the air column between the specimen and the camera distort images through their associated density gradients and lensing effects.[12] Efforts were made to mitigate convective currents coming from the specimen and hotplate, but the incandescent lighting used for imaging proved a sufficient source of interference. A cold light source or a camera orientation that minimizes the effect of convective currents, by positioning it outside the affected area, would reduce the effect. The large scatter in the CTE measurements for Ti is due to the lattice geometry studied, the size of the unit cells, and the short lengths of relatively unstressed material available for measurements taken in the Type D node region.

The analytical model developed in Section II does a good job of predicting lattice thermal expansion. The model uses the simple assumption that the thermal expansion of the type II joint region is a weighted average of the constituent materials (Eq. (21)). In reality, a distinct material interface exists between the constituents, and the thermal response is much more complicated. The analytical CTE prediction of −1.4 ppm/C for the lattice is close to the FE predicted value of −1.1 ppm/C and

Table III.   Measured CTE of Lattices and Constituent Materials (95% Confidence Bounds), and FE Prediction for Lattice using Measured Values (ppm/°C)

	Aluminum (measured)	Titanium (measured)	Lattice (measured)	FE prediction
Exp. 1	26.1 (25.9–26.3)	10.4 (10.0–10.7)	−1.5 (−1.7 to −1.3)	−0.8
Exp. 2	24.9 (24.7–25.1)	10.0 (9.6–10.7)	−0.8 (−0.9 to −0.6)	−0.6
Exp. 3	25.5 (25.4–25.6)	9.1 (8.6–9.6)	−0.6 (−0.7 to −0.5)	−2.6
Exp. 4	24.9 (24.8–25.0)	9.1 (8.7–9.5)	−0.8 (−0.8 to −0.6)	−2.2
Average	25.4	9.7	−0.9	−1.3

sufficiently accurate to help characterize the design space to locate geometries of interest.

CTE measurements performed on a titanium and aluminum lattice using 2D DIC were able to validate the analytical and FE predictive models used in its design. The metal lattice behaved elastically over a temperature range of 175°C, exhibiting consistent negative thermal expansion. The design space identified through the analytical model suggests the ability to realize a family of structures with a wide range of stiffness and thermal expansion properties.

IV.   Stiffness

(1)   Basic Stiffness Results for Pin-Jointed Lattices

(A)   Elastic Properties: The unit cell of the design without spacers, shown in Fig. 12(a), is loaded by a set of forces, parameterized by P, Q, and S. It is presumed that the lattice has been designed in the manner described above, such that the bending stiffness of struts of type I are very low, whereupon their behavior is stretch dominated. In such a situation, the lattice can be analyzed as if it were pin-jointed.

The resulting behavior is isotropic in the plane of the lattice, and stated as

$$\varepsilon_{xx} = \frac{\sigma_{xx}}{E} - \frac{\nu \sigma_{yy}}{E} + \alpha \Delta T \tag{27a}$$

$$\varepsilon_{yy} = \frac{\sigma_{yy}}{E} - \frac{\nu \sigma_{xx}}{E} + \alpha \Delta T \tag{27b}$$

$$\varepsilon_{xy} = \frac{\sigma_{xy}}{2G} \tag{27c}$$

where E, ν, G, and α are the in-plane Young's modulus, the Poisson ratio, shear modulus, and CTE, respectively, of the lattice.

The biaxial stiffness is the ratio of the biaxial stress to the in-plane strain under equibiaxial loading,[5] and can be deduced as

$$S_{\mathrm{b}} = \frac{(\sqrt{3}\cos\theta - \sin\theta)^2}{2\sqrt{3}L_1\left[\frac{1}{E_1 w_1} + \frac{2\sin^2\theta}{3E_2 w_2}(\cos\theta + \sqrt{3}\sin\theta)\right]} \tag{28}$$

Note that this corrects a misprint in the previously published formula in Steeves *et al.*,[5] which is missing the leading 2 in the denominator. The shear modulus is

$$G = \frac{(\sqrt{3}\cos 2\theta + \sin 2\theta)^2}{8\sqrt{3}L_1\left[\frac{1}{E_1 w_1} + \frac{2\sin^2 2\theta}{3E_2 w_2}(\cos\theta + \sqrt{3}\sin\theta)\right]} \tag{29}$$

In these expressions for S_{b} and G, L_1 is used rather than L because L_1/w_1 is the aspect ratio of type I elements.

The Poisson ratio for the lattice may be computed from

$$\nu = \frac{S_{\mathrm{b}} - 2G}{S_{\mathrm{b}} + 2G} \tag{30}$$

and the Young's modulus

$$E = 2G(1 + \nu) \tag{31}$$

Note that, in certain circumstances, the Poisson ratio will be zero or negative. For example, when $E_2 \gg E_1$, the Poisson ratio reduces to

$$\nu = \frac{\cos 2\theta - \sqrt{3}\sin 2\theta}{3 + 2(\sqrt{3}\cos\theta + \sin\theta)\sin\theta} \tag{32}$$

As a consequence, it is zero at $\theta = \pi/12$, and negative for $\theta > \pi/12$.

The introduction of a stiff spacer, as in Fig. 3, has no effect on the elastic properties of the lattice. If the spacer is a stiff component (e.g., composed of a solid plate rather than a set of truss or beam elements), the elastic properties of the lattice are still given by the values in Eqs. (28) and (29). This situation arises because for a stiff spacer, when we neglect its deformation and treat it as rigid, both stress and elastic strain scale in the same way with the size of the spacer. To show this for the case of equibiaxial stress, the loads in Fig. 12(b) are

$$P = (L + \mathrm{H})\sigma_{\mathrm{B}} \tag{33a}$$

$$Q = \frac{\sqrt{3}}{2}(L + \mathrm{H})\sigma_{\mathrm{B}} \tag{33b}$$

where σ_{B} is the applied biaxial stress. Because the spacers are rigid, the forces experienced by the adjacent nodes are the same. The stress are

$$\sigma_{11} = \frac{2Q}{\sqrt{3}L} \tag{34a}$$

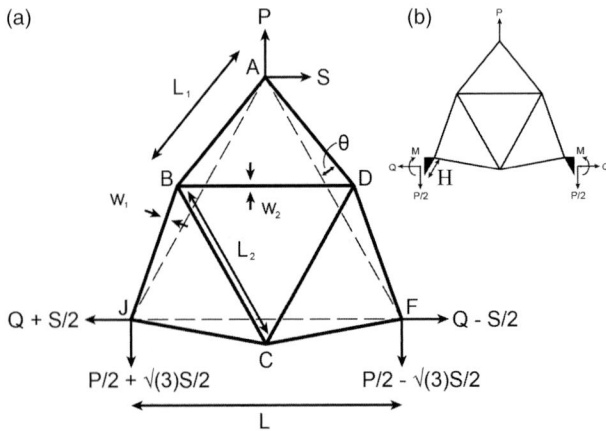

Fig. 12.   Pin-jointed unit cell subject to loads P, Q, and S used in stiffness calculation (a). Spaced lattice unit cell with applied loads P and Q (b).

$$\sigma_{22} = \frac{P}{L} \tag{34b}$$

And the changes in dimensions of the lattice are

$$\Delta u_1 = \varepsilon_{11} L = \frac{1}{E}\left[\frac{2Q}{\sqrt{3}} - \nu P\right] \tag{35a}$$

$$\Delta u_2 = \varepsilon_{22}\frac{\sqrt{3}L}{2} = \frac{\sqrt{3}}{2E}[P - \nu Q] \tag{35b}$$

The strains are

$$\bar{\varepsilon}_{11} = \frac{\Delta u_1}{L+H} \tag{36a}$$

$$\bar{\varepsilon}_{22} = \frac{2\Delta u_2}{\sqrt{3}(L+H)} \tag{36b}$$

Using Eqs. (30), (31), and (32), with (36), the strains become

$$\varepsilon_{11} = \varepsilon_{22} = \frac{1}{E}[\sigma_b - \nu\sigma_b] \tag{37}$$

showing that the strains are not a function of the size of the spacer.

For the case of shear loading for a lattice with spacer with an applied shear stress τ, the forces in Fig. 12(b) are

$$P = (L+H)\tau \tag{38a}$$

$$Q = \frac{-\sqrt{3}(L+H)\tau}{2} \tag{38b}$$

Because the spacer is rigid, the resulting macroscopic strains are

$$\bar{\varepsilon}_{11} = \frac{1}{E}\left[\frac{Q}{\frac{\sqrt{3}(L+H)}{2}} - \nu\frac{P}{L+H}\right] = \frac{1}{E}[-\tau - \nu\tau] \tag{39a}$$

$$\bar{\varepsilon}_{22} = \frac{1}{E}\left[\frac{P}{(L+H)} - \nu\frac{Q}{\frac{\sqrt{3}(L+H)\tau}{2}}\right] = \frac{1}{E}[\tau + \nu\tau] \tag{39b}$$

This indicates that there is no stiffness penalty upon introduction of a stiff spacer.

(2) In-Plane Compression Measurements

In-plane compression experiments have been used to generate stress/strain measurements. The objectives are two-fold: (i) allow calibration of the mechanical robustness and stiffness of representative lattices, and (ii) provide validation data for the ensuing FE calculations. For these purposes, it suffices to fabricate monolithic lattices from 1-mm-thick plates by laser cutting. To probe the yielding and strain-hardening characteristics, one set of lattices has been generated from 304 stainless steel. During testing, out of plane buckling was prevented by constraining the lattice between two 12.7-mm-thick tempered glass plates, bolted to an aluminum frame. The experiments were performed in an MTS™ 810 servohydraulic testing system under displacement control, using a displacement rate of 0.5 mm/min. Images of the lattice were recording every 15 s during the tests using a CCD camera connected to an image correlation system.

Two different type II member geometries were used. The stress–strain behaviors, shown in Fig. 14, reveal robust behavior, characterized by yielding followed by strain hardening. In all cases, yielding occurs in the type I members at critical stress levels in the range, $15 \le \sigma_c \le 18$ MPa. The bright regions in the images (Fig. 13) indicate the occurrence of out-of-plane plastic buckling. Unload–reload measurements reveal hysteresis and decreasing stiffness with increase in plastic strain.

(3) Experimental and FE Results

Stress strain curves generated from FE modeling (FEM) are plotted alongside experimental results in Fig. (14). The 304 stainless steel was modeled by linear elastic response followed by yielding with isotropic hardening. Twenty node biquadratic elements (C3D20R) were used in these calculations. Homogeneous strains were applied to the RVE in two directions. The influence of the glass plates used to confine specimens out of plane was not modeled.

Notable features of the experimental stress–strain curves include a reduction in elastic stiffness with increasing strain past initial yield as evident in the unload–reload regions, and significant hysteresis in these regions. No reduction in stiffness is seen in models restricted to in-plane deformations. Models seeded with imperfections to initiate buckling show the same reduction in stiffness with strain as experiments. Frictional interactions between the lattices and the glass plates used to constrain out-of-plane motion are attributed as the source of hysteresis observed in the experiment. The collapse modes, plastic buckling of type I members oriented most obliquely to the loading direction, and out-of-plane plastic hinging in the same yielded members are accurately captured by the FE results (Fig. 15).

(4) Comparison of Pinned and Bonded Stiffness

To compare pin-jointed analytical and FE-bonded predictions for stiffness, the equations in Section II were used to identify lattice geometries with zero thermal expansion and maximum stiffness. For a given skew angle θ and lattice unit cell length L, Eq. (17) can be used to identify values of w_1 and w_2 that produce maximum stiffness for a given pair of constituent materials. Using the temperature-average properties of Ti and Al (Table Ia), a variety of zero CTE lattices as predicted by Eq. (17) are identified (Fig. 16). The associated aspect ratio of type I members, $\hat{L}_1/w_1$, decreases with increasing skew angle from 20.4 to 5.0 (Figs. 16(a–d)). The stiffness of the analogous pin-jointed structures, those having the same member dimensions, is plotted along with FEM results in Fig. 17, where $V_{f,i}$ is the fraction of solid material of type i.

Fig. 13. The 304 stainless steel lattice compressed along the vertical axis. Bright areas indicate out-of-plane deformation associated with member buckling.

supporting substructures. A low CTE glass, such as Zerodur, used as the type I material will offer a large CTE ratio, λ, when paired with a wide range of other higher expansion materials making available a wide range of achievable CTE (Fig. 5).

Other considerations exist in the design of these systems. These include (i) transient heating effects resulting from mismatches in thermal conductivity between constituents and nonuniform heating, (ii) the net CTE of the lattice is a function of the relative thermal strain between sublattices that may be varying due to transients and inhomogeneous thermal loading, which can affect the response, and (iii) aerodynamic surfaces are often curved surfaces so that nonflat shapes with a tailored thermal expansion coefficient may be desired. The techniques developed and exercised in this paper can be extended to address these issues, and the strength and failure modes of systems can also be addressed.

References

[1]R. S. Lakes, "Cellular Solid Structures with Unbounded Thermal Expansion," *J. Mater. Sci. Lett.*, **15**, 475–7 (1996).

[2]O. Sigmund and S. Torquato, "Composites with Extremal Thermal Expansion Coefficients," *Appl. Phys. Lett.*, **69**, 3203–5 (1996).

[3]L. V. Gibiansky and S. Torquato, "Thermal Expansion of Isotropic Multiphase Composites and Polycrystals," *J. Mech. Phys. Solids*, **45**, 1223–52 (1997).

[4]G. Jefferson, T. A. Parthasarathy, and R. J. Kerans, "Tailorable Thermal Expansion Hybrid Structures," *Int. J. Solids Struct.*, **46**, 2372–87 (2009).

[5]C. A. Steeves, S. L. Lucato, M. Y. He, E. Antinucci, J. W. Hutchinson, and A. G. Evans, "Concepts for Structurally Robust Materials that Combine Low Thermal Expansion with High Stiffness," *J. Mech. Phys. Solids*, **55**, 1803–22 (2007).

[6]J. N. Grima, P. S. Farrugia, R. Gatt, and V. Zammit, "A System with Adjustable Positive or Negative Thermal Expansion," *Proc. R. Soc. A*, **463**, 1585–96 (2007).

[7]C. A. Steeves, C. Mercer, E. Antinucci, M. Y. He, and A. G. Evans, "Experimental Investigation of the Thermal Properties of Tailored Expansion Lattices," *Int. J. Mech. Mater. Des.*, **5**, 195–202 (2009).

[8]M. Danielsson, D. M. Parks, and M. C. Boyce, "Three-Dimensional Micromechanical Modeling of Voided Polymeric Materials," *J. Mech. Phys. Solids*, **50**, 351–79 (2002).

[9]K. Bertoldi, Harvard University, School of Engineering and Applied Sciences (discussion).

[10]Simulia. *ABAQUS 6.9-EF*. Simulia, Providence, RI, 2009.

[11]Mathworks. *MATLAB 2008b*. Mathworks, Natick, MA, 2008.

[12]M. A. Sutton, J. J. Orteu, and H. W. Schreier, *Image Correlation for Shape, Motion and Deformation Measurements*. Springer, Berlin, 2009. □

J. Am. Ceram. Soc., **94** [S1] S55–S61 (2011)
DOI: 10.1111/j.1551-2916.2011.04447.x
© 2011 The American Ceramic Society

Journal

Optimization of Thermal Protection Systems Utilizing Sandwich Structures with Low Coefficient of Thermal Expansion Lattice Hot Faces

Craig A. Steeves[†,‡] and Anthony G. Evans[*,§]

‡Institute for Aerospace Studies, University of Toronto, Toronto, ON, Canada M3H 5T6

§Department of Materials, University of California, Santa Barbara, California 93106

Atmospheric cruise hypersonic vehicles are subject to high viscous heating over large surface areas. Acreage thermal protection systems (TPSs) must be stiff, strong, and light while withstanding large thermal gradients and protecting the cool interior of the vehicle. It is a challenge to design thermal protection to minimize the thermal stresses caused by thermal expansion mismatch. This paper uses a recent concept for low-thermal-expansion periodic lattices to propose a sandwich configuration for acreage TPSs. A key aspect of these concepts is that they can be attached to coll structures without inducing thermal stresses during heating. Sandwich TPSs are analyzed and optimized for minimum mass for required performance characteristics, and compared with an optimized baseline system. For performance requirements relevant to atmospheric hypersonic flight, the sandwich TPSs using low-thermal-expansion periodic lattices are superior to the baseline system for a large range of operating conditions.

I. Background

THERMAL protection systems (TPSs) for the acreage surfaces of hypersonic vehicles are required to protect the vehicle interior from high viscous heating, as well as to provide structural strength and stiffness in order that the vehicle retains its aerodynamic shape.[1,2] For hypersonic cruise vehicles, the acreage surface temperature may be well above 1000°C,[3] which makes thermal expansion mismatch and the resulting thermal stresses a critical issue when a hot structure is connected to a cold underlying vehicle. Furthermore, TPSs must be designed to minimize mass and volume for required thermal and structural performance.

Because of the problems associated with thermal expansion mismatch, it is attractive to use materials with very low thermal expansion. Periodic lattices that exhibit this characteristic have been proposed, among others Lakes,[4] Sigmund and Torquato,[5] Grima *et al.*,[6] and Jefferson *et al.*[7] A recent concept for *relatively stiff* periodic lattices with low or zero net thermal expansion coefficient[8,9] will be utilized to provide a zero thermal expansion stiff layer, which operates at high temperature and can be incorporated as part of a sandwich-type TPS. This lattice geometry is shown in Fig. 1. The behavior and properties of low-thermal-expansion bimaterial lattices are described in Steeves *et al.*[8,9] To summarize, using standard high-temperature aerospace alloys it is possible to build bimaterial lattices that have a low, tailorable net thermal expansion coefficient, and

nearly optimal stiffness for an elastically isotropic lattice attaining a given coefficient of thermal expansion (CTE). This is achieved because the expansion of the individual members is accommodated by bending and rotation. Two parameters govern the thermal expansion behavior: the ratio Σ of the CTE of the two materials and the skewness θ as depicted in the figure. Increasing either Σ or θ decreases the effective CTE of the lattice. Critical to the functioning of this concept is the existence of "stationary nodes" which, if the lattice is designed to have zero thermal expansion, do not move relative to each other upon heating. These nodes can be connected to a cold structure without the generation of thermal stresses. As indicated in Fig. 1 there are type 1 stationary nodes at the intersections of six type 1 low-CTE struts in adjacent unit cells, as well as type 2 stationary nodes at the centers of the high-CTE type 2 material. While there is always solid material at the type 1 unit cell junctions, the space at the center of the type 2 material need not be filled. In addition, there are type 3 stationary nodes in the gap between unit cells (that is, at the center of an equilateral triangle formed by three type 2 stationary nodes), but this node is never filled by solid material and hence cannot be used as a connection point.

The configuration of the lattice unit cells is flexible; three configurations different from that shown in Fig. 1 are shown in Fig. 2. All of the unit cells have the same thermal expansion properties because the locations of the connections between the two materials are identical. Cell (a) is the most structurally efficient. Cell (b) fills the space within the continuous lattice but the two materials are connected at the same location and hence the thermal behavior is unchanged. Cell (c), described by Berger *et al.*,[10] incorporates spacers at the joints between adjacent cells. The spacers slightly increase the net CTE of the lattice for a fixed skewness angle, but permit much higher lattice skewness. By increasing the skewness, the net CTE can be significantly reduced, which greatly exceeds the effect of the uniform thermal expansion of the spacers. This enables an almost fully flat surface with only small slots between the materials. Moreover, the unit cell geometry can be such that at a design temperature, the gaps are fully closed (Berger *et al.*[10] for details).

The design of a TPS with minimum mass for a required thermal resistance and structural strength is discussed herein. Two configurations of systems are optimized and compared: a baseline TPS comprising a thick layer of low-density insulating material attached to a metallic plate, which provides stiffness and strength; and a sandwich-type TPS with low-density insulation sandwiched between two metallic faces, where the hot exterior face of the sandwich is a lattice-type structure with low or zero thermal expansion, shown in Fig. 1.[8] The two TPS configurations are sketched in elevation in Fig. 3; a three-dimensional schematic of the lattice-based TPS with the low-density ceramic foam removed is given in Fig. 4. For the case of the TPS, it is sensible for aerodynamic reasons to use a unit cell configuration shown as Fig. 2(c) to minimize the surface roughness.

A photograph of a prototype sandwich (with the ceramic foam insulation removed for visual access) is shown in Fig. 5. The cold face and the stiffeners will typically be composed of

R. McMeeking—contributing editor

Manuscript No. 28681. Received September 28, 2010; approved January 10, 2011.
This work was supported by the Office of Naval Research through the MURI program Revolutionary Materials for Hypersonic Flight (Contract No. N00014-05-1-0439).
*Member, The American Ceramic Society.
†Author to whom correspondence should be addressed. e-mail: csteeves@utias.utoronto.ca

lattice stiffness–skewness parameter Θ is:

$$\Theta = \frac{\cos\theta\cos^2(\theta+\pi/6)\cos^2(2\theta-\pi/6)}{\cos^2(2\theta-\pi/6)+\cos^2(\theta)\cos^2(\theta+\pi/6)+\cos^2(\theta+\pi/3)\cos^2(\theta+\pi/6)} \tag{10}$$

For a continuous beam loaded by pressure w on stringers spaced at L, the maximum stress induced in the lattice face is:

$$\sigma_{max} = \frac{wL^2}{12tc} \tag{11}$$

This expression implies that the neutral axis of bending is coincident with the cold solid face of the sandwich as described above. The most likely mode of failure for the lattice is for the low-CTE material 1 members to yield, which occurs at an applied axial stress on the lattice of:

$$\sigma = 2\sigma_1\tilde{r}\sin\left(\frac{\pi}{3}-\theta\right) \tag{12}$$

The axial stress in the lattice is due to the bending moment induced by the external pressure; the sandwich lattice face is assumed to carry all the axial loads from bending. The maximum pressure load the sandwich beam can carry before the lattice yields is, therefore:

$$w_{LY} = \frac{24ct\tilde{r}\cos(\pi/6+\theta)\sigma_1}{L^2} \tag{13}$$

It can be ensured that the lattice fails due to yielding by setting the aspect ratio $\tilde{r}$ to be sufficiently large to prevent buckling and by choosing an overall lattice unit cell size such that the thickness of the lattice, determined later, is large enough to prevent buckling for the chosen lattice strut length. The sandwich beam may alternately fail due to shearing of the core; the core shear strength is provided exclusively by the metallic component, which has strength:

$$\tau_c = \frac{v_m\sin^2\beta\sigma_m}{3} \tag{14}$$

where σ_m is the yield strength of the metallic component of the core and β the angle of the core reinforcing struts with respect to the lattice face. This estimate is very conservative because it assumes that the core members that are in compression, buckle at very low loads and hence make no contribution to the core shear strength. Consequently, the maximum pressure load that may be supported by the beam is:

$$w_{CS} = \frac{2v_m\sin^2\beta c\sigma_m}{3L} \tag{15}$$

For any choice of material and geometry, the sandwich will fail in the mode with minimum w for that configuration.

The mass of the sandwich TPS is:

$$M = bLt(a\rho_2+\rho_1) + bLc\rho_c \tag{16}$$

where ρ_1 and ρ_c are the lattice and core densities, respectively. For a lattice that has minimal gaps, the high-CTE material 2 is the dominant component, and hence:

$$\rho_1 \approx \rho_2 \tag{17}$$

The density of the composite core is given by a rule of mixtures:

$$\rho_c = v_m\rho_m + (1-v_m)\rho_f \tag{18}$$

with ρ_m the density of the shear stiffener material, which is typically the high-CTE material 2.

III. Optimization of TPSs

The goal of this analysis is to generate TPS designs that provide minimum mass for given thermal resistance and stiffness or strength. The optimal designs for both the sandwich TPS and the baseline system will be compared with determine the overall minimum mass design. The following nondimensional expressions will be used in the optimizations. Nondimensional performance indices will be denoted by $(\hat{\cdot})$, while nondimensional material parameters by $(\tilde{\cdot})$ and the geometric design parameters will be identified by $(\bar{\cdot})$. Analogous parameters and indices from the sandwich and baseline systems will use the same labels in order to make the comparison clearer.

For the baseline system, the mass index is:

$$\hat{M} = \frac{M}{bL^2\rho_s} \equiv \bar{c}(\bar{t}+\tilde{\rho}) \tag{19}$$

with the nondimensional parameters:

$$\bar{c} = c/L; \quad \bar{t} = t/c; \quad \tilde{\rho} = \rho_f/\rho_s$$

The stiffness index is:

$$\hat{EI} = \frac{EI_{eff}}{bL^3E_s} \equiv \frac{\bar{c}^3\bar{t}^3}{12} \tag{20}$$

The strength index is:

$$\hat{w} = \frac{w}{\sigma_s} \equiv 2\bar{t}^2\bar{c}^2 \tag{21}$$

The thermal resistance index is:

$$\hat{R} = \frac{R}{L/k_s} \equiv \bar{c}\left(\bar{t}+\frac{1}{\tilde{k}}\right) \tag{22}$$

with $\tilde{k} = k_f/k_s$.

The corresponding indices for the sandwich TPS are:

$$\hat{M} = \frac{M}{bL^2\rho_1} \equiv \bar{c}((\tilde{\rho}_1+a)\bar{t}+\tilde{\rho}_c) \tag{23}$$

again $\bar{t} = t/c$ and $\bar{c} = c/L$, while $\tilde{\rho}_1 = \rho_1/\rho_1$ and $\tilde{\rho}_c = \rho_c/\rho_1$. The stiffness index is:

$$\hat{EI} = \frac{EI_{eff}}{bL^3E_1} \equiv 2\sqrt{3}\tilde{r}\Theta\bar{c}^3\bar{t} \tag{24}$$

The two strength indices are, for lattice yielding:

$$\hat{w}_{LY} = \frac{w_{LY}}{\sigma_1} \equiv 24\cos(\pi/6+\theta)\tilde{r}\bar{c}^2\bar{t} \tag{25}$$

and for core shearing:

$$\hat{w}_{CS} = \frac{w_{CS}}{\sigma_1} \equiv 2v_m\sin^2\beta\bar{c}\tilde{\sigma} \tag{26}$$

where $\tilde{\sigma} = \sigma_m/\sigma_1$. The thermal resistance index is:

$$\hat{R} = \frac{R}{L/k_1} \equiv \bar{c}\left((1/\tilde{k}_1+a)\bar{t}+\frac{1}{\tilde{k}_c}\right) \tag{27}$$

where $\tilde{k}_1 = k_1/k_1$ and $\tilde{k}_c = k_c/k_1$.

The design requirements for the TPS, which are parameters of the problem, are $\hat{R}_{req}$, $\hat{EI}_{req}$ and $\hat{w}_{req}$.

(1) Optimization for Stiffness

Examining first the behavior of the baseline TPS, it is instructive to draw contours of the performance indices in geometric $(\bar{c},\bar{t})$ design space, in order to illustrate the relationships between the behavior of the stiffness, resistance, and mass. Figure 6(a) shows

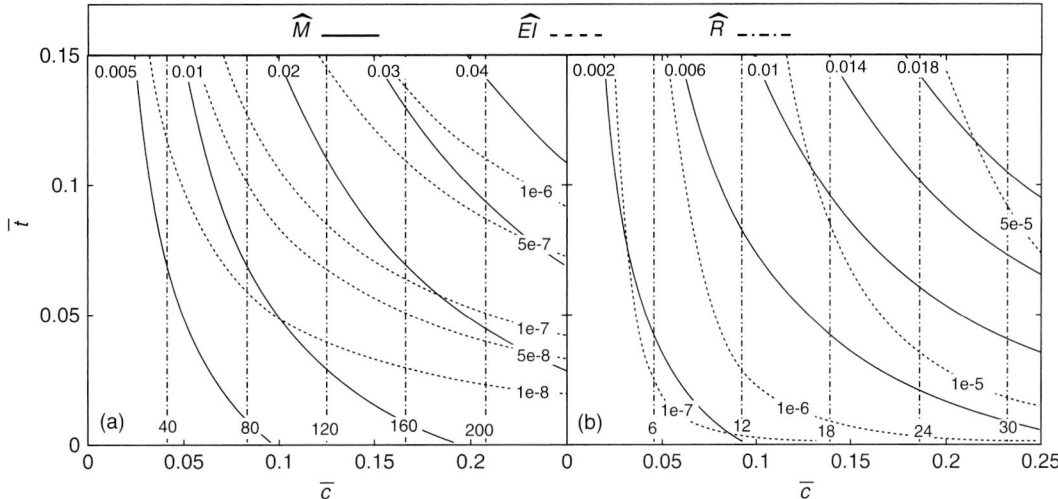

Fig. 6. Contours of $\hat{M}$, $\hat{R}$, and $\hat{E}I$ in $(\bar{c}, \bar{t})$ space (a) for the baseline TPS geometry using aluminum alloy 7075-T6 and ceramic insulating foam LI-900 and (b) for the sandwich TPS geometry using niobium alloy Cb-752, nickel–cobalt alloy NIMONIC PK33 and ceramic insulating foam LI-900 with $v_\mathrm{m} = 0.02$, $\tilde{r} = 0.06$ and $a = 0.25$.

an example of the performance of the baseline TPS comprising a solid face of aluminum 7075-T6 and the ceramic foam insulator LI-900; properties for both materials are found in Table I. In Fig. 6(a), the stiffness indices and the mass indices are strongly dependent upon both $\bar{c}$ and $\bar{t}$, while the resistance index is almost independent of $\bar{t}$. This is because the aluminum has a much higher conductivity (by three orders of magnitude) then the ceramic foam and hence does not contribute to the insulating capability of the baseline TPS.

To minimize mass for required minimum stiffness and resistance indices, it is necessary to find the location of the design point that has the smallest $\hat{M}$ for the required $\hat{E}I_\mathrm{req}$ and $\hat{R}_\mathrm{req}$. In this example, the feasible design region lies above the contour of $\hat{E}I_\mathrm{req}$ and to the right the contour of $\hat{R}_\mathrm{req}$. There are several possibilities for an optimal location. First, if the gradients of the performance indices are parallel to the gradient of the mass index, that is, $\nabla\hat{M} = \lambda\nabla\hat{R}$ or $\nabla\hat{M} = \lambda\nabla\hat{E}I$ (with λ a constant multiplier), optimal design may occur at one of these points provided the other performance requirement is also satisfied. This is not the case in this example: neither of the performance indices have gradients parallel to the gradient of the mass index. Second, if the contours for the desired performance indices, $\hat{w}_\mathrm{req}$ and $\hat{R}_\mathrm{req}$, intersect, the minimum mass design occurs at the intersection point. For the baseline system, this condition is met by:

$$\bar{t} = \frac{\left(12\hat{E}I_\mathrm{req}\right)^{\frac{1}{3}}}{\tilde{k}\left(\hat{R}_\mathrm{req} - \left(12\hat{E}I_\mathrm{req}\right)^{\frac{1}{3}}\right)} \quad (28)$$

with the corresponding

$$\bar{c} = \tilde{k}\left(\hat{R}_\mathrm{req} - \left(12\hat{E}I_\mathrm{req}\right)^{\frac{1}{3}}\right) \quad (29)$$

Recall that this is the baseline TPS, and the appropriate non-dimensionalizations must be used. If the contours of $\hat{R}_\mathrm{req}$ and $\hat{E}I_\mathrm{req}$ do not intersect (this occurs when the stiffness requirement dominates the design), then the optimal design is found at a boundary of the design region; in this case, the maximum $\bar{t}$ permissible will be chosen and the minimum $\bar{c}$ will be found which satisfies both $\hat{R}_\mathrm{req}$ and $\hat{E}I_\mathrm{req}$. Once the optimal pair $(\bar{c}, \bar{t})$ is found, the minimum mass index $\hat{M}_\mathrm{min}$ can be calculated. Trajectories of optimal design will follow either resistance index contours (for fixed required resistance) or stiffness index contours (for fixed required stiffness).

For the sandwich TPS, the optimization process is similar. Figure 6(b) shows the analogous contour plot of the performance indices and the mass index for the sandwich TPS, utilizing a lattice composed of the low-CTE niobium alloy Cb-752 and the high-CTE nickel–cobalt alloy NIMONIC PK33, with LI-900 as the core foam material and NIMONIC PK-33 as the metallic core component. (Note that the two contour plots are not directly comparable because of the different parameters used in the material property normalizations.) Several other geometric parameters must be specified; the net CTE of the lattice is 0 ppm/C, the volume fraction of metal in the core is $v_\mathrm{m} = 0.005$, the slenderness ratio of the lattice struts is $\tilde{r} = 0.05$, and the thickness ratio of the solid face to the lattice is $a = 0.25$. For the sandwich TPS, the gradient of the mass index is parallel to the gradient of the stiffness index at

$$\bar{t} = \frac{\tilde{\rho}_\mathrm{c}}{2(\tilde{\rho}_\mathrm{l} + a)} \quad (30)$$

for all values of $\bar{c}$. For this set of example materials and geometry, the optimal is found at $\bar{t} \approx 0.020$. The value of $\bar{c}$ corresponding to the particular $\hat{E}I_\mathrm{req}$ is:

$$\bar{c} = \left(\frac{\hat{E}I_\mathrm{req}(\tilde{\rho}_\mathrm{l} + a)}{\sqrt{3}\tilde{r}\Theta\tilde{\rho}_\mathrm{c}}\right)^{\frac{1}{3}} \quad (31)$$

Table I. Relevant Material Properties for the Materials Used in the Demonstration Optimization (Unused Data are Omitted)

	Density (kg/m³)	Young's modulus (GPa)	Yield strength (MPa)	Thermal conductivity (W·(m·K)⁻¹)	Thermal expansion (ppm/K)
LI-900	144			0.135	
Al 7075-T6	2810	72	505	130	
Cb-752	9030	110	257	48.7	7.4
NIMONIC PK33	8210	221		25.7	17.2
Toray T300	1760		3500	8.5	

Sources: Cb-752: ATI Wah Chang,[14] PK-33: Special metals,[15] LI-900: Banas.[16]

J. Am. Ceram. Soc., **94** [S1] S62–S75 (2011)
DOI: 10.1111/j.1551-2916.2011.04501.x
© 2011 The American Ceramic Society

Journal —————————————————————————————

Explorations of Hybrid Sandwich Panel Concepts for Projectile Impact Mitigation

Christian J. Yungwirth,[‡] John O'Connor,[§] Alan Zakraysek,[§] Vikram S. Deshpande,[¶] and
Haydn N. G. Wadley[†,‡]

[‡]Department of Material Science & Engineering, School of Engineering and Applied Science, University of Virginia,
Charlottesville, Virginia 22903

[§]Naval Surface Warfare Center, Indian Head, Maryland 20640

[¶]Engineering Department, Cambridge University, Cambridge CB2 1PZ, U.K.

Previous studies have shown that while stainless-steel sandwich panels with pyramidal truss cores have a superior blast resistance to monolithic plates of equal mass per unit area, their ballistic performance is similar to their monolithic counterparts. Here, we explore concepts to enhance the ballistic resistance without changing the volumetric efficiency of the panels by filling the spaces within the core with combinations of polyurethane, alumina prisms, and aramid fiber textiles. The addition of the polyurethane does not enhance the ballistic limit compared with the equivalent monolithic steel plate, even when aramids are added. This poor performance occurs because the polymer is penetrated by a hole enlargement mechanism which does not result in significant projectile deformation or load spreading and engagement of the steel face sheets. By contrast, ceramic inserts deform and erode the projectile and also comminute the ceramic. The ceramic communion (and resultant dilation) results in stretching of both steel face sheets and leads to significant energy dissipation. The ballistic limit of this hybrid is about twice that of the equivalent monolithic steel plate. The addition of a Kevlar fabric to the ceramic hybrid is shown to not significantly change the ballistic limit but does reduce the residual velocities of the debris.

I. Introduction

IT is well known that sandwich panels, with appropriate distribution of mass between the front and back faces and the core, exhibit superior bending stiffness and strength compared with monolithic (solid) plates of the same mass per unit area. Theoretical studies by Fleck and Deshpande[1] and Hutchinson and Xue[2] also predict that sandwich beams with porous cores will have superior shock resistance to monolithic beams. Subsequent experimental studies[3–5] in which edge clamped metallic panels with low-density lattice cores were subjected to high-intensity shock loading (in air and water) have confirmed these predictions and shown significant reductions in sandwich panel deflections compared with monolithic counterparts.

Explosive events in air are often accompanied by high-velocity fragments. Multifunctional protection systems that combine efficient structural load support with air shock and ballistic impact resistance are therefore of considerable interest. Recent experimental studies indicate that metallic sandwich panels with low relative density cellular cores (optimized for structural load support and shock resistance) have approximately the same ballistic performance as monolithic structures of equal areal mass.[6] The study reported here initiates an exploration of concepts that might be used to enhance the ballistic resistance of these structurally efficient sandwich panels.

The ballistic impact resistance of a simple metallic plate of fixed thickness depends upon its density, strength, ductility, and its strain and strain rate hardening characteristics. It can be significantly enhanced by replacing some of the plate mass with a hard ceramic tile placed on the impact side of the bilayer.[7–10] Even lighter alternatives have been proposed where the metal is replaced by a ballistic fiber composite laminate,[11–14] or sandwich panel,[15] or multilayer.[10] All seek to exploit the same defeat mechanism; namely, during impact, the hard ceramic plastically deforms or fractures the projectile, dissipating the projectiles kinetic energy by plastic work within it. This process is usually accompanied by comminution of the ceramic tile and inelastic stretching of the back face, which further reduces the projectile kinetic energy. Recently, Sarva *et al.*[16] have shown that that by restraining the impact-face of ceramic tiles with a membrane of suitable tensile strength, the ballistic efficiency can be improved by as much as 25% for a 2.5% increase in areal density. Deshpande and Evans[17] have shown that this remarkable effect is due to an enhancement of the strength of the ceramic by the extra confinement.

Blast resistant sandwich panels with low relative density cellular cores have significant empty space within the core: filling all or a part of this empty space with ballistic fibers, polymers, or ceramics offer potential opportunities to enhance the ballistic resistance of multifunctional sandwich panels. In order to explore the mechanisms that might be invoked to enhance the ballistic performance of sandwich structures, we experimentally investigate the effects of filling the empty space within a model stainless steel, pyramidal lattice core sandwich panel with polyurethane elastomers, aramid fiber fabrics, and alumina prisms (and combinations of the same). The study investigates the mechanisms of projectile arrest for fillings that span the disparate possibilities of soft, very high elastic strain to failure polymers to very hard, but brittle ceramics, and examines the implications of these fillings on both the panels ballistic limit and spatial extent of damage which influences multihit performance. The results are contrasted with the ballistic penetration mechanisms of a monolithic stainless plate spanning the same range of aerial masses.

F. Zok—contributing editor

Manuscript No. 28899. Received November 12 2010; approved February 11 2010.
The ballistic measurements were supported by the Defense Advanced Research Projects Agency and the Office of Naval Research under Grant number N00014-04-1-0299 (Dr. Leo Christodoulou was its program manager). The analysis work has been performed as part of the Ultralight Metallic Panels with Textile Cores Designed for Blast Mitigation and Load Retention program conducted by the University of Virginia and Cambridge University and funded by the Office of Naval Research (ONR) under Grant number N00014-01-1-1051 (Dr. David Shifler was the program manager).
[†]Author to whom correspondence should be addressed. e-mail: haydn@virginia.edu

II. Ballistic Penetration Mechanisms

There exists a large literature on the mechanics (experiments and modeling) of penetration of monolithic plates made from metallic alloys as well as ceramics. Here, we give a brief review of the key mechanisms of penetration in order to motivate structural concepts that might enhance penetration resistance. In the 1960s and early 1970s, numerous experimental studies were utilized to explore the impact processes and penetration mechanisms in plates, most notably by Hopkins and Kolsky[18] and Goldsmith.[19] A compendium on the mechanics of projectile penetration was published by Backman and Goldsmith[20] and a more recent review was written by Corbett *et al*.[21] Figure 1 depicts the most common failure modes encountered in cases where the projectile strength exceeds that of the target (so that the projectile remains more or less intact). In general these modes can be divided into regimes based on the ductility of the target.

(1) Low Ductility Regime

Three modes of failure are most commonly observed for low ductility target materials such as hardened aluminum and steel alloys as well as ceramics:

(1) *Spall fracture or scabbing (Fig. 1(a))*: A compressive elastic or shock wavefront emanates from the impact site and reflects as a tensile wave from the back or distal plate/air interface. A spall fracture occurs near the back surface when the tensile stress in the reflected (tensile) wave exceeds the tensile strength of the material.

(2) *Plugging (Fig. 1(b))*: Plugging results in a cylindrical slug, nearly the size of the projectile, being ejected from the target. In metals this typically occurs due to large shears at the periphery of the moving plug which result in adiabatic softening and shear failure of the metal. By contrast, in ceramics this type of failure is a result of the cone cracking mechanism.[13,22]

(3) *Radial fracture (Fig. 1(c))*: If the tensile strength of the target is lower than its compressive strength, then a radial fracture behind the initial compressive stress wave occurs due to the tensile radial stresses. This failure mode is often observed in both ceramics and high strength (low ductility) metals.

(2) High Ductility Regime

The two failure modes typically observed here occur during the penetration of high ductility metals (e.g., stainless steels) and polymers.

(1) *Petaling (Figs. 1(d) and (e))*: Petaling, both frontal (Fig. 1(d)) and rear (Fig. 1(e)), is produced by high radial and circumferential tensile stresses after passage of the initial wave near the lip of the penetration. This deformation is the result of bending moments created by the forward motion of the plate material as it is pushed ahead of the projectile and is initiated at inhomogeneities or weaknesses in the target. Petaling is often accompanied by large plastic flows and/or permanent flexure. As the material on the distal side of the plate is further deformed, a star-shaped crack is initiated by the tip of the projectile.[23] Finally, the sectors are rotated back by the ensuing motion of the projectile, often forming three to seven symmetric petals. Rearward petaling commonly occurs from ogival- or conical-shaped noses on projectiles penetrating thin ductile plates while frontal petaling occurs in thicker plates.

(2) *Ductile hole enlargement (Fig. 1(f))*: Ductile hole enlargement is a common mode of failure of thick plates impacted by ogival- or small-angle conical-shaped projectiles.[24] At the beginning of contact, the tip of the projectile begins displacing material radially leading to a radial momentum that continues so that a hole in the target is enlarged along the trajectory of the projectile. Heavily dependent on projectile shape, projectile diameter to target thickness ratio and projectile velocity, ductile hole enlargement is favored over plugging when the thickness of the plate exceeds the projectile diameter.

(1) Concepts to Enhance Ballistic Penetration via Hybrid Constructions

Monolithic systems undergo the penetration mechanisms discussed above and typically offer limited penetration resistance. The ballistic performance is enhanced by combining material systems—so-called hybrid constructions. Some examples of such hybrids and the mechanisms by which they enhance ballistic performance are:

(1) *Ceramic tiles backed by metallic or composite plates*: During penetration, the ceramic comminutes and erodes the projectile, reducing its kinetic energy and importantly resulting in a spreading of the applied load. The remnant energy is dissipated by the deformation of the back plate. The ceramic thus performs two roles as illustrated in Fig. 2(a): (a) it deforms and fragments (erodes) the projectile and (b) it spreads the force of the impacting projectile over a large area of the back plate—this inhibits the local penetration mechanisms of the back plate discussed above.

(2) *Ceramic tiles encased within a metallic or polymer case*: Upon impact the projectile penetrates the front cover easily by one of the modes in Fig. 1. Subsequently, it comminutes the ceramic. This comminuted ceramic dilates and expands, and therefore stretches the outer metallic or polymer fiber casing as illustrated in Fig. 2(b). This mechanism efficiently distributes the load and engages a large fraction of the panel in the energy dissipation process.

Ultrahigh specific strength aramid and polyethylene fibers in the form of ballistic fabrics are also widely used with metal/ceramic protection systems to arrest projectile and ceramic debris.[25] The performance of the above-mentioned systems can typically be enhanced by including a ballistic fabric such as Kevlar on the back face of the hybrid system: this fabric catches any small debris from the impacted target or eroded projectile. Here, we explore concepts for enhancing the ballistic response of sandwich panels by *inserting* various materials within them. The study therefore focuses upon methods for increasing performance without increasing the volume of the structure. Figure 3(a) shows the baseline (empty) sandwich panel

Low Ductility

(a)

High Ductility

(d)

Spall fracture

Petaling (frontal)

(b)

(e)

Plugging

Petaling (rearward)

(c)

(f)

Radial fracture

Ductile hole enlargement

Fig. 1. Sketches of the penetration mode of monolithic plates. The modes are divided into two regimes based on the ductility of the target material.

Fig. 5. Sketch of the punching and folding operation to manufacture the pyramidal truss lattice core.

Table II. **Manufacturer Reported Properties for the Polyurethane (PU 1)**

Property	
Manufacturer	Smooth-On
Product name	PMC-780 dry
Tensile nodulus (MPa)	2.76
Tensile strength (MPa)	6.21
Elongation to break (%)	700
Shore hardness	80 A

Kevlar-Flex-Pro was laser cut into two different designs to accommodate the arrangement of nodal contact points on the front and back face sheets (Figs. 6(a) and (b), respectively). The cut fabric sheets were placed on the inner surfaces of the face sheets before laser welding of the truss cores. Subsequently, the panel was filled with the polyurethane as described above.

(3) Construction of the Ceramic-Filled Panels

Alumina (Al_2O_3) manufactured by CoorsTek (Golden, CO) with the grade designation AD-94 was used to fill the empty truss core sandwich panels. The alumina was diamond saw cut into isosceles triangular cross-section prisms with an apex angle of 70°, base angle of 55°, and base length 3.0 cm. The prisms

were 11.4 cm long. The manufacturers' mechanical property data for AD-94 is listed in Table III. After insertion, the remaining small gaps in the sandwich core were filled with the polyurethane described in Section III(2).

The process for manufacturing the hybrid ceramic-filled lattice panels involved first including the Flex-Pro fabric in the sandwich panel as described in Section III(2). Subsequently, ceramic prisms were inserted into the panel and the gaps filled with polyurethane. The main difference here is that the presence of the Flex-Pro meant that the ceramic prisms were necessarily slightly smaller and had a base length of 2.4 cm, a height of 1.8 cm, an apex angle of 70°, and base angles of 55°.

(4) Mechanical Properties of the Steel and Polyurethane

Tensile specimens of dog-bone geometry were cut from each of the as-received steel sheets. The uniaxial tensile responses of the 304 stainless-steel alloy at an applied strain-rate of 10^{-3} s^{-1} is plotted in Fig. 7(a) using axes of true stress and logarithmic strain. The 304 stainless steel displays a linear hardening post-yield response with a tangent modulus $E_t \approx 1$ GPa. This high strain hardening capacity of the stainless steel also stabilizes the tensile specimens against necking, resulting a very high tensile ductility. The polyurethane used here is well above its glass transition at room temperature ($T_g = -56°C$). The compressive stress–strain curves for this material are shown in Fig. 7(b) for a

Fig. 6. Sketch of the Kevlar fabric cut to accommodate the nodes of the pyramidal core on the (a) front and (b) back face sheets.

Table III.   Manufacturers Data for the Physical Properties of the AD-94 Al_2O_3 Triangular Prisms

Density (g/cm³)	Elastic modulus (GPa)	Flexural strength (MPa)	Tensile strength (MPa)	Compressive strength (MPa)	Fracture toughness (MPa·m$^{1/2}$)	Hardness (GPa)
3.97	303	358	221	2068	4–5	11.5

range of strain rates. These were measured by McShane *et al.*[26] using a split Hopkinson bar set-up. The tangent modulus of the polyurethane increases with strain rate, which is advantageous for the ballistic performance of this polymer.

IV.   Dynamic Test Protocol

The ballistic performance of the above-described panel systems was investigated for projectile impact velocities in the range $225 \leq v_p \leq 1700$ m/s. In addition, for comparison purposes, we also investigated the ballistic performance of monolithic 304 stainless-steel plates of thickness 3 and 10 mm with areal masses $m = 24$ and 80 kg/m, respectively. These plates span the areal masses of the sandwich panel systems and provide a useful basis to compare and rank the performance of the different concepts.

Impact experiments were performed on 25 sandwich plates of each configuration. In all cases, the plates were impacted at zero obliquity and midspan by a spherical, 1020 plain carbon steel projectile of diameter 12.5 mm weighing approximately 8.4 g. Ballistic testing was conducted using a powder gun comprising a breech and a gun barrel as sketched in Fig. 8. The gun propelled plastic sabots carrying the 12.5 mm spherical steel projectiles. An electric solenoid activated a firing pin, which initiated 0.38 caliber blank cartridges (Western Cartridge Company, East Alton, IL). The mixture of solid smokeless propellant IMR 3031, manufactured by IMR (Shawnee Mission, KS), and cotton (Fig. 8) in the breech was ignited by this charge and the expanding propellant gas accelerated the sabot through the gun barrel. The purpose of the cotton was to ensure the ensuing pressure wave remained uniform throughout deflagration of the propellant. The sabot was located within a 25.4 mm bore gun barrel: a series of holes placed along the gun barrel were used to dissipate the shock wave and maintain a smooth acceleration of the sabot until it exited the barrel. The velocity at which the sabot exited the gun barrel was adjusted by selecting an appropriate quantity of gunpowder. The plastic sabot comprised four quarters that, upon mating, surrounded the 12.5 mm diameter spherical projectile. The sabot plugs had an inner diameter of 1.8 cm, an outer diameter of 2.7 cm, a height of 3.5 cm, and weighed 18.6 ± 0.12 g. Separation of the sabot from the projectile by air drag was facilitated by a 40° bevel at the sabot opening. The 12.5 mm diameter spherical steel projectiles weighed $M = 8.42 \pm 0.02$ g and were manufactured by National Precision

Ball (Preston, WA) from 1020 plain carbon steel with an ultimate tensile strength of approximately 375 MPa.[6]

The sample test fixture was located within a blast chamber (Fig. 9). A square, 40 cm long, 2.86-cm-thick steel plate was located one meter from the end of the barrel. It had a 3.8 cm diameter hole located in the center through which the projectile entered the test area. Two pairs of brake screens were used to measure the projectile entry and exit velocities (Fig. 9) and provided impact and exit velocity measurements with a precision of ± 2.0 m/s. The test samples were edge clamped along the top and bottom edges so that the effective span of the plate between the clamped edges was approximately 110 mm while the width of the plate was 120 mm; see Fig. 4. The kinetic energy of the projectile dissipated by the sandwich panel structures depends upon the way in which the projectile interacts with the truss cores. To remove this source of variability in the measurements, the sandwich panels were carefully positioned so that the projectile impacted at the center of the square formed by four nodes of the truss on the face-sheet facing the incoming projectile. The projectile usually impacted the back face sheet at the apex of a truss. High-speed photography was used to observe the dynamic transverse deformation and failure of the plates. An Imacon 200 digital framing camera (Intronix Imaging Technologies, Westlake Village, CA) was used for this purpose; this camera is capable of taking up to 16 frames at a maximum rate of 10^8 frames/s. Interframe times of in the range 4.5–50 µs were used and the exposure time was 300 ns. In addition, the plates were examined after each experiment to understand the failure mechanisms.

V.   Experimental Observations

We proceed to detail the observations in two steps. First we discuss the ballistic performance of the polymer-filled lattices and then contrast this behavior with that of the ceramic-filled sandwich plates. In order to facilitate comparisons between the different systems, the areal masses of the five sandwich systems (including the unfilled pyramidal core sandwich from Yungwirth *et al.*[6]) and two monolithic plates investigated are listed in Table I.

(1)   Polymer-Filled Lattice Plates

The measured projectile exit or residual velocity v_r as a function of the impact velocity v_p is plotted in Fig. 10(a). Full penetration

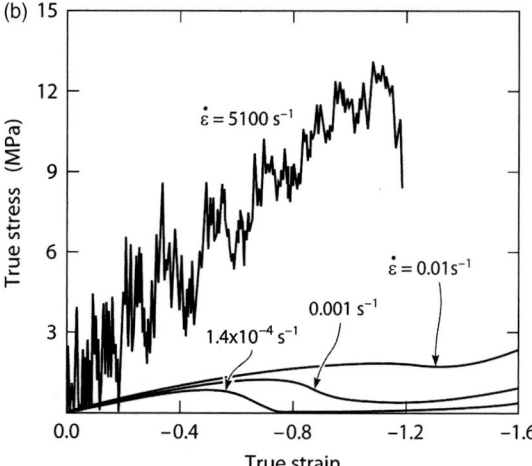

Fig. 7. (a) Measured quasi-static uniaxial stress versus strain curves for the as-received 304 stainless steel and (b) the room temperature compressive response of the polyurethane over a range of strain rates.

Fig. 12. Photographs of the hybrid polymer-filled lattice core sandwich plates impacted at (a) $v_p = 274$ m/s, (b) $v_p = 499$ m/s, (c) $v_p = 731$ m/s, and (d) $v_p = 1181$ m/s. For comparison purposes, the photograph of the unfilled sandwich panel impacted at $v_p = 1206$ m/s is included in (e). The photographs were taken after sectioning the plates along their midplane.

target response. At impact velocities $v_p = 299$ and 371 m/s (Figs. 11(a) and (b), respectively), the projectile penetrated the front face but was arrested by the rear face of the sandwich panels. In these figures we clearly see (i) a shear-off failure of the front face with nearly no spreading of the deformation; (ii) the tensile failure of the nodes connecting the pyramidal core to the rear face near the impact site; (iii) the stretching and bending of the rear face; and (iv) rehealing of the polymer after penetration. This clearly illustrates that the front face and to some extent the polymer-filled core slow the projectile sufficiently so that the mode of deformation/failure of the rear face switches from shear-off on the front face to stretching on the rear face and this engages a significantly larger fraction of the face sheet material. At the higher impact velocities of $v_p = 516$ and 985 m/s (Figs. 11(c) and (d), respectively), the projectile penetrates the rear face sheet, by a petaling mechanism associated with face sheet stretching. Note that, in all cases, penetration through the polymer core seems to occur by a cavity expansion mechanism but the very high elastic recovery strain of the polymer means that the cavity closes behind the penetrating projectile leaving behind an incipient flaw as seen in Fig. 11(c).

The addition of the FlexPro Kevlar fabric into the sandwich core (i.e., the hybrid polymer-filled lattice plates) does not significantly affect the ballistic limit (Fig. 10(a)) but again seems to reduce the residual velocities of the penetrated projectiles compared with the polymer-filled lattice plates. Given that the areal masses of both these systems is approximately the same (Table I), we can conclude that the FlexPro fabric does serve in increasing the energy absorption capacity of the panel (as the deformation of the projectiles is negligible, the kinetic energy of the incoming projectile is primarily dissipated by the deformation of the panel). A sequence of images of the tested and sectioned panels is included in Fig. 12: these images clearly show that the Kevlar fabric does not change the shear-off mode on the front face but undergoes some stretching on the rear face compared with the polymer-filled lattices shown in Fig. 11. This stretching is the likely reason why the addition of the fabric seems to further reduce the residual velocities of the projectile. The presence of a pattern of circular holes in the fabrics (Fig. 6) may also have contributed to the poor performance because this resulted in a fraction of the Kevlar tows having an effective length of only 30 mm. It is worth emphasizing here that placing the Kevlar

Fig. 13. Measurements of the projectile exit velocity v_r as a function of the impact velocity for the ceramic-filled sandwich plates and hybrid ceramic-filled plates. The corresponding data for the 10-mm-thick steel plate is also included.

fabric within the polymer-filled core highly constrained the transverse deformation of the fabric, preventing it from fully stretching and acting in a "catcher mode." This is the primary reason for the rather disappointing performance of these hybrid panels—placing the Kevlar behind the back face of the sandwich panel is expected to give improved performance.

These polymer-filled lattice plates show minimal performance benefits over their monolithic counterparts. We argue that this is due to two main reasons: (i) the polymer filling is sufficiently soft that it does not plastically deform or erode the steel projectile, and (ii) the cavity expansion deformation mode of the polymer does not efficiently distribute the load to the panel faces so the mode of deformation of the polymer-filled sandwich plate

remains unchanged from the empty sandwich plate; compare Figs. 12(d) and (e).

(2) Ceramic-Filled Lattice Plates

The polymer-filled lattice plates show minimal performance benefits over their monolithic counterparts for the reasons outlined above. Filling the sandwich plates with ceramics is expected to address both of these drawbacks.

The measured projectile exit or residual velocity v_r as a function of the impact velocity v_p is plotted in Fig. 13 for both the types of ceramic-filled sandwich panels. The corresponding data for a 10-mm-thick monolithic 304 stainless-steel plate (with areal mass approximately equal to the ceramic-filled lattice panel; see Table I) is included in the figure. The ballistic limit is seen to increase from about 900 m/s for the monolithic plate to about 1600 m/s for the ceramic-filled lattice panels with an equal areal mass.

This increase in the ballistic limit can be rationalized by observing the deformation/failure modes of the sectioned (as-tested) panels. Figures 14 and 15 include photographs of the as-tested 10-mm-thick monolithic steel plate and ceramic-filled lattice panels specimens, respectively. First consider the monolithic plates. At an impact velocity below the ballistic limit ($v_p = 540$ m/s), the projectile is seen to undergo extensive deformation with a small indent forming on the plate (Fig. 14(a)). At just about the ballistic limit ($v_p = 804$ m/s), the projectile again undergoes extensive deformation, but in addition the plate is seen to fail/deform in a plugging type mode wherein a plug is pushed out of the plate directly ahead of the projectile with little deformation away from the impact site (Fig. 14(b)). At higher impact velocities, the projectile penetrates by pushing through a shear plug. These shear plugs leave behind plastic lips at the impact and distal surfaces of the plate.

These deformation modes of the monolithic plate can be contrasted with the corresponding observations for the ceramic-filled lattice plates shown in Fig. 15. At low impact velocities

Fig. 14. Photographs of 10-mm-thick monolithic stainless-steel plate impacted at (a) $v_p = 542$ m/s, (b) $v_p = 804$ m/s, (c) $v_p = 979$ m/s, and (d) $v_p = 1218$ m/s. The photographs were taken after sectioning the plates along their midplane. In (b) the heavily deformed projectile and still attached "plug" can be seen.

Fig. 15. Photographs of the ceramic-filled lattice core sandwich plates impacted at (a) $v_p = 326$ m/s, (b) $v_p = 543$ m/s, (c) $v_p = 1060$ m/s, and (d) $v_p = 1233$ m/s. The photographs were taken after sectioning the plates along their midplane.

($v_p = 326$ and 543 m/s), the impact face sheet deforms by a shear plugging mode with cracking observed in the ceramic prism directly underneath. No other significant residual deformation is observed. This is rationalized as follows: the impacting projectile easily forms a shear plug in the steel face sheet but then undergoes extensive plastic deformation as it encounters the hard ceramic. This ceramic prism is constrained by the surrounding prisms, the steel face sheet and the incompressible elastomer. Recall that Deshpande and Evans[17] have shown that confinement significantly enhances the fracture strength of the ceramic and thus confined ceramic prisms here undergoes limited cracking. This limited cracking combined with the high elastic wave speed in the ceramic enables the impacted ceramic prism to very effectively redistribute the load exerted by the projectile over a significant fraction of the back face sheet and thus we observe little or no deformation of the back face sheet of the panel. At higher impact velocities (Figs. 15(c) and (d)), the deformation is more widespread with more extensive cracking of the ceramic.

This cracking results in dilation (or bulking) of the ceramic, which is resisted by the stretching of the lattice trusses until they fail at their nodes and by the stretching of the steel face sheets of the sandwich panel thereafter.

Like the polymer-filled panels, the addition of the Kevlar FlexPro fabric to the panels does not change the ballistic limit significantly (Fig. 13). This is because the FlexPro does not seem to significantly change the deformation mode of the panels compared with the panels without the Kevlar; contrast Figs. 15 and 16. The Kevlar fabric once again was sufficiently confined within the sandwich that stretching of the fabric (and thus energy dissipation) was inhibited. This resulted in the fabric failing by a shear-off (cutting) mechanism rather than the efficient energy absorbing stretching mode.

In order to gain further insight into these deformation mechanisms a high-speed photographic sequence of the deformation of the hybrid ceramic-filled lattice panel impacted at $v_p = 789$ m/s is included in Fig. 17. The timing of each photograph is marked

Fig. 16. Photographs of the hybrid ceramic-filled lattice core sandwich plates impacted at (a) $v_p = 790$ m/s and (b) $v_p = 1243$ m/s. The photographs were taken after sectioning the plates along their midplane.

Fig. 17. High-speed photographic sequence (exposure time of 300 ns) of the impact of the hybrid ceramic-filled sandwich plate at $v_p = 789$ m/s. Time after the impact of the projectile against the proximal face-sheet is indicated for each frame.

on the figures with $t = 0$ taken as the instant of the impact against the proximal or front face-sheet. Consistent with the basic supersonic flow theory,[27] at $v_p = 789$ m/s (i.e., Mach number $M_\infty \equiv v_p/c \approx 2.25$, where $c = 340$ m/s is the velocity of sound in the undisturbed air), a detached shock wave is formed in front of sphere with a shock wave angle[||] greater than the Mach angle $\omega = \sin^{-1}(c/v_p)$. Penetration of the front face by the projectile results in a debris cloud being ejected backwards. The projectile then enters the sandwich plate and a significant volumetric expansion of the sandwich plate is observed: consistent with the images in Fig. 16, the front face undergoes minimal deformation while the rear face is seen to undergo significant stretching over nearly its entire span. This deformation mode results in the capture of the projectile within the target.

VI. Discussion

The ballistic limit v_c of all the panels investigated is summarized in Fig. 18 as a function of the panel areal masses (sometimes referred to as the specific density m_s). We emphasize that these results are valid for only the spherical, 8 g 1020 plain carbon steel projectile used in this study. Figure 18 clearly illustrates that the monolithic, empty and polymer-filled sandwich panels have an equivalent performance, i.e. the ballistic limit of these panels increases approximately linearly with areal mass at the

rate of 11 m/s for an increase in areal mass of 1 kg/m. We thus conclude that there is no performance benefit in terms of the ballistic limit in utilizing either empty or polymer-filled sandwich panels. This is rationalized by noting that the hydrodynamic stresses generated within the polymer (i.e., inertial stresses of order ρv^2, where ρ is the polymer density and v the impact velocity) are significantly larger than the material strength (Fig. 7) for all velocities near and above the ballistic limit. This implies that the only resistance to penetration provided by the polymer is due to its mass, i.e. no performance gains on a mass basis. The incorporation of layers of a Kevlar fabric within the elastomer-filled lattice had surprisingly little effect upon the ballistic resistance of the panel (Fig. 10) because of its confinement by the elastomer and by the panel's rear face sheet.

By contrast, the ceramic-filled sandwich panels have a significantly higher ballistic resistance compared with their monolithic counterparts and lie well above the $11 \text{ m} \cdot (\text{s} \cdot \text{kg} \cdot \text{m}^2)^{-1}$ line that defines the performance of the monolithic, empty and polymer-filled sandwich plates. In fact the ballistic limit v_c of the ceramic-filled sandwich plates is about twice that of their monolithic counterparts. These performance benefits arise because the ceramic inserts completely change the penetration modes of the sandwich plates by switching-off the plugging deformation mode as discussed above.

This study has shown that the ballistic performance of sandwich structures can be significantly enhanced by using ceramic inserts to fill the empty spaces within the truss core sandwich structures. The ceramic used here is an armor grade alumina

---

[||]The shock wave angle is the angle that the shock wave makes with the direction of motion of the projectile.

Fig. 18. A summary of ballistic limit v_c for the sandwich structures and solid 304 stainless-steel plates that span the areal density range of the hybrid sandwich panels.

ceramic. We note in passing that using higher performance ceramics such as SiC could further enhance the performances reported here. For example C. J. Yungwirth, and H. N. G. Wadley (private communication) have conducted a study wherein the alumina prisms utilized here in the hybrid ceramic-filled sandwich panels were replaced by SiC prisms. The ballistic limit was approximately equal to the equivalent alumina-filled panel. However, the density of SiC was about 18% less than that of alumina and thus in terms of mass efficiency the SiC panels had a superior ballistic performance. We note that the ballistic performance of ceramics could be significantly enhanced (especially for brittle projectiles) by increasing the ceramic confinement.[28,29] Espinosa *et al.*[30] have shown that this inhibits comminution and penetration through the cracked ceramic.[31] While the focus of this study has explored ways of improving the ballistic performance of sandwich panels by inserting materials *within* the structure, significant improvements are also possible by placing the ceramic on the outside of strike face (H.N.G. Wadley *et al.*) or attaching the Kevlar fabric to the rear face of the structure (both at the expense of a reduced volumetric efficiency). We also note that the study has not attempted to optimize the mass distribution amongst the various components of the system. Using the data and dimensionless parameter approach of Cuniff,[32] it can be shown that a monolithic panel of Kevlar 29 composite with an areal density of 80 kg/m^2 has a ballistic limit several hundred meters per second higher than that of the ceramic-filled system. Allocating a larger fraction of the panels mass to a Kevlar-based textile attached to the rear of the structure is therefore anticipated to result in significant performance improvements.

In this study we have demonstrated that filling empty spaces within sandwich panels with ceramic inserts enhances their bal-

listic performance. It remains to be determined what effect these fillings have on the performance of sandwich panels in terms of their blast resistance. If core compressibility is an essential requirement for the superior blast performance of sandwich panels, multi-layer concepts (e.g., see Fig. 19) that combine a compressible core with a hard core might be required in order to endow the structures with this multifunctional capability. This is a topic for future investigations.

VII. Concluding Remarks

The ballistic performance of stainless-steel pyramidal core sandwich panels with the empty spaces filled with polyurethane or alumina prisms was investigated. The measurements show that addition of the polyurethane does not enhance the ballistic limit compared with the equivalent monolithic steel plate, but it reduces the residual velocity of the penetrated projectiles. This poor performance of the polymer-filled lattice is due to the fact that the polymer is penetrated by a hole enlargement mechanism which does not spread the load and engage the steel sandwich casing adequately, i.e. the failure mode of the polymer-filled sandwich panel remains unchanged from that of the empty sandwich panel. By contrast, the ceramic inserts comminute and dilate while eroding the projectile. This results in the stretching of the steel face sheets and significant energy dissipation. The ballistic limit of this structure is about twice that of the equivalent monolithic steel plate. The addition of a Kevlar fabric within the systems is shown to not significantly change the ballistic limit but help reduce the residual velocities of the penetrated projectile and panel debris. This is attributed to the fact that the Kevlar fabric is significantly constrained within the sandwich and is unable to flex and stretch and thus fails by its weak shear-off (cutting) mechanism.

Acknowledgments

The views, opinions, and/or findings contained in this article are those of the authors and should not be interpreted as representing the official views or policies, either expressed or implied, of the Defense Advanced Research Projects Agency or the Department of Defense. Distribution Statement "A" (Approved for Public Release, Distribution Unlimited).

References

[1]V. S. Deshpande and N. A. Fleck, "One-Dimensional Response of Sandwich Plates to underwater Shock Loading," *J. Mech. Phys. Solids*, **53** [11] 2347–83 (2005).
[2]J. W. Hutchinson and Z. Xue, "Metal Sandwich Plates Optimized for Pressure Impulses," *Int. J. Mech. Sci.*, **47** [4–5] 545–69 (2005).
[3]L. F. Mori, S. Lee, Z. Y. Xue, A. Vaziri, D. T. Queheillalt, K. P. Dharmasena, H. N. G. Wadley, J. W. Hutchinson, and H. D. Espinosa, "Deformation and Fracture Modes of Sandwich Structures Subjected to underwater Impulsive Loads," *J. Mech. Mater. Struct.*, **2** [10] 1981–2006 (2007).
[4]H. N. G. Wadley, K. P. Dharmasena, D. T. Queheillalt, Y. Chen, P. Dudt, D. Knight, K. Kiddy, Z. Xue, and A. Vaziri, "Dynamic Compression of Square Honeycomb Structures during Underwater Impulsive Loading," *J. Mech. Mater. Struct.*, **2** [10] 2025–48 (2007).
[5]K. P. Dharmasena, D. T. Queheillalt, H. N. G. Wadley, P. Dudt, Y. Chen, D. Knight, A. G. Evans, and V. S. Deshpande, "Dynamic Compression of Metallic Sandwich Structures during Planar Impulsive Loading in Water," *Eur. J. Mech.-A/Solids*, **29** [1] 56–67 (2010).
[6]C. J. Yungwirth, H. N. G. Wadley, J. O'Connor, A. Zakraysek, and V. S. Deshpande, "Impact Response of Sandwich Plates with a Pyramidal Lattice Core," *Int. J. Impact Eng.*, **35** [8] 920–36 (2008).
[7]M. L. Wilkins, C. F. Cline, and C. A. Honodel, "Fourth Progress Report of Light Armour Program"; Report UCRL 50694, Lawrence Radiation Laboratory, University of California, 1969.
[8]M. L. Wilkins, "Mechanics of Penetration and Perforation," *Int. J. Eng. Sci.*, **16** [11] 793–807 (1978).
[9]M. Mayseless, W. Goldsmith, S. P. Virostek, and S. A. Finnegan, "Impact on Ceramic Targets," *J. Appl. Mech.*, **54** [2] 373–8 (1987).
[10]B. A. Gama, T. A. Bogetti, B. K. Fink, C. Yu, T. D. Claar, H. H. Eifert, and J. W. Jr. Gillespie, "Aluminum Foam Integral Armor: A New Dimension in Armor Design," *Comput. Struct.*, **52** [3–4] 381–95 (2001).
[11]C. Navarro, M. A. Martinez, R. Cortes, and V. Sanchez-Galvez, "Some Observations on the Normal Impact on Ceramic Faced Armours Backed by Composite Plates," *Int. J. Impact Eng.*, **13** [1] 145–56 (1993).
[12]I. S. C. Benloulo and V. Sanchez-Galvez, "A New Analytical Model to Simulate Impact onto Ceramic/Composite Armours," *Int. J. Impact Eng.*, **21** [6] 461–71 (1998).

Fig. 19. Sketch of a multilayer sandwich structure comprising a sandwich core with ceramic inserts as the top layer and the empty sandwich panel as the bottom layer. This structure is expected to have multifunctional capabilities in terms of both blast and ballistic performances.

[13]M. J. Normandia, J. C. LaSalvia, W. A. Gooch, and J. W. McCauley, "Protecting the Future Force: Ceramics Research Leads to Improved Armor Performance," *Amptiac Q.*, **8** [4] 21–7 (2004).

[14]P. M. Cunniff, "Dimensionless Parameters for Optimization of Textile Based Body Armor Systems"; pp. 1303–10 *Proceedings of the 18th International Symposium on Ballistics*, San Antonio, TX, 1999.

[15]H. Senf, E. Strassburger, and H. Rothenhausler, "Investigation of Bulging during Impact in Composite Armor," *J. Phys. IV Fr.*, **7** [C3] 301–6 (1997).

[16]S. Sarva, S. Nemat-Nasser, J. McGee, and J. Isaacs, "The Effect of Thin Membrane Restraint on the Ballistic Performance of Armor Grade Ceramic Tiles," *Int. J. Impact Eng.*, **34** [2] 277–302 (2007).

[17]V. S. Deshpande and A. G. Evans, "Inelastic Deformation and Energy Dissipation in Ceramics: A Mechanism-Based Dynamic Constitutive Model," *J. Mech. Phys. Solids*, **56**, 3077–100 (2008).

[18]H. G. Hopkins and H. Kolsky, "Mechanics of Hypervelocity Impact of Solids"; *Proceedings of the 4th Hypervelocity Impact Symposium APGC-TR-60-39*, Eglin Air Force Base 1, 1960.

[19]W. Goldsmith, *Impact*. Arnold, New York, 1960.

[20]M. Backman and W. Goldsmith, "The Mechanics of Penetration of Projectiles into Targets," *Int. J. Eng. Sci.*, **16**, 1–99 (1978).

[21]G. Corbett, S. Reid, and W. Johnson, "Impact Loading of Plates and Shells by Free-Flying Projectiles: A Review," *Int. J. Impact Eng.*, **18**, 131–230 (1996).

[22]F. C. Frank and B. R. Lawn, "On the Theory of Hertzian Fracture," *Proc. R. Soc.*, A**299**, 291–306 (1967).

[23]G. Corbett, S. Reid, and W. Johnson, "Impact Loading of Plates and Shells by Free-Flying Projectiles: A Review," *Int. J. Impact Eng.*, **18**, 131–230 (1996).

[24]M. Backman and W. Goldsmith, "The Mechanics of Penetration of Projectiles into Targets," *Int. J. Eng. Sci.*, **16**, 1–99 (1978).

[25]B. A. Cheeseman and T. A. Bogetti, "Ballistic Impact into Fabric and Compliant Composite Laminates," *Compos. Struct.*, **61** [1–2] 161–73 (2003).

[26]G. J. McShane, C. Stewart, M. T. Aronson, H. N. G. Wadley, N. A. Fleck, and V. S. Deshpande, "Dynamic Rupture of Polymer–Metal Bilayer Plates," *Int. J. Solids Struct.*, **45** [16] 4407–26 (2008).

[27]H. W. Liepmann and A. Roshko, *Elements of Gas-Dynamics*. Dover Publications Inc., New York, 1985.

[28]G. Hauver, P. Netherwood, R. Benck, and L. Kecskes, "Ballistic Performance of Ceramic Targets"; *Proceedings of the Army Symposium on Solid Mechanics*, Plymouth, MA, 1993.

[29]F. Malaise, J. Y. Tranchet, and F. Collombet, "An Experimental Investigation of Ceramic Block Impenetrability to High Velocity Long Rod Impact"; *Proceeding of the 6th International Conference on Mechanical and Physical Behaviour of Materials under Dynamic Loading*, Cracow, Poland, 2000.

[30]H. D. Espinosa, N. S. Brar, G. Yuan, Y. Xu, and V. Arrieta, "Enhanced Ballistic Performance of Confined Multi-Layered Ceramic Targets Against Long Rod Penetrators through Interface Defeat," *Int. J. Solids Struct.*, **37** [36] 4893–914 (2000).

[31]B. A. Gailly and H. D. Espinosa, "Modeling of Failure Mode Transition in Ballistic Penetration with a Continuum Model Describing Microcracking and Flow of Pulverized Media," *Int. J. Numerical Methods Eng.*, **54** [3] 365–98 (2002).

[32]P. M. Cuniff, "Dimensionless Parameters for Optimization of Textile-Based Body Armor Systems"; *18th International Symposium on Ballistics*, San Antonio, TX, November 15–19, 1999. □

J. Am. Ceram. Soc., **94** [S1] S76–S84 (2011)
DOI: 10.1111/j.1551-2916.2011.04601.x
© 2011 The American Ceramic Society

journal

In-Plane Compression Response of Extruded Aluminum 6061-T6 Corrugated Core Sandwich Columns

Russell Biagi, Jae Yong Lim, and Hilary Bart-Smith[†]

Mechanical and Aerospace Engineering Department, University of Virginia, Charlottesville, Virginia, 22904, U.S.A.

The compression response of extruded aluminum 6061-T6 corrugated core sandwich columns is investigated. Analytical equations that predict the collapse load are used to generate failure mechanism maps. From these maps dominant failure mechanisms can be identified as a function of various geometric parameters and material properties. Experimental testing and numerical simulations are performed to test the fidelity of the analytical predictions. Fabrication of the sandwich panels involves extrusion of an aluminum billet through a specially designed die. To create longer columns the extruded panels are joined together using friction stir welding (FSW). Studies of the thermo-mechanically affected zone (TMAZ) show that the hardness within these regions drops by approximately 50%. This significantly influences the observed failure load and failure mechanism and hence heat treatment post welding is required to ensure uniformity of properties. Good agreement is achieved between the predictions and experiment. Lastly, optimal designs are calculated based on the analytical analysis and results compared with hat-stiffened and truss core sandwich columns.

I. Introduction

R ECENT studies of the column response of truss core sandwich columns demonstrate that they can compete with current industry standard hat-stiffened panels.[1] With advances in fabrication methods, there are many alternatives for the lightweight core structure, which necessitates the need to characterize the panel response. In this spirit, the following study attempts to characterize the column response of aluminum alloy 6061-T6 corrugated core sandwich columns compressed perpendicular to the corrugations (Fig. 1). This study employs a new fabrication process that has recently been developed to create sandwich panels.[2] This process uses extrusion in conjunction with friction stir welding to create aluminum alloy 6061-T6 corrugated core sandwich structures.

The properties of the core play an integral role in the overall behavior of the sandwich structure, governing the capacity of the whole structure (i.e., sandwich beam bending). For this reason, much of the recent research has focused on the compressive and shear response of various core topologies and of sandwich beams in bending (for example).[3–6] Moreover, several of the core options present opportunities for additional functionality, such as thermal transport and energy absorption.[7–12] However, the in-plane compressive response has received limited attention in the literature. As metallic core sandwich structures are being considered for various new

applications, such as energy absorbent ship hulls, in-plane compression is a loading scenario that will inevitably be encountered and must be addressed. This may be the exclusive loading direction, or may occur in conjunction with out-of-plane loading. In either case, understanding the in-plane loading response is essential in order for sandwich structures to achieve widespread use in next generation multifunctional applications. The present work focuses on corrugated core sandwich structures loaded perpendicular to the corrugations. This orientation allows for greater exploitation of multifunctional capabilities. To date, there has not been a comprehensive study of the competing failure modes of a corrugated core sandwich column under this loading direction. This work attempts to address this deficiency.

The outline of the article is as follows. Section II describes the manufacturing process to create the corrugated core sandwich columns and testing done to characterize the base material. Section III presents analytical equations that predict the collapse load for competing failure modes as a function of geometric parameters and material properties. Mechanism maps—where non-dimensional geometrical parameters represent the axes—are generated that highlight the dominant failure mode. Section IV presents finite element simulations to compare with the analytical predictions. Using the mechanism maps presented in Section III, columns are designed to fail by predetermined mechanisms and the results from experiments are compared with the predictions. These results are presented in Section V. Section VI presents minimum mass designs for the corrugated core columns and comparisons are made with hat-stiffened and pyramidal core designs. Implications and conclusions are presented in Section VII.

II. Manufacturing

Aluminum columns were created using an extrusion process, producing the core and faceplates in a single continuous piece. The manufacturing procedure will be presented along with the experimental protocol for parent material characterization and sandwich column compression.

Aluminum alloy 6061 is used to manufacture extruded corrugated core columns in the same fashion described by Queheillalt *et al.*[2] This alloy was chosen based on a variety of considerations including its availability, economy compared to other alloys, ability to be extruded into high tolerance corrugated structures, and it can also be easily machined and heat-treated. A 17.8 cm diameter, 300 ton direct extrusion press was used to force the aluminum billet through a regular prismatic structure die at 482°C. This process creates corrugated structures that are 3.05 m long, parallel to the corrugations. The die was designed by Cellular Materials International Inc. (Charlottesville, VA) and the extrusion process was performed by Mideast Aluminum Inc. (Mountaintop, PA). The extruded panels were then solutionized, water quenched and heat-treated to the T6 condition. A cross section of an extrusion is shown in Fig. 2.

The relative density for this extruded core structure is given as

F. Zok—contributing editor

Manuscript No. 28960. Received November 22, 2010; approved April 02, 2011.
[†]Author to whom correspondence should be addressed. e-mail: hb8h@virginia.edu

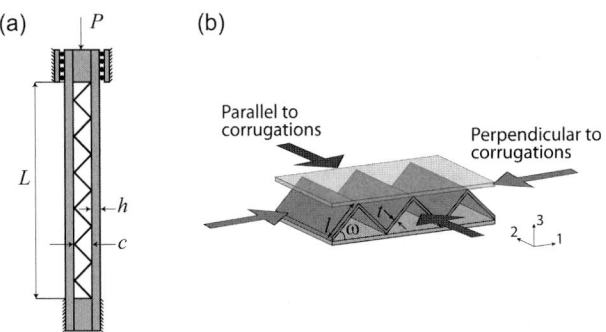

Fig. 1. (a) Corrugated core column compressed perpendicular to the corrugations with fixed end conditions. (b) Illustration of the in-plane loading direction for a corrugated core sandwich structure.

Fig. 2. (a) Schematic and (b) photographic representation of the cross section of an extruded aluminum corrugated sandwich structure.

$$\bar{\rho} = \frac{2}{\sin(2\omega) + (2t/l)} \frac{t}{l} \tag{1}$$

where l is the ligament length, t is the ligament thickness and ω is the core angle.[2] The measured relative density $\bar{\rho} = 26.2\% \pm 0.1\%$ and core angle $\omega = 61.4° \pm 0.1°$ were very close to the target geometry of $\bar{\rho} = 25\%$ and $\omega = 60°$. The geometry of the core, and therefore $\bar{\rho}$, was constrained by the dimensions of the die available. The specific dimensions were chosen to produce consistent extrusions with high geometric tolerances.

The extrusions were initially 3.05 m long parallel to the corrugations, but only four cells wide. The in-plane column compression experiments carried out in this study load the structures perpendicular to the corrugations (see Fig. 1). To create columns of varying length, 203 mm long prepared sections were joined together using a friction stir welding process (Friction Stir Link Inc., Waukesha, WI). The span length of the designed column determines the total number of sections to be joined. Those sections placed in the interior of the column length were modified by removing the solid vertical material on both outer edges, whereas the first and last sections retain the outer vertical material to enable clamping during the compression experiments (Fig. 3). This resulted in a continuous corrugation along the length of the panel. The face sheets of each whole panel were then machined to a desired thickness, h. Three, 2-inch wide columns were water jet cut from each panel.

Friction stir welding is a process where two metals are joined through frictional heat and mechanical deformation induced by a rotating tool traveling along the line of contact between the pieces.[13–15] The tool heats the metal, but does not melt it, and mechanically mixes the softened material through severe plastic deformation, creating a weld by solid-state plastic flow at an elevated temperature.[14,15] A 17-mm

Fig. 3. Illustration of the friction stir weld process. Adapted from figure in[2].

diameter welding tool rotating at 1000 rpm with a traverse speed of 400 mm/min is used to join the Al 6061-T6 corrugated extrusion segments. This process creates two zones that can affect the material properties of the original alloy. The area of material that undergoes plastic deformation from the rotating tool is referred to as the thermomechanically affected zone (TMAZ).[13] The area adjacent to this region, known as the heat affected zone (HAZ), experiences elevated temperatures that can affect the heat treatment of the material.[15] Friction stir welding is an attractive joining technique because it avoids many of the difficulties and defects associated with traditional liquid-phase welding such as porosity in the weld region, filler metal composition, and compatibility and environmental and safety issues.[13,15] This process is depicted in Fig. 3.

As stated above, the material properties can be affected by the friction stir welding process. It was found that the material properties within the weld region were significantly degraded from those of the bulk material, which can influence the column response. To correct this, the columns were subsequently re-tempered to the T6 condition to ensure uniform material properties throughout the corrugated columns. Comments on the manufacturing process will be discussed in Section VII.

Following ASTM standard E8[16], tension tests were carried out on Al 6061-T6 specimens machined directly from the face of a corrugated extrusion. The Young's modulus, E, and 0.1% offset yield stress, σ_y, were found from the stress–strain response (Fig. 4). The specimen was oriented such that the tensile loading direction was parallel to the direction of extrusion. It is anticipated that there will be a slight difference between the response of the material in the direction of the extrusion and the response of the material perpendicular to the extrusion direction. However, it is assumed that this discrepancy will be minimized as a result of the two heat treatments performed subsequent to the extrusion process. The aluminum 6061-T6 shows very little strain hardening after yield, behaving similarly to an elastic-perfectly plastic solid with Young's modulus $E = 75$ GPa and yield stress $\sigma_y = 290$ MPa.

To quantify the effect of friction stir welding on the joint material, hardness tests were performed in the bulk material and across a weld joint, perpendicular to the weld direction. Measurements were taken both before and after the re-temper stage of the manufacturing process and will be referred to as pre-temper hardness and post-temper hardness herein. Rockwell B hardness tests were carried out in accordance with ASTM standard E18[17] using a Wilson Instrument

Fig. 4. Uniaxial tensile stress versus strain response for Al 6061-T6.

Fig. 5. Hardness distribution across a friction stir weld joint.

Fig. 6. Failure modes identified for a corrugated core column compressed perpendicular to the corrugations.

Rockwell hardness tester (Model OUR-a; Milford, CT) with a 1.588-mm (1/16 in) diameter ball indenter and 980N (100 kgf) major test load. A separation of 5 mm was held between sampling points. The pre-temper and post-temper hardness of the bulk aluminum material is reported as the average of 21 measurements, taken in a region far from the weld. The average pre-temper hardness and post-temper hardness were calculated to be 63 ± 0.7HRB and 67 ± 0.5 HRB, where uncertainties are reported as the standard error of the mean with 95% confidence. Figure 5 shows the measured hardness distribution across the weld. The width of the TMAZ marked is equal to the diameter of the welding tool, $d_t = 17$ mm, that plastically deformed the material. It is clear that the material properties in this region are degraded from the welding process. The material transitions back to the bulk properties through the HAZ, which extends ~$d_t/2$ from the TMAZ edge. This profile is consistent with the typical hardness of a metallic friction stir weld in which the weld center is slightly harder than the weld edges due to a recrystallized fine-grained microstructure referred to as the weld nugget.[13,15] The post-temper measurements show consistent hardness across the weld region, indicating uniform material properties over the length of the corrugated column.

III. Analytical Predictions

Failure modes are identified for a corrugated core column in compression and analytical predictions are presented. The predictions are used to construct failure mechanism maps that are a function of column geometry, from which specific experimental geometries were chosen. The mechanical behav-

ior of sandwich structures subject to in-plane compression is dependent on the overall panel geometry, core topology, and parent material properties. Columns with fixed end conditions are loaded perpendicular to the corrugations (Fig. 1). The corrugated column of span, L, consists of two face sheets of thickness, h, and core with ligament length, l, thickness, t, and core angle, ω (Fig. 1). The core and face sheets are composed of the same isotropic material. The failure mechanisms for a corrugated core column compressed perpendicular to the corrugations, composed of a strain hardening material, have been identified as: (i) macro elastic buckling, (ii) macro plastic buckling, (iii) elastic face wrinkling, (iv) plastic face wrinkling, and (v) elastic core shear failure. These can be seen in Fig. 6. Following the analysis of Côté *et al.*[1] analytical predictions governing the sandwich column failure are presented.

(1) Macro Buckling

Euler buckling, characterized by bending of the structure, and core shear failure are two modes of macroscopic failure that have been identified. Rather than being separate mechanisms, the bending and shear buckling modes are coupled. This can readily be seen in a simplified formula where, assuming thin face sheets with negligible bending stiffness about their own axis, the critical sandwich buckling load is given as[18–20]

$$\frac{1}{P_{cr}} = \frac{1}{P_E} + \frac{1}{P_S}, \tag{2}$$

where P_E and P_S are the Euler buckling load and shear buckling load, respectively. Equation 2 matches Engesser's expression for the critical buckling load of an elastic column taking into account shear deformation[21,22] but is valid only for sandwich structures with thin face sheets. The general formula for the critical macro elastic bucking load of a sandwich column with thick faces is given by Allen[18] and Zenkert[20] to be

$$P_{cr} = \frac{(2k^4\pi^4 D_f D_0/L^4) + (k^2\pi^2 D/L^2)}{(k^2\pi^2 D_0/L^2) + S} \tag{3}$$

with the flexural rigidities for a corrugated core sandwich defined as

$$D_0 = \frac{Ebh(h + l\sin(\omega))^2}{2(1 - v^2)}$$

$$D_f = \frac{Ebh^3}{12(1 - v^2)} \tag{4}$$

$$D = 2D_f + D_0$$

where E is the Young's modulus and v is the Poisson's ratio of the material and the geometric properties are as defined above. Fixed column end conditions are assumed with buckling coefficient $k = 2$. The contribution of the core to the overall flexural rigidity has been neglected. When loaded in-plane, perpendicular to the corrugations, the deformation of the core is bending dominated and has a negligible contribution to the overall flexural rigidity of the sandwich structure. Note that plane strain conditions are assumed and a factor of $(1-v^2)$ is introduced to the expressions.

The shear rigidity, S, of the sandwich column is assumed to be set by the shear rigidity of the corrugated core, G_{31}, only[1], neglecting the shear stiffness of the faces. Following the analyses of Deshpande and Fleck[4] and Côté *et al.*[23], the shear rigidity in terms of the core geometry is given as

$$S = G_{31}bl\sin(\omega) = \frac{Ebt\sin(\omega)\sin^2(2\omega)}{2(1-v^2)(\sin(2\omega)+2(t/l))}. \qquad (5)$$

The preceding equations for macro buckling are valid if the material remains elastic, or if

$$P_{cr} < \frac{4bh\sigma_y}{\sqrt{3}} \qquad (6)$$

where $2\sigma_y/\sqrt{3}$ is the material yield stress assuming plane strain conditions. If this criterion is not satisfied, macro buckling is governed by the plastic response of the material. The macro plastic buckling load is calculated using Eq. 3, with the flexural rigidities modified using the Shanley tangent modulus assumption.[24] The elastic plane-strain modulus, $E/(1-v^2)$, is replaced with the tangent modulus, E_t, where $E_t \equiv d\sigma_s/d\varepsilon_s$ is the tangent modulus of the plane strain true tensile stress versus logarithmic strain curve of the parent material, evaluated at $\sigma_s = P_{cr}/2bh$. In practice, an iterative process is used to obtain the macro plastic buckling failure load. Equation 3 is evaluated at various E_t along the material data curve until $P_{cr}/2bh = \sigma_s$. It is assumed that small transverse deflections take place at the bifurcation load and the core remains elastic with shear rigidity given by Eq. 5. If the material is composed of an elastic-perfectly plastic material, the column is unable to support additional load if the elastic limit is reached and the appropriate failure is face yielding. Face yielding occurs when the stress in the face sheets equals the plane strain yield stress of the material:

$$P_{FY} = \frac{4bh\sigma_y}{\sqrt{3}}. \qquad (7)$$

(2) Face Wrinkling

The face wrinkling response is a short wavelength instability, set by buckling failure of the face between points of attachment to the core. Dictated by the core geometry, the wrinkling span length is $2l\cos(\omega)$. For a column with face sheets of thickness h, the elastic wrinkling load is

$$P_{FW} = \frac{k_1^2\pi^2bE}{24\cos^2(\omega)(1-v^2)}\frac{h^3}{l^2} \text{ if } \frac{h}{l} < \sqrt{\frac{96\cos^2(\omega)(1-v^2)\sigma_y}{\sqrt{3}k_1^2\pi^2E}} \qquad (8)$$

If the inequality in Eq. 9 is not met, the stress in the faces exceed the elastic limit of the material and the plastic face wrinkling load is given by

$$P_{FW} = \frac{k_1^2\pi^2bE_t}{24\cos^2(\omega)}\frac{h^3}{l^2} \text{ if } \frac{h}{l} \geq \sqrt{\frac{96\cos^2(\omega)(1-v^2)\sigma_y}{\sqrt{3}k_1^2\pi^2E}} \qquad (9)$$

where, as above, E_t is the tangent modulus of the plane strain true stress versus true strain curve of the material evaluated at $\sigma_s = P_{FW}/2bh$ An iterative process is again used to obtain the plastic face wrinkling failure load with Eq. 9

evaluated at various points along the material data curve until the wrinkling stress equals the material stress in the faces; $P_{FW}/2bh = \sigma_s$. Due to the geometry of the core/node interface of the extruded column, the end constraint is chosen to be built-in, with $k_1 = 2$. If the parent material is elastic-perfectly plastic and cannot support additional load beyond yield, the column fails by face yielding given by Eq. 7.

(3) Failure Mechanism Maps

Using the above analytical expressions, failure mechanism maps for the in-plane compression of aluminum 6061-T6 corrugated core columns were constructed. These give a visual representation of failure, making it easier to identify the operative collapse mode (i.e., the mode associated with the lowest critical load P) for a given column geometry. These maps are developed as a function of non-dimensional geometric parameters, h/l and L/l, for a sandwich structure with constant core ligament slenderness ratio, t/l, core angle, ω and given material response. The active failure mode boundaries are generated by evaluating the minimum normalized collapse load $\bar{P} \equiv P/(\sigma_y bl\sin(\omega))$ over a range of h/l and L/l. The failure mechanism map for the fabricated core construct is presented in Fig. 7. Lines of constant normalized critical failure load, $\bar{P}$, and normalized mass, $\bar{M} = M/\rho bl^2$, are indicated on the maps. Although not shown in this map, it is possible to construct columns whose geometries ($\bar{\rho} < \sim 4.6\%$) allow core shear to become an active mechanism in the buckling regime.[25] Experimental geometries were selected to probe the failure regimes along lines of constant force and are compared to analytical predictions and finite element simulations. The measured column geometries tested are indicated on the map. Column geometries that lie in the macro elastic buckling region could not be tested due to specimen length limitations in existing experimental facilities.

(4) Transverse Loading

It is instructive to consider the response of the panel loaded in the transverse direction. For the corrugated core column, this loading scenario is parallel to the corrugations (Fig. 1). For an elastic-perfectly plastic material, the failure modes identified are: (i) macro elastic buckling, (ii) local elastic face buckling, (iii) local elastic core buckling, and (iv) material yielding. The macro elastic buckling failure of a corrugated core column of length, L, and width, b, compressed parallel to the corrugations is taken to be described by the critical

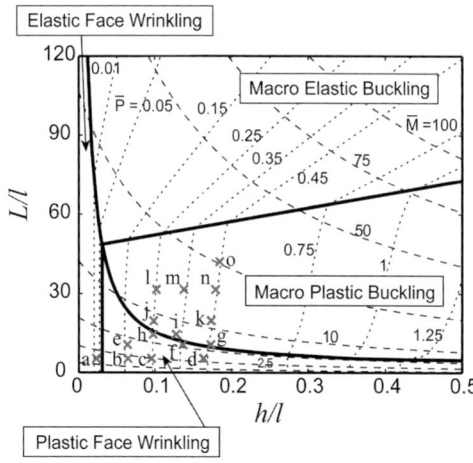

Fig. 7. Failure mode map for the measured Al 6061-T6 extruded corrugated core geometry with $t/l = 0.1488$ and $\omega = 61.4°$. The experimentally tested column geometries are marked on map.

sandwich buckling load given in Eq. 3. When loaded in this orientation, the core deformation is stretch dominated and the core rigidity, D_c, cannot be neglected in the overall flexural rigidity of the sandwich structure as it was when loaded perpendicular to the corrugations. The core flexural rigidity is expressed as

$$D_c = \frac{Ebtl^2\sin^2(\omega)}{12(l\cos(\omega) + t/\sin(\omega))} \quad (10)$$

and the total flexural rigidity is now

$$D = 2D_f + D_0 + D_c \quad (11)$$

with D_0 and D_f given in Eq. 4.

Following Côté *et al.*,[23] the shear rigidity is obtained as

$$S = \frac{Ebt\sin^3(\omega)}{(1+v)(\sin(2\omega) + (2t/l))}. \quad (12)$$

Local face buckling failure is the short wavelength plate buckling that can occur along the length of the column, between points of attachment of the core and faceplates. The width of the plate is set by the core geometry to be $2l\cos(\omega)$ Following the analysis of Budiansky[26] for the local plate failure in a hat-stiffened panel, the local elastic failure load can be calculated. The critical elastic face buckling load is

$$P_{FB} = \frac{K\pi^2 Eh^2 A}{48(1-v^2)l^2\cos^2(\omega)} \quad \text{if} \quad \frac{h}{l} < \sqrt{\frac{48(1-v^2)\cos^2(\omega)\sigma_y}{K\pi^2 E}}, \quad (13)$$

where K is the buckling coefficient set by the boundary conditions and the aspect ratio $L/(2l\cos(\omega))$. It is assumed that $L >> 2l\cos(\omega)$ and the face buckles in a pattern of half-sine waves of length equal to plate width $2l\cos(\omega)$ along L.[22,26] Therefore, simply supported conditions are estimated on the loaded edges.[27] Due to the geometry of the extruded construction, the unloaded edges are assumed to be fixed giving $K=7$.

Similarly, the local elastic core failure is short wavelength buckling along the length of a core web with width, l. The critical local elastic core load is expressed as

$$P_{FB} = \frac{K\pi^2 Et^2 A}{12(1-v^2)l^2} \quad \text{if} \quad \frac{t}{l} < \sqrt{\frac{12(1-v^2)\sigma_y}{K\pi^2 E}} \quad (14)$$

as above, it is assumed that the unloaded edges are fixed and the loaded edges are simply supported, giving $K=7$.

The elastic limit is set by the plane strain material yield stress as

$$P_{cr} < \frac{2}{\sqrt{3}}\sigma_y A \quad (15)$$

where the cross-sectional area is

$$A = \frac{bt}{\cos(\omega) + t/l\sin(\omega)} + 2bh \quad (16)$$

If the critical load exceeds the elastic limit, the critical plastic failure load is

$$P_y = \frac{2}{\sqrt{3}}\sigma_y A \quad (17)$$

IV. Numerical Simulations

Numerical simulations were performed for the in-plane loading of corrugated core sandwich columns, compressed perpendicular to the corrugations. Abaqus-Standard finite element software (Simulia of Dessault Systemes, Providence, RI) was used to for all simulations. Four node, first order continuum plane-strain elements with reduced integration (Abaqus CPE4R) were used with the measured material properties from the uniaxial tensile test results in Fig. 4. These elements prevent shear and volumetric locking that can occur with fully integrated elements in bending.[28] Loading was simulated using a general static analysis with a displacement boundary condition applied to one end of the column. The nodes at the column ends were fixed from all other displacements and all rotations to simulate built-in end conditions. No initial imperfection was introduced into the column. Lack of symmetry about the column centroid was enough to cause a preferential buckling direction. Therefore, the simulations are an upper bound on the column failure, to be compared with the analytical predictions, which are also derived under the assumption of "perfect" column geometry.

V. Comparison with Experimental Results

Extruded aluminum corrugated core columns, designed to probe the macro plastic buckling and face wrinkling regimes, were fabricated and tested. Column designs a–d (Fig. 7) were constructed from a single extrusion, in which no friction stir welding was necessary. All other geometries required friction stir welding to produce columns of necessary length, L, as described in Section II. Three specimens were tested for each geometry, except point *o* with two viable samples available for testing. The measured results, compared to the analytical and finite element predictions, are summarized in Table I. Slight variations in column geometry existed due to the fabrication process: specifically with respect to face sheet thickness. As such, the analytical and numerical predictions are calculated using the average measured face thickness of each column.

(1) Plastic Face Wrinkling

Various column geometries were manufactured to probe the plastic face wrinkling regime (b–f, h; Fig. 7). The measured compressive response for a specimen from sample geometry *c* ($h=2.12$ mm, $L=117$ mm) is given in Fig. 8. After initial loading there is a nonlinear material response to a peak load of 65.46 kN, followed by post-buckling softening. Face wrinkling occurs concurrently in both faces as seen in Fig. 8(b). Both the analytical and numerical peak load are within 7% of the measured value, showing very good agreement. Recall geometry *c* did not require friction stir welding. Figure 9 presents the plastic face wrinkling loading response for a specimen with geometry *f*, a geometry requiring friction stir welding, which exhibits a similar response to sample *c*. It should be noted that the face wrinkling failure did not occur at the weld locations.

(2) Macro Plastic Buckling

Macro plastic buckling is the predicted failure mechanism for eight sample geometries manufactured and tested (samples g, i–o). The measured response of a specimen from sample point *o* ($h=4.08$ mm, $L=934$ mm) is given in Fig. 10. The load increases linearly with displacement followed by an increasing nonlinear response up to a peak load of 88.75 kN. After the peak load, post-buckling softening occurs. Photographs of the deformation history show global buckling failure in a shape consistent with a column having clamped end conditions (Fig. 10(b)). The analytical and numerical predictions for

Table I. Summary of the Observed and Predicted Failure Loads and Failure Mechanisms for the Column Geometries Tested. For All Columns, the Predicted Failure Load and Mechanism are Reported Using the Average Measured Face Thickness, h_{avg}. For the Columns that Exhibit Face Wrinkling Failure, the Face Wrinkling Load Using the Minimum Measured Face Thickness, h_{min}, is also reported

			Failure load (kN/mm)				Failure mode		
			$P_{predicted}$						
Pt	L/l	h/l	h_{avg}	FW(h_{min})	$P_{simulation}$	$P_{measured}$	Predicted	Simulation	Observed
	5.3	0.023	0.157	0.112	0.140	0.090	EFW	EFW	EFW
	5.3	0.024	0.186	0.154	0.165	0.092	EFW	EFW	EFW
a	5.3	0.023	0.165	0.136	0.148	0.102	EFW	EFW	EFW
	5.3	0.065	0.912	0.885	0.897	0.759	PFW	PFW	PFW
	5.3	0.065	0.905	0.879	0.89	0.743	PFW	PFW	PFW
b	5.3	0.064	0.892	0.865	0.869	0.768	PFW	PFW	PFW
	5.3	0.096	1.370	1.353	1.377	1.27	PFW	PFW	PFW
	5.3	0.096	1.372	1.360	1.377	1.29	PFW	PFW	PFW
c	5.3	0.096	1.367	1.360	1.369	1.24	PFW	PFW	PFW
	5.3	0.163	2.402	2.395	2.467	2.27	PFW	PFW	MPB/PFW
	5.3	0.164	2.423	2.381	2.467	2.41	PFW	PFW	MPB/PFW
d	5.3	0.165	2.430	2.353	2.474	2.42	PFW	PFW	MPB/PFW
	10.5	0.064	0.886	0.8449	0.869	0.648	PFW	PFW	PFW
	10.5	0.064	0.886	0.8047	0.869	0.688	PFW	PFW	PFW
e	10.5	0.064	0.892	0.8115	0.876	0.635	PFW	PFW	PFW
	10.6	0.136	1.982	1.877	2.000	1.910	PFW	PFW	PFW
	10.5	0.137	2.005	1.919	2.021	1.860	PFW	PFW	PFW
f	10.5	0.137	1.997	1.919	2.015	1.882	PFW	PFW	PFW
	10.5	0.174	2.556	2.515	2.603	2.482	MPB	MPB	MPB/PFW
	10.6	0.173	2.553	2.500	2.589	2.441	MPB	MPB	MPB/PFW
g	10.6	0.172	2.527	2.437	2.568	2.388	MPB	MPB	MPB/PFW
	14.5	0.098	1.407	1.326	1.406	1.129	PFW	PFW	PFW
	14.5	0.099	1.416	1.312	1.413	1.138	PFW	PFW	PFW
h	14.5	0.097	1.385	1.251	1.385	1.113	PFW	PFW	PFW
	14.5	0.128	1.841	1.732	1.865	1.660	MPB	MPB/PFW	PFW
	14.5	0.127	1.834	1.704	1.851	1.640	MPB	MPB/PFW	PFW
i	14.5	0.128	1.847	1.746	1.865	1.621	MPB	MPB/PFW	PFW
	19.8	0.098	1.387	1.101	1.393	1.119	MPB	MPB/PFW	PFW
	19.8	0.099	1.401	1.142	1.407	1.117	MPB	MPB/PFW	PFW
j	19.8	0.098	1.393	1.128	1.393	1.118	MPB	MPB/PFW	PFW
	19.8	0.173	2.457	2.479	2.488	1.765	MPB	MPB	PFW @ weld
	19.8	0.174	2.469	2.486	2.508	1.754	MPB	MPB	PFW @ weld
k†	19.8	0.173	2.461	2.463	2.495	1.743	MPB	MPB	PFW @ weld
	31.6	0.101	1.381	1.251	1.384	1.050	MPB	MPB	PFW
	31.6	0.102	1.399	1.230	1.402	1.082	MPB	MPB	PFW
l	31.6	0.102	1.400	1.203	1.402	1.041	MPB	MPB	MPB/PFW
	31.5	0.138	1.888	1.725	1.894	1.594	MPB	MPB	PFW
	31.6	0.139	1.901	1.711	1.900	1.570	MPB	MPB	MPB/PFW
m	31.6	0.138	1.881	1.704	1.888	1.568	MPB	MPB	MPB/PFW
	31.6	0.179	2.456		2.456	2.261	MPB	MPB	MPB
	31.6	0.179	2.466		2.463	2.230	MPB	MPB	MPB
n	31.6	0.179	2.469		2.469	2.246	MPB	MPB	MPB
	42.1	0.185	2.484		2.452	2.043	MPB	MPB	MPB
o	42.1	0.184	2.467		2.434	1.969	MPB	MPB	MPB

†Specimens did not undergo re-tempering process after friction stir welding.

peak load agree within 1.5%. The measured failure load is 20% lower than predicted.

VI. Minimum Weight Design

Corrugated core sandwich structures are optimized such that the geometric parameters minimize the panel mass for a given critical load. Comparisons are made with optimized hat-stiffened panels, which are considered an industry standard for axial load bearing applications and widely used. The corrugated sandwich panels are also compared with pyramidal core sandwich structures, which are a competing sandwich design. Non-dimensional parameters are used in the optimization allowing various designs to be compared.

The non-dimensional loading index is defined as $\bar{P}^* \equiv P_{cr}/\sigma_y bL$ and the non-dimensional panel mass index is given as $\bar{M}^* \equiv M/\rho bL^2$. The panel mass $\bar{M}^*$ is minimized for a given $\bar{P}^*$ subject to the possible critical failure loads as constraints. For all panels, the minimum mass is found numerically using an extensive search method in which the non-dimensional column geometries are varied over the design space with the minimum mass configuration being calculated at each loading index.

The optimization carried out assumes an elastic-perfectly plastic material with the modulus and yield stress of Al 6061-T6, defined from the tension test described earlier. This is a reasonable assumption as the material response of Al 6061-T6 demonstrates very little strain hardening after yield (Fig. 4).

Fig. 8. (a) The measured loading response for a specimen from sample geometry *c* with the finite element and analytical predictions included. (b) Photographs illustrating the deformation history at points marked in (a).

Fig. 9. (a) The measured loading response for a specimen from sample geometry *f* with the finite element and analytical predictions included. (b) Photographs illustrating the deformation history at points marked in (a).

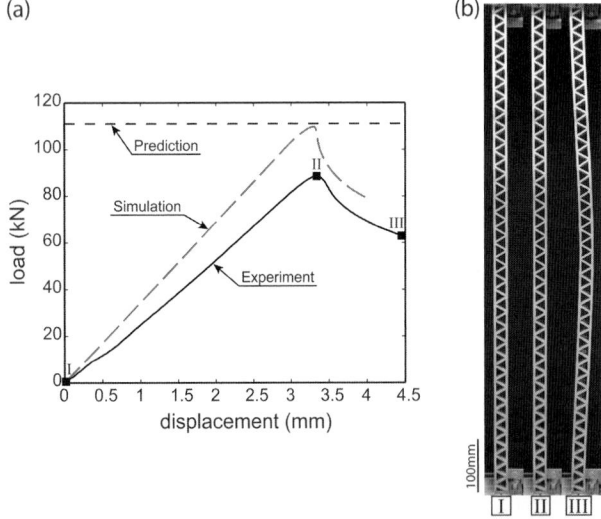

Fig. 10. (a) The measured loading response for a specimen from sample geometry *o* with the finite element and analytical predictions included. (b) Photographs illustrating the deformation history at points marked in (a).

The extruded corrugated core column mass is expressed as

$$M = 2bL\rho\left(h + \frac{t}{2\cos(\omega) + (2t/l\sin(\omega))}\right). \quad (18)$$

The overall length, L, of the load-bearing column is typically a design constraint and is not varied in this optimiza-

tion. In addition, the core angle, ω, is held constant ($\omega = 60^\circ$) so as to correspond to the fabricated extrusions. For a corrugated core column compressed perpendicular to the corrugations, the non-dimensional geometric variables, h/L, t/L, and l/L, are sought (with ω held constant) that minimize the mass index $\bar{M}^*$ at a given load $\bar{P}^*$, subject the critical failure load constraints given in Section III for an elastic-perfectly plastic material.

Optimization of a hat-stiffened panel follows the analysis set forth by Budiansky[26] for a simplified geometry that has square stiffeners of length w spaced at regular intervals, w. Compressed in the transverse direction, perpendicular to the stiffeners, this panel behaves as a monolithic plate. The mass optimization for an extruded pyramidal core column follows the analysis of Côté *et al.*,[1] with only a slight variation to the core relative density due to the extruded geometry (see[2]). The pyramidal core is constrained to a square base, and thus displays transverse isotropy. Therefore, the in-plane loading response of column is the same in the two orthogonal directions. The optimized loading results are presented in Fig. 11. Figure 12 illustrates the corresponding non-dimensional geometric parameters for a corrugated core column loaded perpendicular to the corrugations and a pyramidal core column.

VII. Discussion

(1) Column Compression

The results presented clearly demonstrate that the analytical predictions accurately predict the column failure for both face wrinkling and macro buckling. The peak load and failure mechanism agreement between the predictions and experiments, particularly along the predicted failure boundary, confirm that fixed face wrinkling boundary conditions are an appropriate assumption. Also, for all geometries tested, the analytical predictions are in excellent agreement with the finite element simulations. The discrepancies seen in the failure response highlight the importance of both the precision and sequence of the fabrication process used to construct extruded sandwich panels. It is essential that all stages of the fabrication be rigorously monitored to produce reliable panels, thus allowing analytical and numerical tools to be implemented. Variations in column geometry can cause degradation in failure load and initiate unanticipated failure mechanisms. The friction stir welding process is an attractive and capable manufacturing method to join sandwich panel sections. However, the panels must be manufactured in a way that minimizes imperfections. For example, large panels should be tempered immediately following the friction stir welding, prior to any machining, where all panel geometries (e.g., face thickness and panel width) are greatest, thereby reducing the likelihood of warping. Then the faces should be machined to the desired thickness and columns extracted. It should also be highlighted that, as expected, precise fabrication is vital to the performance of the sandwich panel. The extrusion-friction stir welding process can produce corrugated panels with very few defects if closely monitored. However, small variations in panel geometry can have significant influence on performance.

Looking more closely at some of the results there are some samples designed in the MPB region that exhibit a failure mechanism discrepancy (Table I). Face wrinkling was observed to be the active failure mode or occur in conjunction with global buckling. This is reasonable for the samples with geometries that lie close to failure boundaries (geometries *g*, *i*, and *j*), where the plastic face wrinkling and macro plastic buckling failure loads are approximately equal. Slight imperfections (geometric or experimental) can activate plastic face wrinkling mechanism prior to macro buckling. However, the observed face wrinkling response is not reasonable for samples designed to lie well within the macro plastic buckling region, away from failure boundaries (such as geometries *l*

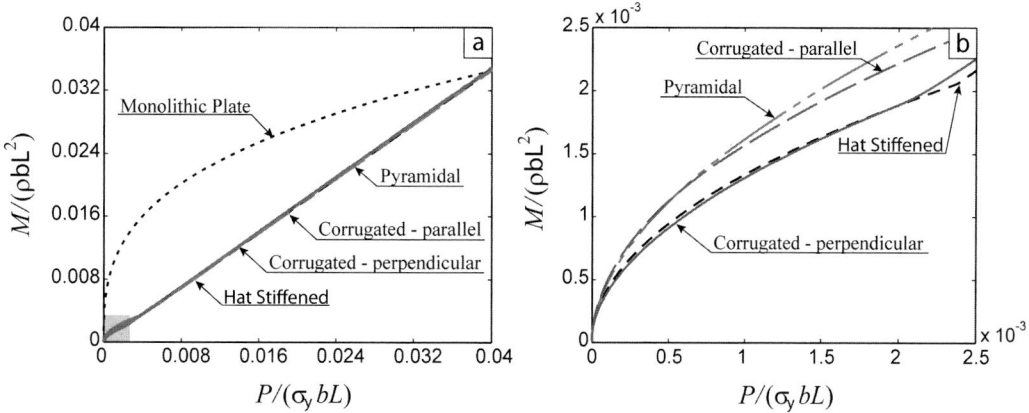

Fig. 11. (a) Failure load comparison for minimum mass optimized corrugated core sandwich columns ($\omega=60°$), pyramidal core sandwich columns ($\omega=60°$) and hat-stiffened panels. The sandwich panels are fabricated using an extrusion process. The corrugated core sandwich columns are optimized for loading perpendicular to the corrugations and parallel to the corrugations. The hat-stiffened panels are optimized for axial loading and transverse loading. All panels are composed of elastic-perfectly plastic material with $\sigma_y = 290$ MPa and $v = 0.33$.

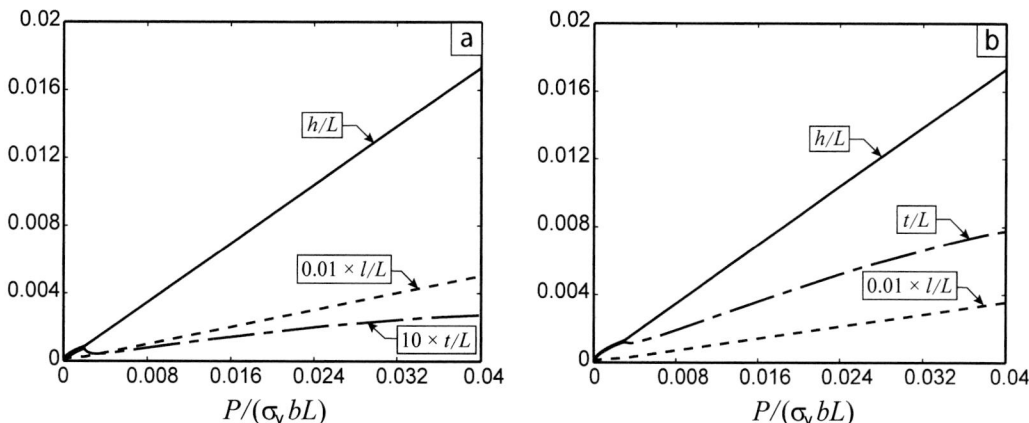

Fig. 12. The optimal geometric design parameters, corresponding to the mass optimization presented in Figure 11, for a (a) corrugated core sandwich column compressed perpendicular to the corrugations and (b) pyramidal core sandwich column, made of an elastic-perfectly plastic material.

and m). Examining these columns further shows flaws in the manufacturing process that result in slight variations in the face sheet geometry. The cause of these variations is primarily attributed to the welding process that was used to join individual extruded columns. This process resulted in specimens that were not perfectly flat prior to machining the faces to a desired thickness. While the exterior face surfaces were milled flat, the inner face surfaces retained the original offset resulting in local variations in face sheet thickness. The face wrinkling load, proportional to h^3 (Eq. 9), is sensitive to variations in face sheet thickness. In certain instances, the critical load for face wrinkling using the thinnest face sheet thickness measured can become less than the macro plastic buckling load calculated using the average face sheet thickness. Thus, localized face wrinkling becomes an active failure mechanism prior to macro buckling. The minimum measured face thickness was sufficient to activate plastic face wrinkling prior to the designed macro plastic buckling mode for several column geometries tested (g, i, j, l, m). In most instances, the calculated plastic face wrinkling load in the thinnest section was within 10% of the macro plastic buckling load using the average face thickness. Using the minimum face thickness, the predicted plastic face wrinkling load is in excellent agreement with the measured failure load and correctly captures the operative failure mechanism. The peak loads agree within 8% for all columns except point l, which is within 16%. This analysis can be applied to the columns designed in the plastic

face wrinkling region as well, bringing the agreement between the analytical and measured peak loads even closer—in many cases within 5%. For the columns that are observed to fail exclusively by macro plastic buckling, reasonable to very good agreement is seen between the analytical and measured peak loads (8%–20%), in spite of the potential manufacturing flaws present. Variations in geometry that result from manufacturing flaws can significantly alter both the predicted failure load and mechanism and must be mitigated to properly use the predictive tools (i.e., failure mode maps). However, when appropriate parameter values are used, the analytical predictions accurately capture the column response.

To show the detrimental effect that friction stir welding can have on the column response, sample k was tested without undergoing heat treatment post weld. These columns were designed with a geometry that placed them well within the macro plastic buckling region. However, the active failure mode observed was face wrinkling, which occurred at a friction stir weld joint. Upon examination of the tested columns, it was found that failure occurred at each of the friction stir weld joints. It should be noted that the minimum face sheet thickness was not sufficient to activate plastic face wrinkling prior to macro buckling as discussed above. These tests demonstrate how the friction stir welding process can have a significant affect on the column failure, dictating overall column response. This must be accounted for, through additional processing to prevent performance degradation.

II. Delamination Mechanics and Examples of TBC Delaminations

At temperatures representative of the highest temperatures experienced by coatings in service it is generally believed that stresses in the topcoat and the TGO are relaxed due to creep of the constituent materials. With this assumption, the largest in-plane stresses in the topcoat and TGO are compressive and they occur during cool down due to thermal expansion mismatch between the ceramic layers and the metal substrate. Thus, while the degradation processes primarily occur at high temperatures, it is generally believed that the critical conditions for delamination occur during cool down. The coatings must be able to withstand the stresses at the lowest temperatures when the turbines are cool. For this reason, the relevant toughness is believed to be "room temperature" toughness, and tests to measure the delamination toughness of TBC systems have invariably been conducted at room temperature.

A variety of simplified models are considered in this paper, none of which account for either the bond coat or the TGO. The model results presented here are intended to illustrate basic ideas; they will need to be embellished in quantitative applications to account for additional layers such as the TGO. In the models considered here, the Young's modulus and Poisson's ratio of the coating are, E_C and v_C, and those of the substrate are E and v. For plane strain, the two Dundurs elastic mismatch parameters are

$$\alpha = \frac{\bar{E} - \bar{E}_C}{\bar{E} + \bar{E}_C} \text{ and } \beta = \frac{1}{2}\frac{\mu(1 - 2v_C) - \mu_C(1 - 2v)}{\mu(1 - v_C) + \mu_C(1 - v)} \quad (1)$$

with $\bar{E}_C = E_C/(1 - v_C^2)$, $\bar{E} = E/(1 - v^2)$, $\mu_C = E_C/[2(1+v_C)]$ and $\mu = E/[2(1+v)]$.

It is assumed that the interface plane is the weak link between the coating and the substrate and that the delamination crack stays within this plane. Kinking of an interface crack out of the interface region defeats the purpose of the test, and conditions that encourage the crack to stay within the interface will be discussed later in the paper. Denote the toughness of the controlling failure plane between the coating and the substrate (interface or interfacial layer) in its current state by $\Gamma(\psi)$ in units J/m^2 with the convention defined in Fig. 1 having the coating above the substrate. For most interfaces, the toughness depends on the mode mix as measured by ψ and as depicted in

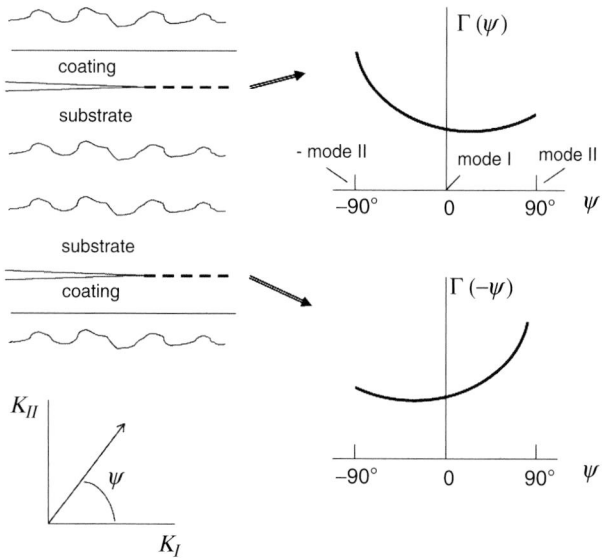

Fig. 1. Schematic of coating/substrate interface toughness as a function of mode mix. The convention in this paper defines the toughness function, $\Gamma(\psi)$, such that the coating lies above the substrate. When the coating lies below the substrate the toughness is $\Gamma(-\psi)$ when subject to the mode mix, ψ. The toughness function $\Gamma(\psi)$ can be symmetric in ψ but generally it should be assumed to be asymmetric.

Fig. 2. Two examples of edge delamination cracks for a coating experiencing a tensile stress due to thermal expansion mismatch with the substrate. (A) An edge delamination emanating from an edge of the substrate. (B) An edge delamination emerging from a coating edge at an interior point of the substrate or possibly an open vertical sinter crack. For tensile stress, the crack is open with roughly equal components of mode I and mode II. For compressive stress, the crack is a mode II delamination, and the above results apply approximately if friction is neglected. The results reveal the inherently greater resistance to delamination initiation for coatings terminating at an edge compared with a coating terminating at an interior edge.

Fig. 1. If $\beta = 0$, $\psi = \tan^{-1}(K_{II}/K_I)$, where K_I and K_{II} are the conventional mode I and II stress intensity factors. With the convention adopted in this paper (c.f., Fig. 1), the interface toughness for situations in which the coating lies below the substrate is $\Gamma(-\psi)$. The toughness function, $\Gamma(\psi)$, is not necessarily symmetric in ψ and, specifically, the negative mode II toughness, $\Gamma(-90°)$, is not necessarily the same as the positive mode II toughness, $\Gamma(90°)$. If $\beta \neq 0$, the definition of ψ is slightly more complicated and is given in Appendix A. With G as the energy release rate (units of J/m^2) and ψ as the measure of the mode mix of the interface crack subject to the present loading, the condition for incipient advance of the delamination in the interface plane is

$$G = \Gamma(\psi) \quad (2)$$

The emphasis in this paper is on tests to measure $\Gamma(\psi)$ for a wide range of ψ relevant to coating delamination failures.

Edge delaminations are among the most common TBC delaminations observed on components where spallations occur. Two types of edge delaminations are illustrated in Fig. 2. In these examples, the coating is taken to be of uniform of thickness, t, and is bonded to an infinitely thick substrate. The film has a uniform thermal expansion mismatch with the substrate such that away from its edge the film is subject to a uniform equi-biaxial stress, σ. The steady-state, or long crack, energy release rate limit is independent of the delamination length:

$$G_{SS} = \frac{\sigma^2 t}{2\bar{E}_C} \quad (3)$$

If the thermal mismatch conditions produce tensile stress in the coating, the crack is completely open with mix of mode I and II given by $\psi \cong 55°$, depending somewhat on the elastic mismatch.[7] However, if the stress in the coating is compressive, as it would be for the TBC under cool down, the crack faces contact each other and the examples in Fig. 2 are *mode II (or near-mode II) delamination cracks* with $\psi \cong -90°$.[‡] Friction between the faces of the crack is neglected in these results.

[‡]The results in Fig. 2 have been taken from Yu *et al* [7] where a coating under tension due to a uniform thermal expansion mismatch is analyzed. The tensile loading produces an open crack. The results are only approximately valid for the compressive case due, in part, to neglect of contact that occurs between the crack faces. However, when friction is neglected, the energy release rate is only slightly affected by contact and the results in Fig. 2 are approximately applicable. An example where friction is taken into account in mode II delamination is considered later in this paper and a coating under compression is analyzed in Balint and Hutchinson.[8]

For an edge delamination emerging from an edge of the substrate in Fig. 2(A), the energy release rate, G, requires a delamination length, a, at least several times the coating thickness to approach the steady-state limit, even for a mismatch with $\alpha = 0.5$ $(\bar{E}_C/\bar{E} = 1/3)$ which is representative of some TBC systems. By contrast, an edge delamination emanating from an interior edge of the coating as in Fig. 2(B) attains the steady-state limit when a is a small fraction of the coating thickness for all mismatches. An interior edge delamination arises, as illustrated, where the coating terminates abruptly away from the substrate edge or, for example, at an open vertical sinter crack in the coating. An interior coating edge is more susceptible to delamination initiation than a coating that extends all the way to the substrate edge. The latter has some built in protection against the initiation due to the extra compliance of the substrate edge that lowers the stress in the coating in that vicinity.

Edge delaminations can initiate from corners and from air holes in a substrate. As for delaminations emanating from a substrate edge, the local stress distribution in the coating will be affected by substrate compliance at such features. If the substrate thickness is comparable to that of the coating the compliance will be increased. A reduction in stored strain energy in the coating at potential initiation locations adds to the protection against delamination.

The mode mix of the delamination depends on the distribution of the compressive stress through the thickness of the coating. A uniform stress distribution, or a stress distribution which is more compressive at the surface than at the interface, produces mode II. If the stress is sufficiently more compressive just above the coating's interface than at its surface, then the crack may open.[9] For example, if the stress in the coating vanishes in a layer of thickness kt below the surface and is uniform compression in the remaining layer, then a mode I component will exist if $k < 0.449$; otherwise mode II prevails. This example assumes no elastic mismatch between the coating and substrate, but elastic mismatch is less important than the details of the stress distribution.

III. Tests for Measuring Delamination Toughness

Measuring the delamination toughness, $\Gamma(\psi)$, of TBC coatings has been challenging, especially so given the importance of a range of mixed mode toughness in applications. This section begins by citing several tests that have been used to measure delamination toughness and some representative results. The fracture mechanics approach to determining TBC durability is illustrated in conjunction with the tests.

Vastinonta and Beuth[10] used a conical brale C indenter to induce a circular delamination in a 100-μm-thick electron beam deposited TBC system. The indenter is pushed into the substrate through the topcoat and TGO. The indenter forces an outward radial plastic flow of the substrate under the coating that decays with distance from the central axis of symmetry. The substrate motion induces additional compression of the coating in the radial direction and increases the elastic energy density stored in the TBC in the vicinity of the indent. The indenter also initiates a delamination edge where it pushes through the coating. The delamination spreads axisymmetrically until the driving force falls below the toughness, i.e., until $G = \Gamma(\psi)$ can no longer be met. A difficult aspect of this test is the analysis of the delamination crack problem required to generate both G and ψ as a function of the radius delamination and the indenter depth. A sophisticated elastic-plastic finite element calculation is required. In addition, the induced radial compressive stress can be high enough to buckle the coating, further complicating the determination of G and ψ. The loading is a mix of mode I and mode II (with $\psi > 0$ if buckling does not occur) that varies with crack radius.

Jones *et al.*[11] sliced a planar section of a electron beam deposited TBC system on a turbine blade substrate. Then these authors used a focused ion beam to cut a broad notch under the

coating system creating a trilayer bridge consisting of the bond coat, the TGO and the topcoat. The bridge was supported (effectively clamped) at its ends by the uncut substrate and coating system. A concentrated load normal to the surface was applied to the center of the bridge. In the first series of tests, the load was applied to the underside of the bridge pushing upward. This load produces a moment distribution in the bridge, which at sufficient force causes the topcoat to develop a vertical crack above the load that runs from the surface to the interface. This crack branches into the weak interface and spreads as a delamination crack. The geometry and preparation of the sample allow for clear visualization of the various stages, including the stable advance of the crack tip as the bridge is pushed upward. This is a highly sophisticated test that is not likely to be used routinely.

This in situ test allows for the interrogation of engine hardware but is a highly sophisticated test and is not likely to be used routinely. It also requires a detailed finite element analysis to obtain G and ψ. Plastic deformation of the metal bridge components can occur. The mode I component is somewhat larger than the mode II component when the load is applied under the bridge. To avoid cracking in brittle coatings and to obtain toughness data under conditions closer to mode II, Jones, Manning and Hemker (unpublished work) carried out a second set of tests in which they created a narrow open notch in the center of the top coat to the interface of interest and loaded the specimen by pushing down from above the bridge. This direction of loading adds to the compression in the coating from the thermal expansion mismatch and therefore more closely mimics the delamination conditions experienced in service. Kagawa and colleagues have developed two related tests, the barb test[12] and the push-out test,[13] to measure delamination toughness of TBCs. In each test, the coating is subject to additional compression by forcing it, but not the substrate, against a hard block. The force at which the coating delaminates is used to determine the critical energy release rate and the associated toughness. These tests add compression to any residual compression in the coating, but they impose a delamination displacement and shearing stress on the interface in the opposite direction from that experienced by a delamination in service driven by thermal expansion mismatch (i.e., $\psi > 0$ rather than $\psi < 0$). The energy release rate and mode mix of the barb test has been analyzed by finite element methods[14] with the finding that $\psi \cong 60°$. Like the aforementioned bridge test, the barb and push-out tests require highly refined specimen preparation and sophisticated testing which are likely to limit their use for routine toughness testing. Nevertheless, Kagawa and colleagues[13] have collected an extensive data set showing how the toughness

Fig. 3. Delamination toughness of an EB-PVD TBC on a NiCoCrAlY bond coat as a function of thermal exposure plotted as a function of the thickness of the TGO from Tanaka *et al.*[13] The upper band of data is for isothermal exposure and the lower band is for cyclic thermal exposure. The TGO thickness increases with thermal exposure time in accord with the markers on the upper band of data.

The *mode I double-cantilever beam specimen* in Fig. 9(B) has

$$K_I = 2\sqrt{3} P_I a h^{-3/2}, \quad K_{II} = 0, \quad G = \frac{P_I^2 a^2}{B} \quad (13)$$

Superposition of the above gives for the mixed mode double-cantilever beam specimen

$$G = \frac{P_I^2 a^2}{B} + \frac{3}{4}\frac{P_{II}^2 a^2}{B}, \quad \tan\psi = \frac{\sqrt{3} P_{II}}{2P_I} \quad (14)$$

The accuracy of both contributions to G can be improved by including extra terms of order h/a, as has been done in Hutchinson and Suo.[9] Based on beam theory predictions, $\Delta_{II} = P_{II}(w^3 + 3a^3)/12B$ and $\delta = 2P_I a^3/3B$, such that one can rewrite (14) as

$$G = 108\frac{B\Delta_{II}^2 a^2}{(w^3 + 3a^3)^2} + \frac{9}{4}\frac{B\delta^2}{a^4},$$
$$\tan\psi = \frac{4\sqrt{3} a^3 \Delta_{II}}{(w^3 + 3a^3)\delta} \quad (15)$$

The crack will be open if $\delta > 0$ and crack growth is expected to be stable under prescribed displacements.

In the literature of laminated composites, double-cantilever specimens are also called end-notched specimens and they have been widely used to the measure mixed mode toughness dependence with various loadings including those outlined above. An important recent development is a loading device for carrying out a mixed mode double-cantilever test by applying unequal moments to the two layers.[17] This system is capable of applying the full range of mode mix and it has the advantage that steady-state crack growth occurs under constant applied moments. This system holds great promise for TBC testing if it could be scaled down to an appropriate size.

The *Brazil nut specimen* has been used successfully to measure a substantial range of mixed mode toughness for specific interfaces[18] and for composite laminates.[19] It is possible this specimen could also be adapted to measure TBC coating delamination toughness by bonding one half of a circular metallic disk to the surface of the coating and the other half to the underside of the substrate. Unlike the test just described, the Brazil nut test does not exhibit a steady-state and crack growth can occur unstably. The range of energy release rates achievable in the Brazil nut test with a specimen having a radius on the order of 1 cm is sufficient to drive most coating delaminations. Friction is also a concern in this test.[20]

(2) Specimens with a Coating Layer

For all of the tests discussed above, the effect of the elastic mismatch of the coating layer on the mode mix of the interface crack can be estimated simply if the coating is thin compared with the thickness of the substrate and the stiffener. The delamination crack lies on the interface between the coating and the substrate. An important distinction is whether the coating lies above or below the substrate, as has been noted in Fig. 1. With the convention employed in this paper, the interface toughness for a coating lying above the interface is $\Gamma(\psi)$, while that for a coating lying below the interface is $\Gamma(-\psi)$. As in the previous section, the discussion which follows assumes that the bending stiffness of the coating/stiffener bilayer, B, is the same as that of the lower substrate layer, $\bar{E}h^3/12$. With this choice, all of the above expressions for the energy release rate are the same, to the accuracy to which they hold.[§] It will also be assumed that the

---

§Some of the energy release rates may vary slightly from those quoted if one undertook more accurate plane strain calculations. It is strongly recommended that finite element calculations of final specimen geometries and properties be performed. Such calculations also reveal when the crack begins to interact with the specimen ends. In this paper, to simplify the discussion, we have taken the bending stiffness of the bilayer to be the same as that of the lower layer, and we have not directly accounted for the TGO. These effects, and contributions from residual stress in the coating, can be included if one is prepared to compute the energy release rate and the mode mix for the specimen.

Table I. The Phase Angle Shift, $\omega(\alpha,\beta)$ in Degrees, for Combinations of the Elastic Mismatch Parameters[21]

β	α								
	−0.8	−0.6	−0.4	−0.2	0.0	0.2	0.4	0.6	0.8
−0.4	2.2	3.5							
−0.3	3.0	4.0	3.3	1.4					
−0.2	3.6	4.1	3.4	2.0	−0.3	−3.3			
−0.1	4.0	4.1	3.3	2.0	0.1	−2.3	−5.5	−10.8	
0.0	4.4	3.8	2.9	1.6	0.0	−2.1	−4.7	−8.4	−14.3
0.1		2.3	1.1	−0.5	−2.3	−4.5	−7.4	−11.6	
0.2				−1.3	−3.0	−4.9	−7.3	−10.5	
0.3						−5.8	−7.8	−10.4	
0.4								−11.1	

coating is sufficiently thin and/or compliant that the elastic energy released associated with the residual stress in the coating can be ignored, as discussed in conjunction with Fig. 7.

The mode mix for the crack on the interface between the coating and the substrate, as measured by ψ, can be estimated using a general asymptotic relationship for cases with $t/h \ll 1$.[21] Denote the mode mix for any of the specimens in the absence of a coating (i.e., with $t = 0$, or, equivalently, with $E_C = E$ and $\nu_C = \nu$) by ψ^0, and denote the mode mix for the crack on the interface between the coating and the substrate under the same loading by ψ. The relation between the two measures is

$$\psi = \psi^0 \pm \omega(\alpha, \beta) \quad (16)$$

where the "+" applies if the coating lies below the substrate and the "−" applies if the coating lies above the substrate and where $\omega(\alpha,\beta)$ is given in Table I.[¶] For thin coatings ($t/h \ll 1$), this asymptotic result does not depend on t. The result (16) is based on an elasticity analysis of a crack on a bimaterial interface which, for $\beta \neq 0$, neglects the consequences of crack face interpenetration behind the crack tip on the assumption that the interpenetration is subsumed within the fracture process zone. Thus, (16) is limited to cases in which interpenetration does not invalidate application of the solution. If $\beta \neq 0$, the predicted zone of interpenetration increases as the loading becomes dominantly mode II, and thus (16) should be used with caution near mode II.

To illustrate the effect of elastic mismatch between the coating and the substrate for the specimens in Fig. 6, suppose $\bar{E}_C/\bar{E} = 1/4$ and $\nu_C = \nu = 0.3$: then, $\alpha = 3/5$, $\beta = 0.17$ and, from Table I, $\omega = -7.3°$. If the coating lies above the substrate, by (16), phase angle of the *tension edge-delamination specimen* increases from $\psi^0 = 49.1°$ to $\psi = 56.4°$; that of the *four-point bend specimen* increases from $\psi^0 = 40.9°$ to $\psi = 48.2°$; while mode II component of the *inverted four-point bend specimen* decreases by $\psi^0 = -60°$ to $\psi = -52.7°$. The mode II component of the *compression edge-delamination specimen* is also predicted to decrease from $\psi^0 = -90°$ to $\psi = -82.7°$, but this should only be considered as a trend rather than a quantitative estimate for reasons mentioned above. In general, the shift in mode mix due to the compliance of the coating layer is relatively small, as these results illustrate, except for coatings that are exceptionally stiff or compliant.

(3) Interchanging the Substrate and Stiffener

For any of the specimens considered, by turning the trilayer upside down and creating a notch down to the interface in the substrate and having a single continuous stiffener, one can access $\Gamma(\psi)$ in the opposite sign range of ψ. The interchange in Fig. 4 from (A) to (B) illustrates precisely what this entails. The

---

¶This table is reproduced from Suo and Hutchinson.[21] In Suo and Hutchinson,[21] the thin layer was taken to lie below substrate and thus "+" was used. Here, the convention of the earlier reference in defining the Dundurs parameter was followed wherein $\alpha > 0$ for systems with coatings more compliant than substrates. The reader should be alert to the fact that the usual convention has $\alpha > 0$ when the material above the interface crack is stiffer than the material below it.

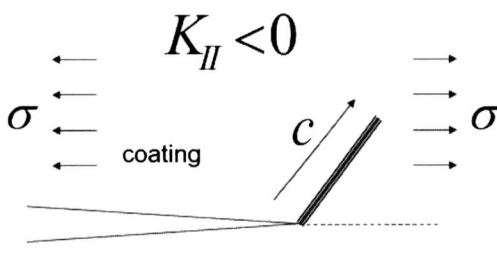

Fig. 13. Conventions related to kinking of an interface crack into the coating. For a coating lying above the substrate, as shown, a negative K_{II} encourages kinking into the coating. If the coating lies below the substrate a positive K_{II} encourages kinking into the coating. A tensile residual stress in the coating ($\sigma > 0$) increases the likelihood of kinking into the coating while a compressive residual stress ($\sigma < 0$) discourages kinking into the coating.

discussion, which follows continues to assume the interface of interest is that between the substrate and the coating. As noted in Fig. 1, inverting the substrate/coating bilayer changes the toughness from $\Gamma(\psi)$ to $\Gamma(-\psi)$ if the mode mix on the interface does not change. Let ψ^0 be the reference mode mix of the specimen under the particular loading in the absence of the coating as defined above. If the coating lies above the substrate (as in Fig. 4(A)), then $\psi = \psi^0 - \omega$ and the toughness is $\Gamma(\psi^0 - \omega)$. If the coating lies below the substrate (as in Fig. 4(B)), then $\psi = \psi^0 + \omega$ and the toughness is $\Gamma(-\psi^0 - \omega)$.

The possibility of interchanging the roles of the substrate and the stiffener in this manner has several advantages. The most obvious advantage is that it potentially provides a method to generate data for both positive and negative mode II contributions. Less obvious is the advantage that can accrue by changing the position of the coating relative to the substrate in helping to suppress kinking of the crack out of the interface. As discussed in the next subsection, the sign of the mode II stress intensity factor and the location of the coating relative to the interface can have a strong influence on the propensity of the crack to remain in the interface, especially if the interface toughness is comparable to that of the coating. Thus, a delamination crack that tends to stray from the interface for one substrate/coating/stiffener configuration might remain in the interface when the roles of the substrate and stiffener are interchanged in the manner suggested. In other words, if it is difficult to measure $\Gamma(\psi)$ for loadings with one sign of ψ, it may be possible to measure the toughness for the opposite sign of ψ. For toughness functions, which are symmetric in ψ, this would suffice. Generally, however, $\Gamma(\psi)$ need not be symmetric in ψ, as the set of data for an epoxy/glass interface illustrates.[22] For some interfaces $\Gamma(\psi)$ does appear to be symmetric.[19]

(4) Propensity for a Delamination Crack to Remain in or Near the Interface

Success in measuring delamination toughness requires that the path followed by the crack in a test is similar to the path followed by the crack in the application of interest. If the interface between the coating and the substrate has low toughness compared with the coating itself (and compared with the substrate, which for the discussion here will be assumed to be very tough), then crack propagation initiated in the interface is likely to remain in the interface for all loading conditions. Depending on the actual system, this interface might lie between the coating and the TGO or between the TGO and the bond coat. For systems without a TGO, this interface would lie between the coating and the bond coat. For some coating systems, delaminations occur within the coating but just above the coating/substrate interface (i.e., just above the TGO if one is present or above the bond coat if not). This may be a consequence of loading conditions which continually drive the crack toward the interface, as

discussed below, or it may be due to the existence of a very thin interfacial layer of less tough material. Residual stress in the coating can also play a role in determining whether the crack stays in or near the interface. A residual compressive stress in the coating acts to discourage cracks from turning into the coating, while tensile stress encourages deviations away from the interface or the low-toughness interfacial layer. A few brief observations related to these effects close out this section.

With the coating above the substrate and with the tip at the right end of the interface crack as in Fig. 13, a positive K_{II} ($\psi > 0$) would promote kinking of the crack downward into the substrate were it not too tough. Thus, if a crack is propagating in the coating just above the interface, a positive K_{II} will tend to cause it to hug the interface. Conversely, a negative K_{II} ($\psi < 0$) promotes upward kinking into the coating. If the coating lies below the interface the situation reverses: a positive K_{II} ($\psi > 0$) promotes downward kinking into the coatings while a negative K_{II} ($\psi < 0$) keeps the crack in or near the interface. Quantitative conditions based on the relative toughness of the interface to that of the coating are available for assessing the likelihood of kinking out of the interface when the sign of K_{II} promotes kinking into the substrate.[23]

The discussion thus far suggests that a test with the sign of K_{II} favoring kinking is less likely to succeed in delivering delamination data than one with the opposite sign of K_{II}. However, other factors must clearly be in play because some relatively brittle TBC coating systems appear to have a fracture path in the coating just above the interface with the TGO, or just above the bond coat if no TGO is present, due to compressive edge delaminations and buckle delaminations having negative mode II ($\psi = -90°$), or nearly so. These are conditions under which kinking into the coating should be most likely to occur. One possibility is that the thin layer of the coating material just above the interface has significantly lower toughness than the coating itself due to chemical or microstructural differences. Microcracking along, or just above, the interface under thermal exposure, which is thought to produce the degradation of toughness seen in Figs. 3 and 5, would be an example. Another possibility is that the in-plane compressive stress in the coating suppresses any tendency for cracks to wander away from the interface or the low-toughness layer. Quantitative mechanics results are also available to assess the role of residual stress.[23] The relevant dimensionless parameter is

$$\eta = \frac{\sigma \sqrt{c}}{\sqrt{E^* G}} \tag{17}$$

Here, σ is the residual stress, c is the putative kinked crack length (see Fig. 13), and $\bar{E}^* \cong 2(1/\bar{E} + 1/\bar{E}_C)^{-1}$. A compressive residual stress significantly reduces any tendency to kink into the coating if η is larger in magnitude than about 0.5, and, conversely, a residual tension of this level strongly promotes kinking. Assuming $c \cong t/10$ and representative values for the other quantities in (17), one concludes that compressive residual stress in the coating can indeed play a role in suppressing kinking even when the sign of K_{II} favors kinking into the coating.

V. Conclusions

Experimental measurement of mixed mode delamination toughness as a function of thermal history is an essential element of TBC durability assessment. Edge delaminations are among the most common types of TBC failures, and, therefore, test methods to measure mode II, and near-mode II, delamination toughness must be developed. To date, most toughness data acquired for TBC systems has fallen within the range of mode mix $0 \leq \psi < 60°$, using the convention adopted in this paper. Several mixed mode tests have been surveyed here which, in principle, could generate data over the entire range of mode mix, although serious obstacles to their implementation may exist. The compression edge-delamination specimen in Fig. 6 closely mimics edge

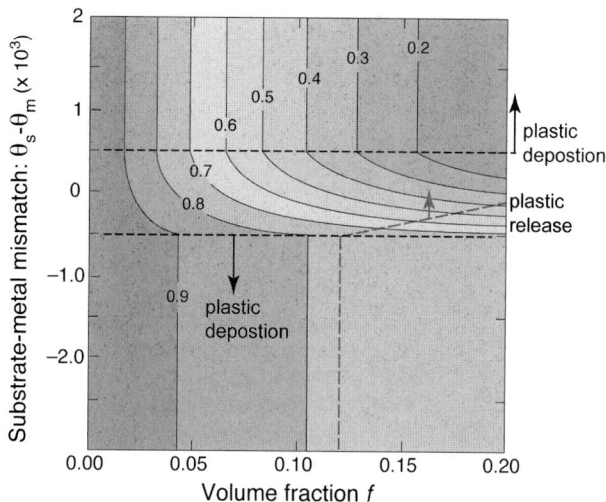

Fig. 5. Contours of G/G_o as a function of coating/metal mismatch and metal layer fraction, for $\Delta\theta_{sc} = -1.5 \times 10^{-3}$ and $\sigma_Y \sim 85$ MPa. The elastic modulus ratio of the coating to the metal is ~ 0.15. The vertical axis can be reinterpreted as $\Delta\theta_{cm} = \Delta\theta_{sm} - \Delta\theta_{sc} = \Delta\theta_{sm} + 1.5 \times 10^{-3}$.

Fig. 6. Contours of G/G_o as a function of coating/metal mismatch and yield stress, for $\Delta\theta_{sc} = -1.5 \times 10^{-3}$, and $f = 0.1$. The elastic modulus ratio of the coating to the metal is ~ 0.15. The vertical axis can be reinterpreted as $\Delta\theta_{cm} = \Delta\theta_{sm} - \Delta\theta_{sc} = \Delta\theta_{sm} + 1.5 \times 10^{-3}$.

The red dashed line indicates the boundary between elastic recovery during debonding (below the line) and plastic straining during debonding (above the line).

A striking feature in Fig. 5 is that the reduction in ERR is independent of the level of substrate/metal mismatch $\Delta\theta_{sm}$ when yielding occurs *both* during cooling from deposition and during debonding. Examining the expression for G in this regime, Eq. (35) indicates that the ERR is only a function of the substrate/metal mismatch through the the plastic strain increment term, $\Delta\varepsilon_p^r$. The combination of Eqs. (37) and (39) implies that $\Delta\varepsilon_p^r \propto \Delta\theta_{cm} - \Delta\theta_{sm}$ when yielding has occurred both during cool-down and debonding. Because $\Delta\theta_{ij} = \theta_i - \theta_j$, the plastic strain increment during release $\Delta\varepsilon_p^r$ scales as $\Delta\theta_{cs}$. This quantity is held fixed in Fig. 5, such that the ERR is independent of $\Delta\theta_{sm}$ whenever yielding has occurred during both steps. A key point is that the *total* plastic strain in the metal layers will scale with $\Delta\theta_{cm}$, but a portion of this misfit is offset by plastic strains generated during cool-down.

As shown at the top of Fig. 5, when $\Delta T_i < 0$ (i.e., cooling from deposition) and $\Delta\theta_{sc} < 0$, the reductions in ERR are the largest when $\alpha_m > \alpha_s > \alpha_c$. In this case, the stress in the metal while adhered is tensile, and equal to the yield stress above the dashed black line. After release, the stress in the metal is still tensile, which keeps the ceramic in compression, implying a smaller change in elastic strain energy in the ceramic during debonding. Moreover, plastic strains are generated during debonding, which further dissipates energy that would otherwise drive cracking. In the top of Fig. 5, the ERR reduction due to plastic work is much larger than that illustrated in Fig. 4, comprising nearly half of the total reduction. At the bottom of Fig. 5, the relative CTE's are $\alpha_s > \alpha_c > \alpha_m$ (for $\Delta T_i < 0$ and $\Delta\theta_{sc} < 0$). In this region, compressive plastic strains are generated during cooling from deposition, which partially accommodate the ceramic/metal mismatch and lead to smaller plastic strains during debonding and hence smaller reductions in ERR.

Similar behavior is evident in Fig. 6, which plots normalized ERR (G/G_o) contours as a function of substrate/metal mismatch ($\Delta\theta_{sm}$) and yield stress (σ_Y), for fixed layer fraction ($f = 0.1$). Once again, the dashed lines correspond to transitions in behavior, regarding yielding during cool-down and debonding. The maximum reductions in ERR correspond to property combinations for which the metal experiences tensile yielding both after cooling from deposition and during debonding. A critical feature of this plot is that there are property combinations for which the ERR is *increased* relative to the uniform coating, which occurs when the metal experiences compressive yielding during cool-down and tensile yielding during debonding. The

ERR can be large for high levels of yield stress because the elastic strain energy in the metal in the adhered state can be significant, and because plastic straining is prohibited during debonding.

It is worth emphasizing that the above discussion is for $\Delta\theta_{sc} < 0$ and $\Delta T_i < 0$; naturally, different conclusions are reached if the sign of the substrate/coating CTE mismatch changes or the system is heated above its assembly temperature.

VI. Summary

A model has been developed to investigate the effects upon the driving force for delamination of coating architecture and the thermomechanical properties for multilayers of two different materials. For elastic systems and layers that experience the same relative change in temperature, the ERR is minimized by the minimum value of an objective function (from Eq. (9)):

$$F = fE_m(\alpha_s - \alpha_m) + (1 - f)E_c(\alpha_s - \alpha_m)$$

This expression may find utility for identifying favorable combinations of constituent properties and the volume fraction of the embedded layers that reduce the delamination driving force. We note that optimal solutions may not always include the use of an elastic–plastic layer; a ceramic multilayer with carefully chosen CTEs and layer fractions can have a delamination driving force of zero (at least theoretically). The fundamental underlying mechanism is that the strain energy in the adhered state should be close to that in the release state, because the strain energy density is transferred from one phase to the other during debonding.

For elastic–plastic-embedded layers with $\Delta\theta_{sc} < 0$ and $\Delta\theta_{sm} < 0$, the greatest benefit is reached when yielding occurs during release, as this dissipates energy that would otherwise have been released during crack advance. For layers with equivalent deposition temperatures (i.e., all layers experience the same temperature change from deposition), this occurs when

$$|\alpha_s - \alpha_m| \geq \varepsilon_Y$$

$$|\alpha_c - \alpha_s| \geq \frac{fE_m\varepsilon_Y}{(1 - f)E_c}$$

The beneficial effect of plastic straining during release is most effective when these conditions are met, and the sign of $\alpha_s - \alpha_m$ is opposite to $\alpha_s - \alpha_c$. In this scenario, the compressive stress in the ceramic layer after deposition is maintained after debonding, thus decreasing the release of strain energy during debonding. Moreover, plastic strains generated during debonding are greater, leading to futher reductions. For the properties considered here, the reduction in the relative ERR (i.e., that compared with a uniform coating) is roughly twice the volume fraction of the metal; for example, for the Pt/YSZ system, 50% reductions of the elastic ERR are achieved with $f \sim 0.2$.

Finally, it is interesting to consider generalization of the central concept of an inelastic layer with alternative dissipative mechanisms. The underlying strategy would be to design systems for which the embedded layers experience additional damage during release; the work associated with this damage subtracts from the overall energy available to drive debonding. For example, it may be possible to design composite coatings comprising entirely brittle layers, in which the embedded layers experience microcracking (i.e., tunnel cracks in the embedded layers) or undergo martensitic transformations that are triggered only when the layers are released. Naturally, this would involve slightly different micromechanics than that outlined above, and the scaling factors associated with microcracking or phase change might not be as favorable as those associated with the strongly dissipative mechanism of metal plasticity, and the benefits might be limited to a single loading cycle. At this point, it is difficult to fully grasp the full range of possibilities afforded by metal toughening of ceramic coatings; for this, the remarkable insights and vision of of A. G. Evans are sorely missed.

Acknowledgments

H. N. G. W. is grateful for support of this research by the Office of Naval Research, under grant N00014-03-1-0297 (Dr. David Shiffler, Program Manager). M. R. B. is grateful for the support of the National Science Foundation under grant CMII0800790.

References

[1]R. A. Miller, "Current Status of Thermal Barrier Coatings—An Overview," *Surf. Coat. Technol.*, **30**, 1–11 (1987).

[2]C. Levi, "Emerging Materials and Processes for Thermal Barrier Systems," *Curr. Opin. Solid State Mater. Sci.*, **8**, 77–91 (2004).

[3]S. Stecura, "Advanced Thermal Barrier System Bond Coatings for Use on Nickel-, Cobalt- and Iron-Base Alloy Substrates," *Thin Solid Films*, **136**, 241–56 (1986).

[4]A. G. Evans, D. R. Mumm, J. W. Hutchinson, G. H. Meier, and F. S. Pettit, "Mechanisms Controlling the Durability of Thermal Barrier Coatings," *Prog. Mater. Sci.*, **46**, 505–53 (2001).

[5]K. N. Lee, D. S. Fox, and N. P. Bansal, "Rare Earth Silicate Environmental Barrier Coatings for SiC/SiC Composites and S_3N_4 Ceramics," *J. Eur. Ceram. Soc.*, **25**, 1705–15 (2005).

[6]I. T. Spitsberg, D. R. Mumm, and A. G. Evans, "On the Failure Mechanisms of Thermal Barrier Coatings with Diffusion Aluminide Bond Coatings," *Mater. Sci. Eng.*, **394**, 176–91 (2005).

[7]T. Xu, M. Y. He, and A. G. Evans, "A Numerical Assessment of the Propagation and Coalescence of Delamination Cracks in Thermal Barrier Systems," *Interface Sci.*, **11**, 349–58 (2003).

[8]M. R. Begley, D. R. Mumm, A. G. Evans, and J. W. Hutchinson, "Analysis of a Wedge Impression test for Measuring the Interface Toughness Between Films/Coatings and Ductile Substrates," *Acta Mater.*, **48**, 3211–20 (2000).

[9]H. Zhao, Z. Yu, and N. G. Wadley, "The Influence of Coating Compliance on the Delamination of Thermal Barrier Coatings," *Surf. Cout. Technol.*, **204**, 2432–41 (2010).

[10]A. G. Evans, M. Y. He, A. Suzuki, M. Gigliotti, B. Hazel, and T. M. Pollock, "A Mechanism Governing Oxidation-Assisted Low-Cycle Fatigue of Superalloys," *Acta Mater.*, **57**, 2969–83 (2009).

[11]V. K. Tolpygo and D. R. Clarke, "Surface Rumpling of a (Ni, Pt)Al Bond Coat Induced by Cyclic Oxidation," *Acta Mater.*, **13**, 3283–93 (2000).

[12]D. S. Balint and J. W. Hutchinson, "An Analytical Model of Rumpling in Thermal Barrier Coatings," *J. Mech. Phys. Solids*, **53**, 949–73 (2005).

[13]D. D. Hass, P. A. Parrish, and H. N. G. Wadley, "Electron Beam Directed Vapor Deposition of Thermal Barrier Coatings," *J. Vac. Sci. Technol.*, **16**, 3396–401 (1998).

[14]J. W. Hutchinson and Z. Suo, "Mixed Mode Cracking in Layered Materials," *Adv. Appl. Mech.*, **29**, 63–181 (1992).

[15]C. Subramanian and K. N. Stafford, "Review of Multicomponent and Multilayer Coatings for Tribological Applications," **165**, 85–95 (1993).

[16]W. Y. Lee, D. P. Stinton, C. C. Berndt, F. Erdogan, Y. D. Lee, and Z. Mutasim, "Concept of Functionally Graded Materials for Advanced Thermal Barrier Coating Applications," *J. Am. Ceram. Soc.*, **79**, 3003–12 (1996).

[17]S. Widjaja, A. M. Limarga, and T. H. Yip, "Modeling of Residual Stresses in a Plasma-Sprayed Zirconia/Alumina Functionally Graded-Thermal Barrier Coating," *Thin Solid Films*, **434**, 216–27 (2003).

[18]Z. Yu, H. Zhao, and H. N. G. Wadley, "The Vapor Deposition and Oxidation of Pt/YSZ Multilayers," *J. Am. Ceram. Soc.*, submitted (2010).

[19]Y. Wei and J. W. Hutchinson, "Nonlinear Delamination Mechanics for Thin Films," *J. Mech. Phys. Solids*, **45**, 1137–59 (1997).

[20]A. Kelly, R. J. Stearn, and L. N. McCartney, "Composite Materials of Controlled Thermal Expansion," *Compos. Sci. Technol.*, **66**, 154–9 (2006).

[21]Y. N. Ioginov, A. V. Yermakov, L. G. Grohovskaya, and G. I. Studenak, "Platinum Metals," *Review*, **4**, 178–84 (2007). □

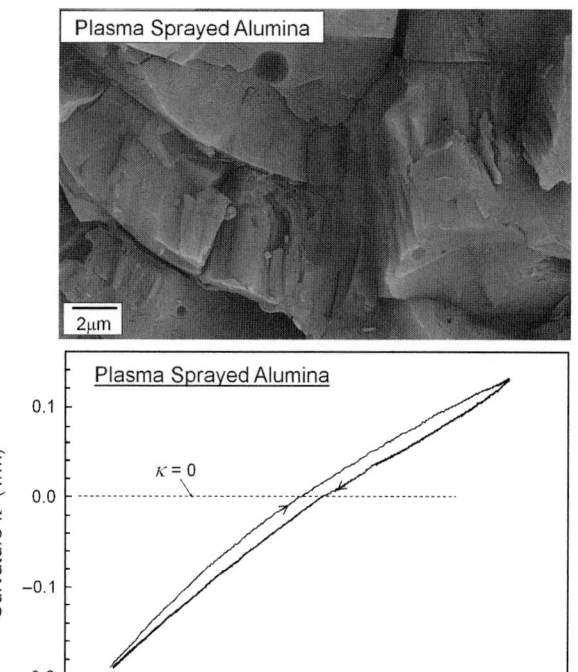

Fig. 12. SEM micrograph of a fractured plasma-sprayed alumina coatinga. Unlike YSZ, it has fewer intrasplat boundaries and cracks. Measured curvature–temperature during thermal cycle is shown below.

IV. Three-Point-Bend Test Results

Thermal cycling offers robust test data, is relatively easy to apply and enables exploration of wide ranging materials and process conditions. However, because in this test, coatings are constrained by the metallic substrate, the stress impositions on the coatings are always associated with thermal mismatches. In fact, the interpretations of measured curvature data require the knowledge of coupling conditions of coating and substrate. In order to isolate the response of plasma-sprayed materials, free-standing specimens have been prepared as described earlier and subject to mechanical loading using three-point-bending.

Two free standing coatings, YSZ and alumina (thickness 1.6 and 1.27 mm, respectively), were examined under three-point-bend test. These samples were prepared with same parameters as described earlier (YSZ-B and alumina). For the given condition, due to higher melting temperature the YSZ coating was more porous than the plasma-sprayed alumina. Measured midspan load–deflection records are shown in Fig. 13. Note that cyclic loading unloading was performed with successive load amplitude to increase until failure. Several observations can be made.

(1) The loading as well as unloading curves are not linear, and the nonlinearity increases with applied load until fracture. However, unlike the thermal cycle test, the through-thickness stress varies from tension to compression and the behavior reflects the net effect of both.

(2) The YSZ coating exhibits greater hysteresis than that of aluminum consistent with the thermal cycle results shown in Figs. 5 and 12. As verification, a sintered alumina plate of similar dimension was tested which displayed an ideally elastic response with no measurable nonlinearity or hysteresis.

(3) Unlike the curvature–temperature response, the unloading curves did not return back to its initial state, i.e. coatings had residual displacement after each cycle. This is due to the fact that the loading starts at zero stress and not at a residual compressive stress in the thermal cycle test. The residual stress in the specimen was removed as the substrate was separated.

(4) At failure, alumina specimen exhibits sudden drop in the load, which represents more brittle nature of denser alumina

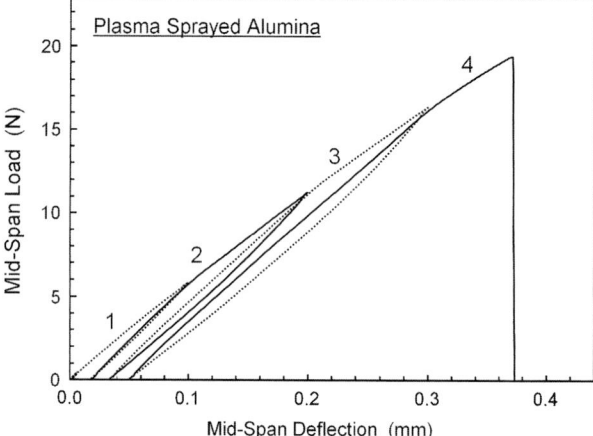

Fig. 13. Cycled load–deflection measurements under three-point-bend condition for shown for plasma-sprayed YSZ and alumina specimens. Amplitudes are increased after each cycle. Schematic of predicted behavior for nacre[32] is shown in the inset.

coating. Note that both *ND* and *HD* of YSZ-B are significantly larger than those of alumina (see Figs. 7 and 10), which imply greater density of interfaces and microcracks.

The anelastic responses are in many ways analogous to those reported and modeled for natural composites such as nacre. In fact our observations are very similar to those predicted by Evans *et al.*[32] for nacreous composites as shown inset of Fig. 13. They had speculated that in the case of nacre, frictional effects will likely result due to sliding across interfaces. However, there are differences. The sprayed coatings do not contain any organic layer at the interfaces and the internal friction and nonlinearity are purely geometric effects. Secondly, nacre is a much more ordered material compared with the chaotic assembly of the plasma-sprayed system. Nevertheless, the underlying similarities in the operating mechanisms are noteworthy.

V. Summary and Conclusions

In thermal-sprayed materials, their unique geometrical/microstructural features allow mechanical responses to be anelastic, namely, nonlinear and hysteretic. In ceramics, the former introduces a quasi-ductile response in otherwise brittle solids, and the latter is associated with energy-dissipative process. The mechanisms behind these characteristics are attributed to the large numbers of embedded interfaces and microcracks, resulting from the rapid quenching process. These geometrical attributes collectively produce the anelastic behavior through opening/closing and sticking/sliding of interfaces.

In this paper, we present strategies and results in controlling the anelastic response. First the anelasticity is quantitatively identified by three parameters, elastic modulus (E), ND, and HD which allow for comparison among the various sprayed samples.It was shown that for the plasma-sprayed yttria-stabilized zirconia, all of the anelastic parameters are sensitive to processing conditions, and the results indicate that it is feasible to tune the process for controlled introduction of anelasticity.

It is further demonstrated that the hysteretic response relates both to the dimensionality of the interfacial separations (intersplat porosity) as well as the roughness of the constituent splat surface. As such it is feasible to suppress the hysteretic behavior while maintaining the nonlinear response. Extrinsic factors such as chemical modification of interfaces will also affect the anelastic processes in these layered materials.

The observed anelastic behavior is similar to those predicted for natural brick and mortar structures by Evans and colleagues, although, compositionally, sprayed materials do not generally contain any polymeric interfacial binder. As such some of the anelastic attributes are intrinsic to layered architecture resulting from processing. The results point to strategies for imparting compliance and toughness to thermostructural coatings via microstructural engineering.

Acknowledgments

This paper is dedicated to Professor Tony Evans, who appreciated the significance of the initial observations some years ago and encouraged our group to further develop the idea. Tony's comment "people have been trying to make synthetic analogues of natural materials for some time, you guys just have it !" will continue to inspire our work in this arena. Authors are also thankful to Dr. K. Shinoda and Dr. Y. Tan for their assistance in processing some of the plasma-sprayed samples and to Prof. C. M. Weyant for his review of the manuscript and suggestions. We are also grateful to Dr. C. Johnson for identifying the issues of thermal gradient in our initial test, which led to our improvement of our experimental approach.

References

[1]P. Fauchais, "Formation of Plasma-Sprayed Coatings," *J. Therm. Spray Technol.*, **4** [1] 3–6 (1995).

[2]H. Herman, S. Sampath, and R. McCune, "Thermal Spray: Current Status and Future Trends," *MRS Bull.*, **25** [7] 17–25 (2000).

[3]A. Kulkarni, Z. Wang, T. Nakamura, S. Sampath, A. Goland, H. Herman, J. Allen, J. Ilavsky, G. Long, J. Frahm, and R. W. Steinbrech, "Comprehensive Microstructural Characterization and Predictive Property Modeling of Plasma-Sprayed Zirconia Coatings," *Acta Mater.*, **51** [9] 2457–75 (2003).

[4]S. Kuroda, T. Dendo, and S. Kitahara, "Quenching Stress in Plasma-Sprayed Coatings and its Correlation with the Deposit Microstructure," **4** [1] 75–84 (1995).

[5]J. Matejicek and S. Sampath, "Intrinsic Residual Stresses in Single Splats Produced by Thermal Spray Processes," *Acta Mater.*, **49** [11] 1993–9 (2001).

[6]R. Mcpherson, "The Relationship between the Mechanism of Formation, Microstructure and Properties of Plasma-Sprayed Coatings," *Thin Solid Films*, **83** [3] 297–310 (1981).

[7]R. Mcpherson, "The Structure and Properties of Plasma Sprayed Alumina Coatings," *Ceram.-Silik.*, **35** [3] 273–82 (1991).

[8]S. Sampath, "Thermal Sprayed Ceramic Coatings: Fundamental Issues and Application Considerations," *Int. J. Mater. Prod. Technol.*, **35** [3–4] 425–48 (2009).

[9]J. Ilavsky, A. J. Allen, G. G. Long, S. Krueger, C. C. Berndt, and H. Herman, "Influence of Spray Angle on the Pore and Crack Microstructure of Plasma-Sprayed Deposits," *J. Am. Ceram. Soc.*, **80** [3] 733–42 (1997).

[10]T. Nakamura, G. Qian, and C. C. Berndt, "Effects of Pores on Mechanical Properties of Plasma-Sprayed Ceramic Coatings," *J. Am. Ceram. Soc.*, **83** [3] 578–84 (2000).

[11]Z. Wang, A. Kulkarni, S. Deshpande, T. Nakamura, and H. Herman, "Effects of Pores and Interfaces on Effective Properties of Plasma Sprayed Zirconia Coatings," *Acta Mater.*, **51** [18] 5319–34 (2003).

[12]A. Kulkarni, A. Vaidya, A. Goland, S. Sampath, and H. Herman, "Processing Effects on Porosity-Property Correlations in Plasma Sprayed Yttria-Stabilized Zirconia Coatings," *Mater. Sci. Eng.A—Struct. Mater. Prop. Microstruct. Process.*, **359** [1–2] 100–11 (2003).

[13]Y. Tan, J. P. Longtin, and S. Sampath, "Modeling Thermal Conductivity of Thermal Spray Coatings: Comparing Predictions to Experiments," *J. Therm. Spray Technol.*, **15** [4] 545–52 (2006).

[14]W. Chi, S. Sampath, and H. Wang, "Ambient and High-Temperature Thermal Conductivity of Thermal Sprayed Coatings," *J. Therm. Spray Technol.*, **15** [4] 773–8 (2006).

[15]A. Cipitria, I. O. Golosnoy, and T. W. Clyne, "Sintering Kinetics of Plasma-Sprayed Zirconia TBCs," *J. Therm. Spray Technol.*, **16** [5–6] 809–15 (2007).

[16]W. G. Chi, S. Sampath, and H. Wang, "Microstructure-Thermal Conductivity Relationships for Plasma-Sprayed Yttria-Stabilized Zirconia Coatings," *J. Am. Ceram. Soc.*, **91** [8] 2636–45 (2008).

[17]Y. Tan, J. P. Longtin, S. Sampath, and H. Wang, "Effect of the Starting Microstructure on the Thermal Properties of As-Sprayed and Thermally Exposed Plasma-Sprayed YSZ Coatings," *J. Am. Ceram. Soc.*, **92** [3] 710–6 (2009).

[18]J. Matejicek and S. Sampath, "In situ Measurement of Residual Stresses and Elastic Moduli in Thermal Sprayed Coatings—Part 1: Apparatus and Analysis," *Acta Mater.*, **51** [3] 863–72 (2003).

[19]J. Matejicek, S. Sampath, D. Gilmore, and R. Neiser, "In Situ Measurement of Residual Stresses and Elastic Moduli in Thermal Sprayed Coatings—Part 2: Processing Effects on Properties of Mo Coatings," *Acta Mater.*, **51** [3] 873–85 (2003).

[20]S. Sampath, V. Srinivasan, A. Valarezo, A. Vaidya, and T. Streibl, "Sensing, Control, and In Situ Measurement of Coating Properties: An Integrated Approach Toward Establishing Process–Property Correlations," *J. Therm. Spray Technol.*, **18** [2] 243–55 (2009).

[21]T. Nakamura and Y. J. Liu, "Determination of Nonlinear Properties of Thermal Sprayed Ceramic Coatings Via Inverse Analysis," *Int. J. Solids Struct.*, **44** [6] 1990–2009 (2007).

[22]Y. Liu, T. Nakamura, V. Srinivasan, A. Vaidya, A. Gouldstone, and S. Sampath, "Non-Linear Elastic Properties of Plasma-Sprayed Zirconia Coatings and Associated Relationships with Processing Conditions," *Acta Mater.*, **55** [14] 4667–78 (2007).

[23]Y. J. Liu, T. Nakamura, G. Dwivedi, A. Valarezo, and S. Sampath, "Anelastic Behavior of Plasma-Sprayed Zirconia Coatings," *J. Am. Ceram. Soc.*, **91** [12] 4036–43 (2008).

[24]E. F. Rejda, D. F. Socie, and T. Itoh, "Deformation Behavior of Plasma-Sprayed Thick Thermal Barrier Coatings," *Surf. Coat. Technol.*, **113** [3] 218–26 (1999).

[25]S. R. Choi, D. M. Zhu, and R. A. Miller, "Mechanical Properties/Database of Plasma-Sprayed ZrO_2—8 wt% Y_2O_3 Thermal Barrier Coatings," *Int. J. Appl. Ceram. Technol.*, **1** [4] 330–42 (2004).

[26]F. Kroupa, "Nonlinear Behavior in Compression and Tension of Thermally Sprayed Ceramic Coatings," *J. Therm. Spray Technol.*, **16** [1] 84–95 (2007).

[27]N. Tassini, S. Patsias, and K. Lambrinou, "Ceramic Coatings: A Phenomenological Modeling for Damping Behavior Related to Microstructural Features," *Mater. Sci. Eng. A—Struct. Mater. Prop. Microstruct. Process.*, **442** [1–2] 509–13 (2006).

[28]P. J. Torvik, "A Slip Damping Model for Plasma Sprayed Ceramics," *J. Appl. Mech.-Trans. ASME*, **76** [6] 061018, 8pp (2009).

[29]M. Kachanov, I. Tsukrov, and B. Shafiro, "Effective Moduli of Solids with Cavities of Various Shapes," *Appl. Mech. Rev.*, **47** [1] 151–74 (1994).

[30]B. R. Lawn and D. B. Marshall, "Nonlinear Stress–Strain Curves for Solids Containing Closed Cracks with Friction," *J. Mech. Phys. Solids*, **46** [1] 85–113 (1998).

[31]G. Dwivedi, T. Wentz, S. Sampath, and T. Nakamura, "Assessing Process and Coating Reliability Through Monitoring of Process and Design Relevant Coating Properties," *J. Therm. Spray Technol.*, **19** [4] 695–712 (2010).

[32]A. G. Evans, Z. Suo, R. Z. Wang, I. A. Aksay, M. Y. He, and J. W. Hutchinson, "Model for the Robust Mechanical Behavior of Nacre," *J. Mater. Res.*, **16** [9] 2475–84 (2001).   □

Fig. 3. Characteristic R-line luminescence from a TGO. The sharp line labeled Ar is from an argon discharge lamp used to calibrate frequency.

stresses or elastic strain energy that motivate failure. To provide this information, we have developed an imaging system for measuring the spatial distribution of the local strains in the TGO, by PSLS through the TBC.[6] In essence, the method uses the TGO as a strain sensor of the damage in its vicinity to monitor the stress re-distribution on thermal cycling.

(1) Piezospectroscopy

The physical basis of the piezospectroscopy technique has been described in detail previously so is only briefly summarized here.[10–14] The thermally grown oxide is aluminum oxide and contains trace concentrations of Cr^{3+} incorporated during its growth at high temperature. The Cr^{3+} ions luminesce when excited optically, producing a very sharp emission doublet at 694 nm, the R_1 and R_2 lines as illustrated in Fig. 3. The radiative and nonradiative transitions following the excitation of electrons from the ground state are shown in Fig. 4 with the R-lines resulting from emission from the split E2 state. Because the energy of the $2E \rightarrow 4A_2$ transition responsible for the R-line emission is sensitive to the overlap of the $3d^3$ orbitals, and consequently the distance between the Cr^{3+} ions and the surrounding O^{2-} ions, the wavelength of the R lines is sensitive to strain.[15] Measurement of the wavelength can thus be used to nondestructively monitor strain.[11] Although the stress dependence of the shift of the R_1 and R_2 lines is slightly different and is also dependent on both the crystallographic orientation of the alumina[11] as well as on polarization,[12,16] the frequency shift reduces to being proportional to the trace of the stress tensor, and hence the mean stress, σ_m

$$\bar{\Delta v} = \frac{\Pi_{ii}}{3}\langle(\sigma_{11} + \sigma_{22} + \sigma_{33})\rangle = \Pi_{ii}\sigma_m \qquad (1)$$

for a polycrystalline alumina having a grain size that is much smaller than the probe size and where the grains are randomly oriented.[11,17] The trace of the piezospectroscopic coefficient tensor, Π_{ii}, has a value of 7.61 $(cm \cdot GPa)^{-1}$ for the R_2 line so by inverting Eq. (1), the mean stress can be obtained from the measured change in frequency from its' stress-free value. The intensity of the R-line luminescence is not dependent on stress and so variations in intensity due to, for instance, variations in TBC thickness, local optical scattering lengths or incident illumination, do not affect the value of the measured stress.

(2) Imaging System

Until recently, the stress distribution in the TGO could only be monitored point by point moving a focused laser beam across a coating and subsequently forming maps pixel by pixel, a laborious process ill-suited to evaluation of coatings on thermal cycling. In the imaging system shown schematically in Fig. 5, an area of interest on a coating is illuminated by a uniform laser beam and the luminescence is collected by a CCD camera, after filtering through a wavelength-tunable filter. A series of images, each at a slightly different narrow wavelength band, is recorded, thus mapping the R_1 and R_2 spectra for each pixel on the CCD. From this a map of the strain distribution is assembled by analyzing the spectra by software. Details of the algorithm for curve fitting and extracting the strain are described elsewhere.[18,19] The tunable filter is a motorized Fabry–Perot filter that only passes a very narrow band of the luminescence as illustrated schematically in Fig. 3. The observations described in this contribution were formed using a 1 W frequency doubled Nd:YAG laser (532 nm) and a low noise CCD camera (Santa Barbara Imaging Group, Goleta, CA). The mean stress is taken to be over the volume of the resolution element, i.e., the volume of the TGO corresponding to the measurement over an area projected onto a pixel of the CCD.

IV. Observations and Analysis

Images of the spatial distribution of the mean stress in four different coatings at the different fractions of thermal cycles are reproduced in Fig. 6. The values of the stresses are color coded (note that the stress-scale is adjusted in each case to optimize the color contrast). Each data point in the stress image corresponds to approximately 16 μm spatial resolution, such that the total imaged area is about 4.4 by 5.6 mm. Most measurements allow determination of the stress, except in a few locations where either luminescence is too weak or too strong. These areas are colored white. Apart from the evident variations in color there are no obvious spatial correlations in the stress variations that might indicate the size or extent of damage or its evolution. Analysis of the frequency distribution of the stresses extracted

Fig. 4. Schematic of the energy levels of Cr^{3+} in alumina, the decay paths and the energy shift caused by a compressive stress.

Fig. 5. Schematic of multispectral imaging system in which a TBC-coated sample is uniformly illuminated with an argon ion laser beam and the luminescence imaged onto a CCD through a variable wavelength tunable Fabry–Perot filter.

Fig. 6. Piezospectroscopic images of four coatings at different fractions of their average life. The color coding corresponds to the mean stress levels shown on the vertical color stripes on the right hand side of each panel. The numbers on the left hand ordinate and horizontal are pixels, each 16 μm in size, so the images correspond to areas of 4.4 mm × 5.6 mm.

from the images shown in Fig. 6 is more revealing. This data, Fig. 7, indicates that the average value of the mean stress decreases with thermal cycling, consistent with the release of elastic strain energy density. Furthermore, the stress variation over the imaged area more closely fits a Gaussian rather than a Weibull distribution, although this data is for the overall stresses rather than those in the immediate vicinity of the damaged regions. A Gaussian distribution indicates that stress values across the image can be interpreted as independent or at least of short lateral order, i.e., with little correlation between different subregions. This lack of correlation is also evident from an overall featureless two-dimensional Fourier transform (not shown).

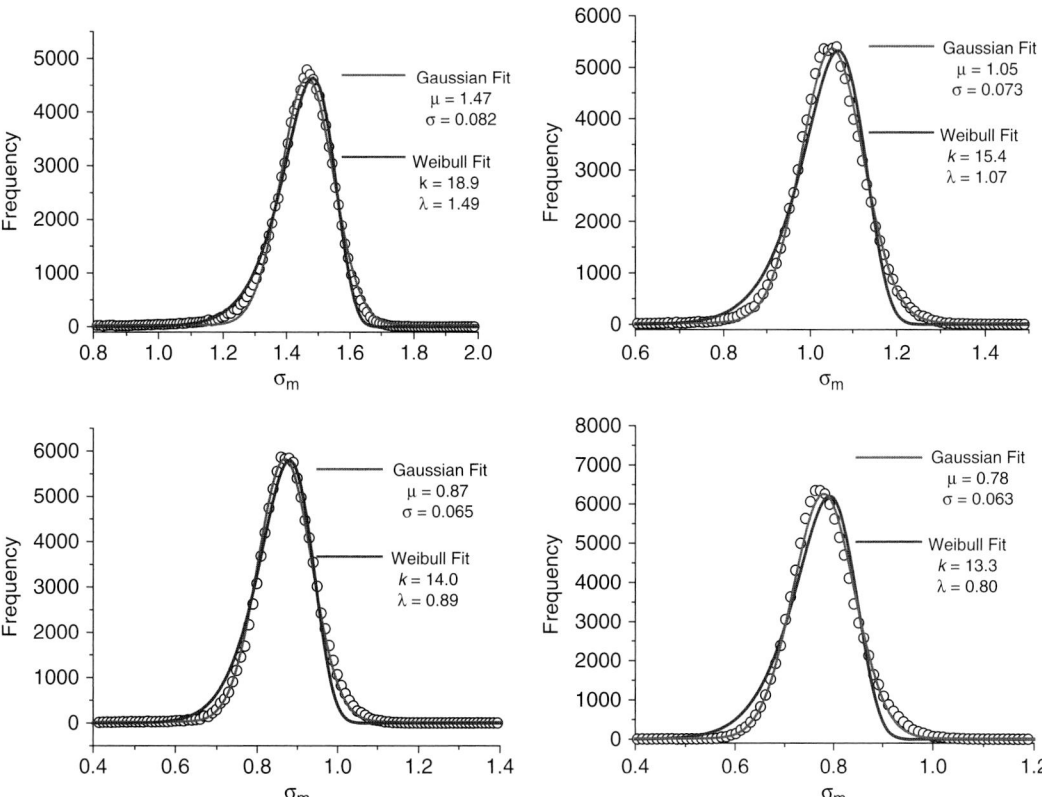

Fig. 7. Distribution functions of the stresses in the images shown in Fig. 6 as a function of the mean stress.

of the bond-coat surface produced by a combination of rumpling and swelling that occur with thermal cycling. Rumpling causes local bending of the TGO, creating out-of-plane strains and stresses, which not only decrease the mean stress in the TGO but also can produce TGO cracks by local tension or tearing. An example of this form of cracking is shown at B in Fig. 2. Inhomogeneous swelling, in particular, of the bond-coat from preferential nickel diffusion into the bond-coat from the underlying superalloy will cause nonuniform spatial displacements of the surface, promoting rumpling. This is believed to be responsible for separations of the type shown at D in Fig. 2. If the TBC is thick enough and remains in contact with the TGO, it suppresses rumpling. However, preexisting separations at the TBC/TGO interface or local regions where there are gaps in the TBC columns, either as a result of shadowing or very small grains, would result in less constraint and allow local rumpling and hence the development of out-of-plane tensile stresses. In some cases, as at location D in Fig. 2(d), the local TBC constraint is lost as a divot of TBC, a small wedge shaped grain formed at the intersection of two larger misoriented columns, appears to have been pulled away by the differential bond-coat motion.

Our finding that TBCs fail by a progressive nucleation, growth and linkage process until linked damaged regions extend across an area corresponding to the macroscopic buckling condition, provide some insight into why the thermal cycle life is sensitive to so many thermal cycling conditions.[5,23] Long hot times combined with slow heating and cooling rates and relatively low maximum temperature leave the bond-coat surface relatively flat and unchanged roughness and so the stresses normal to the interfaces are small and few local separations can occur. The lives are consequently relatively long. Under these conditions, it is speculated that processes that lower the interface fracture toughness cause failure such as might occur by long-range diffusion of species, for instance, S, from the underlying superalloy. Under these conditions, failure tends to occur once a critical thickness of the TGO is formed by oxidation and separation occurs preferentially by delamination along the bond-coat/TGO interface. The life can then be reasonably well predicted by oxidation life codes, such as COATLIFE,[24,25] embodying a critical thickness based on attaining a critical strain energy in the TGO and TBC. At the other extreme is rapid, short cycling at high temperatures, above about 1100°C. The lives under these conditions do not fit the oxidation life codes as has been remarked by Chan *et al.*,[24] being both much shorter and having a much larger statistically variability. These conditions, more typically encountered in military engines, are known to produce pronounced rumpling of bare PtNiAl aluminide coatings and, as illustrated in this work, several different types of interface damage but few bond-coat/TGO delaminations. Experiments and models indicate that at temperatures of 1100°C and above, the surface displacements are dependent on not only temperature but heating and cooling rates in the thermal cycles, too.[5,23,26] Consequently, there remains an outstanding problem of how to relate thermal cycle life to both maximum temperature and thermal cycle time, reconciling Eqs. (2) and (3) introduced above.

The complexity of the damage observed makes it difficult to describe failure in terms of detailed crack driving forces and fracture resistances. Not only does the propagation of a single crack seem an inappropriate description for the failure until the latest stages of the failure but also the separations are not strictly coplanar. Furthermore, although the luminescence shift is a measure of the mean stress it does not provide any information about the thickness of the TGO and hence is not a direct measure of the elastic strain energy, U_{SE}. Neither does it explicitly include the strain energy in the TBC. Nevertheless, as fracture of the TGO does not decrease its elastic modulus, the stored elastic strain energy in the TGO is proportional to the square of the mean stress. The TGO thickens by oxidation and so its thickness increases approximately parabolically with time. Consequently, the total elastic strain energy will be a slowly varying function of

time. As the fracture resistance is unlikely to change very rapidly with thermal cycling, the critical condition for fracture at which the total strain energy area density equals the fracture resistance is likely to be poorly defined and variable. Consequently, one might expect that the failure life exhibits a large statistical variation about a mean value determined by the thermal cycling conditions. Furthermore, as described by Hutchinson in a companion article, failure (as well as buckling) occurs under mixed mode conditions, so the statistical variability is expected to be even greater than just discussed. Further complicating any detailed energy balances is the fact that the TBC sinters with time increasing its elastic modulus and with its strain energy. This emphasizes the importance of knowing where the separations lie; if they lie largely above the TGO, it's the strain energy in the TBC that provides the driving force for failure, whereas if the separations are principally at the alloy/TGO interface, then the strain energy in both the TGO and TBC must be included.

Finally, the imaging system used in this work remains a prototype. Further improvements in the optics as well as image processing hardware and software, currently being developed, can be expected to yield faster imaging of the stresses and make it possible to follow the damage evolution in even greater detail. For instance, the instrumentation should enable images of the residual stresses to be generated after each thermal cycle of the same test object, so that the growth and coalescence of individual separation events can be followed with cycling. The ability to generate large databases of residual stress data also allows better estimates of variance in residual stress data between samples for a given fractional life, in comparison with the current results based on a few samples. With better knowledge of the conditions for delamination coalescence, the image system promises to provide nondestructive monitoring of the health of individual blades and vanes. This would also facilitate comparison of our laboratory scaling findings to those on engine hardware as well as other types of coatings, including overlay coatings.

Acknowledgments

The authors are grateful to Ken Murphy of Alcoa Howmet Research Center for providing the coatings studied in this work. The authors are also grateful for support from the Office of Naval Research (D. R. C., V. K. T.) and the Air Force Research Laboratory (B. H., while at MetroLaser).

References

[1]T. Editors, "Britain's Lonely High-Flier"; The Economist, 2009

[2]A. G. Evans and J. W. Hutchinson, "On the Mechanics of Delamination and Spalling in Compressed Films," *Int. J. Solids Struct.*, **2**, 455–66 (1984).

[3]M. D. Thouless, H. M. Jensen, and E. G. Liniger, "Delamination from Edge Flaws," *Proc. Roy. Soc.*, **A447**, 271–9 (1994).

[4]J. W. Hutchinson and Z. Suo, "Mixed Mode Cracking in Layered Materials," *Adv. Appl. Mech.*, **29**, 63–191 (1992).

[5]V. K. Tolpygo and D. R. Clarke, "Rumpling of CVD (Ni,Pt)Al Diffusion Coatings Under Intermediate Temperature Cycling," *Surf. Coat. Technol.*, **203**, 3278–85 (2009).

[6]R. J. Christensen, D. M. Lipkin, D. R. Clarke and K. Murphy, "Nondestructive Evaluation of the Oxidation Stresses Through Thermal Barrier Coatings Using Cr^{3+} Piezospectroscopy," *Appl. Phys. Lett.*, **69** [24] 3754–6 (1996).

[7]B. Heeg and D. R. Clarke, "Optical Measurement of the Thermal Diffusivity of Intact Thermal Barrier Coatings," *J. Appl. Phys.*, **104**, 113119, 7pp (2008).

[8]J. I. Eldridge, C. M. Spuckler, and R. E. Martin, "Monitoring Delamination Progression in Thermal Barrier Coatings by Mid-Infrared Reflectance Imaging," *Int. J. Appl. Ceram. Technol.*, **3** [2] 94–104 (2006).

[9]V. K. Tolpygo and D. R. Clarke, "Morphological Evolution of Thermal Barrier Coatings Induced by Cyclic Oxidation," *Surf. Coat. Technol.*, **163–164**, 81–6 (2003).

[10]Q. Ma and D. R. Clarke, "Stress Measurement in Single-Crystal and Polycrystalline Ceramics Using their Optical Fluorescence," *J. Am. Ceram. Soc.*, **76** [6] 1433–40 (1993).

[11]Q. Ma and D. R. Clarke, "Piezospectroscopic Determination of Residual-Stresses in Polycrystalline Alumina," *J. Am. Ceram.c Soc.*, **77** [2] 298–302 (1994).

[12]S. H. Margueron and D. R. Clarke, "The Use of Polarization in the Piezospectroscopic Determination of the Residual Stresses in Polycrystalline Alumina Films," *Acta Mater.*, **54** [20] 5551–7 (2006).

[13]B. Heeg and D. R. Clarke, "Non-Destructive Thermal Barrier Coating (TBC) Damage Assessment Using Laser-Induced Luminescence and Infrared Radiometry," *Surf. Coat. Technol.*, **200** [5–6] 1298–302 (2005).

[14]J. A. Nychka and D. R. Clarke, "Damage Quantification in TBCs by Photo-Stimulated Luminescence Spectroscopy," *Surf. Coat. Technol.*, **146**, 110–6 (2001).

[15]B. Henderson and G. F. Imbusch, *Optical Spectroscopy of Inorganic Solids.* Claredon Press, Oxford, 1989.

[16]D. J. Gardiner, M. Bowden, S. H. Margueron and D. R. Clarke, "Use of Polarization in Imaging the Residual Stresses in Polycrystalline Alumina Films," *Acta Mater.*, **55**, 3431–56 (2007).

[17]D. R. Clarke and D. J. Gardiner, "Recent Advances in Piezospectroscopy," *Int. J. Mater. Res.*, **98**, 1–8 (2007).

[18]B. Heeg and J.B Abbiss, "Piezospectroscopic Imaging with a Tunable Fabry-Perot Filter and Tikhonov Reconstruction," *Opt. Lett.*, **32** [7] 859–61 (2007).

[19]J. B. Abbiss and B. Heeg, "Imaging Piezospectroscopy," *Rev. Sci. Instrum.*, **79**, 123105, 9pp (2008).

[20]M. Wen, E. H. Jordan, and M. Gell, "Evolution of Photostimulated Luminescence of EB-PVD (Ni,Pt)Al Thermal Barrier Coatings," *Mater. Sci. Eng.*, **A398**, 99–107 (2005).

[21]V. K. Tolpygo, D. R. Clarke, and K. S. Murphy, "The Effect of Grit Blasting on the Oxidation Behavior of a Platinum-Modified Nickel-Aluminide Coating," *Metall. Mater. Trans. A-Phys. Metall. Mater. Sci.*, **32** [6] 1467–78 (2001).

[22]G. Lee, A. Atkinson, and A. Selcuk, "Development of Residual Stress and Damage in Thermal Barrier Coatings," *Surf. Coat. Technol.*, **201** [7] 3931–6 (2006).

[23]V. K. Tolpygo and D. R. Clarke, "Temperature and Cycle Time Dependence of Rumpling in Platinum-Modified Diffusion Aluminide Coatings," *Scri. Mater.*, **57**, 563–6 (2007).

[24]K. Chan, S. Cheruvu, and R. Viswanathan, "Development of a Thermal Barrier Coating Life Model"; pp. 591–5 in *ASME Turbo Expo 2003*. ASME, Atlanta, 2003.

[25]T.A Cruse, S. E. Stewart, and M. Ortiz, "Thermal Barrier Coating Life Prediction Model Development," *J. Eng. Gas Turbines Power*, **110**, 610–6 (1988).

[26]D. S. Balint and J. W. Hutchinson, "An Analytical Model of Rumpling in Thermal Barrier Coatings," *J. Mech. Phys. Solids*, **53** [4] 949–73 (2005). ☐

Fig. 2. The optical image on the top (a) is showing the actual configuration used in the measurements comprising a columnar TBC deposited onto a Ni-based alloy and the cut out section below the bond coat. The grid used for the digital image correlation is shown in the lower left (b) where a vertical and interfacial crack has been formed. The geometry is shown in the lower right (c).

interpretation then requires FE solution. In the end, asymmetric loading emerged as the method of choice for the current study.

In this method, a dominant interface crack was generated from the precrack by relocating the load point off-center. This mode of loading caused an interface crack to form and propagate toward the load point in a stable manner. The ensuing interface crack path is shown in SEM images of the cracks in Fig. 4(a). It penetrates the TGO and then follows the interface with the bond coat. The tip region comprises a series of *en echelon* microcracks traversing the TGO (Figs. 4(b) and (c)), indicative of a mode II mechanism, as elaborated later. To prevent the load/unload/load sequence from influencing the interface at the crack tip, all subsequent tests were conducted in consecutive load steps without unloading in between.

The experimental crack propagation events for two different beams are summarized in Table I. The results from these series of critical events are the basis for the FE-based fracture toughness calculation described in the subsequent section. For beam 1, the aforementioned precrack through the top coat formed at a distance of $x_0 = 29$ µm away from the geometric center of the

beam. The crack proceeded from the TBC surface at the critical load and arrested at the interface between top and bond coat. In the next step the loading tip was offset to $c = 36$ µm from the center of the beam and then loaded again. The loading was conducted in steps to $F = 4.9$ N and after unloading the analysis showed that an interfacial crack had nucleated at a load of $F_c = 3.1$ N. During this event (not listed in Table I), the interface crack had propagated to a length of $a = 29$ µm and extended beyond the position of the loading tip. This was the only time this occurred in all experiments. After nucleating the interface crack, the load tip was again offset to a new location ahead of the interface crack tip at a position $c = 78$ µm. After loading to a maximum force of $F = 3.7$ N, the analysis revealed that the crack had extended after applying a critical force of $F_c = 3.3$ N to a length of $a = 50$ µm (not listed in Table I). Continuing the experiment, the tip was then offset to $c = 171$ µm relative to the beam center. Increasing the load in a stepwise fashion led to several consecutive crack propagation events for this configuration. As documented in Table I, the crack propagated by amounts of $a = 16, 18, 16, 14,$ and 3 µm when the critical loads

Table I. Event table

Crack event #	Schematic	x_o (μm)	a (μm)	c (μm)	Critical load, P_c (N)	Propagation, Δa (μm)	Interfacial toughness, Γ (J/m^2)	Phase angle, ψ (°)
Beam 1: $L = 1.1$ mm; $W = 550$ μm; $h_{bc} = 36$ μm; $h_{tbc} = 112$ μm; $E_{bc} = 155$ GPa, $E\|_{tbc} = 20$ GPa; $E_{tbc}^{\perp} = 180$ GPa								
1		29	60	171	2.3	16	27.1	17.9
2		29	76	171	3.0	18	42.5	18.0
3		29	94	171	3.5	16	53.6	18.1
4		29	110	171	4.1	14	67.8	18.9
5		29	124	171	5.1	3	95.2	20.5
Beam 2: $L = 1.1$ mm; $W = 550$ μm; $h_{bc} = 51$ μm; $h_{tbc} = 112$ μm; $E_{bc} = 155$ GPa, $E\|_{tbc} = 20$ GPa; $E_{tbc}^{\perp} = 180$ GPa								
1		0	14	50	3.2	6	51.1	18.9
2		0	20	50	3.5	7	63.8	19.1

of $F_c = 2.3$, 3, 3.5, 4.1, and 5.1 N were reached (crack events 1–5, beam 1).

The second sample (beam 2) was treated in a similar fashion and the crack propagation and critical loads and corresponding fracture energy and phase angle values as calculated by FEA can be found in Table I. First, the beam was center loaded till a crack was induced in the top coat at the very center of the beam with $x_0 = 0$ μm. In a second step, the load was applied at an off center position $c = 50$ μm and the load was increased. As a result, the crack channeled to the interface and extended and to a length of $a = 14$ and 20 μm at corresponding critical loads of $F_c = 3.2$ and 3.5 N (crack events, beam 2).

IV. Energy Release Rates and Interface Toughness

The mechanical response of the bilayer beam has been analyzed by the FE method using the commercial code ABAQUS Standard. Only the bilayer beam and part of the support material was simulated in the analysis and the boundary conditions were assumed as rollers. No energy exchanges are allowed between the modeled specimen and the rest of the ring. The finite-element model and the mesh are depicted in Fig. 5. The eight-node biquadratic plane strain elements with reduced integration were used. At the crack tip focused elements were used and the mid-size node parameter is chosen to be 0.25. In these simulations the

Fig. 3. The resulting full-field displacement u_{xx} in the unloaded (a) as well as cracked and loaded (b)–(d) is shown in the upper images. The 3D deformation plot (e) shows how the crack tip can be measured from the full-field displacement data.

TBC is assumed aniostropic in material property: the in-plane modulus, E_{Π}, is considered to be 30 GPa according to an elasticity study of the same material,[8] and the out-of-plane modulus, $E_{\perp}$, is assumed to be 150 GPa (compared with $E_{\perp} \approx 200$ GPa for dense YSZ). Poisson's ratio of TBC is considered to be 0.2. The bond coat is considered to be elastic/plastic with power law hardening having the stress/strain curve obtained in Kim *et al.*[9] (Young's modulus $E = 155$ GPa, Poisson's ratio $v = 0.25$, yield strength $\sigma_Y = 750$ MPa, and strain hardening exponent $n = 0.2$).

(1) Basic Features

ABAQUS calculations based on the appropriate specimen dimensions and material properties allows the energy release rate and mode mixity to be determined as a function of the crack length. The energy release rate, i.e. the *J*-integral, is calculated in ABAQUS for the interface crack as the loading increases. The results are presented in the normalized coordinates suggested by beam theory,[5] $G_p E_{bc} h_{bc}^3 / (P_c)^2$ as a function of a/L (Fig. 6). Note that, at fixed load, the energy release rate decreases as the crack extends, *reaffirming the stable nature of this test configuration.* The stress fields and mode mixity of an interfacial crack has

been shown to have an oscillatory character in a bimaterial specimen. These oscillations were observed in the FE simulations and the associated mode mixity varied with distance from the crack tip.[1] At a distance of 1 μm, which corresponds with the plastic zone size in the FE simulation, the mode mixity was calculated to be 20°. The mode mixity increased to approximately 40° at a distance of 8–10 μm. This larger value approaches the mode mixity that is traditionally reported for four-point flexure measurements, but is still slightly smaller[1,2] for the other methods that have been used to measure TBC interfacial toughness.[9–12]

(2) Experimental Findings

Specific values of fracture resistance, Γ_R, have been calculated for each of the critical crack growth events reported in the previous section. The calculated *J*-integral at the critical load is considered as the interface toughness. These fracture resistance values are included in Table I, and inspection of the data indicates that the measured toughness values fall into a rather broad range (25–95 J/m²), mimicking perhaps the tortuosity of the TGO/bond coat interface. It is worth noting that the measured

Fig. 4.   Scanning electron micrographs of the region near the center of a test specimen reveal both, precrack (a) and delamination (b), (c) and (d) at the interface between the TGO and the bond coat.

fracture resistance appears to rise with crack length in both data sets. This rise cannot be attributed to changes in mode mixity and may be indicative of a rising resistance with crack extension. This could be explained if the crack may arrest at selectively tougher features on the tortuous interface. Alternatively, the role of permanent plastic deformation in the bond coat has not been completely ruled out. It is also possible that the TBC was

damaged when the sample was initially loaded. No evidence of damage was noted, but it is possible that undetected damage resulted in the lower value of interfacial toughness for the first event in Table I. Given this uncertainty, we prefer not to ascribe any significance to the variation in toughness with mode mixity and to represent our measurement with the mean of all measurements. The mean values of the toughness for each individual

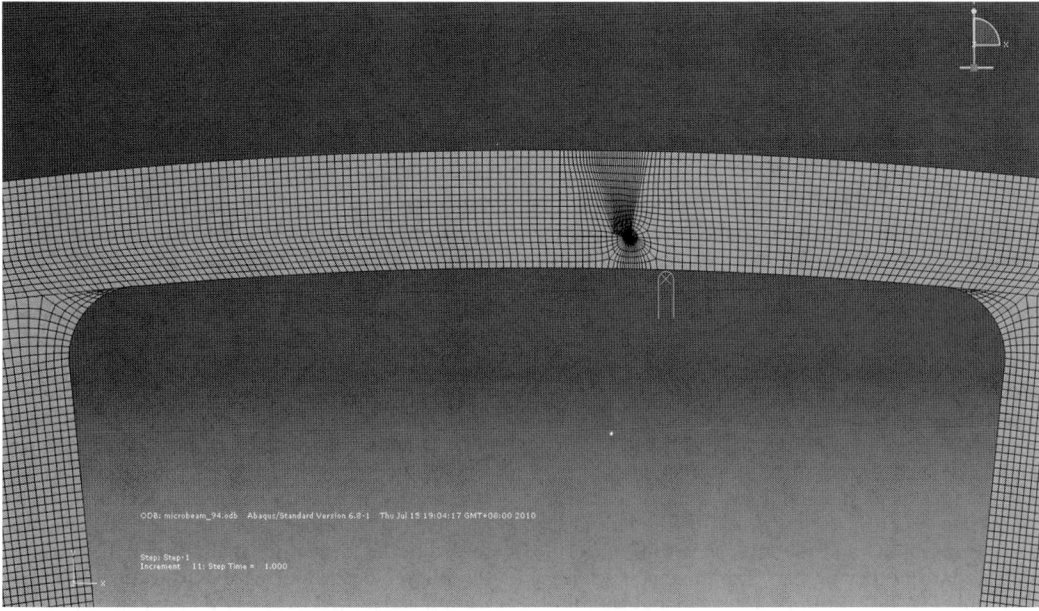

Fig. 5.   The finite-element model and the mesh, with the refined mesh close at the crack tip.

J. Am. Ceram. Soc., **94** [S1] S128–S135 (2011)
DOI: 10.1111/j.1551-2916.2011.04539.x
© 2011 The American Ceramic Society

journal

Thermomechanical Fatigue Damage Evolution in a Superalloy/Thermal Barrier System Containing a Circular Through Hole

Makoto Tanaka,[‡,§] Christopher Mercer,[¶,‖] Yutaka Kagawa,[†,‡,¶] and Anthony G. Evans[‖]

[‡]Research Center for Advanced Science and Technology, The University of Tokyo, 4-6-1 Komaba, Meguro-ku, Tokyo, Japan

[§]Materials Research and Development Laboratory, Japan Fine Ceramics Center, 2-4-1 Mutsuno, Atsuta-ku, Nagoya 456-8587, Japan

[¶]National Institute for Materials Science, 1-2-1 Sengen, Tsukuba, Ibaraki, 305-0047, Japan

[‖]Materials Department, University of California Santa Barbara, Santa Barbara, CA 93106

The results of an investigation of thermomechanical fatigue (TMF) on a superalloy specimen, with an applied thermal barrier coating (TBC) and a circular through hole, are presented. Tensile loads were applied in phase with increasing temperature. Damage evolution in the form of cracks develops in the TBC adjacent to the hole. These cracks run perpendicular to the loading axis. Stress mapping of the thermally grown oxide (TGO) using luminescence spectroscopy determined an increase in compressive residual stress with increasing TMF cycling. Scanning electron microscopy examination, following cross sectioning, determined the TBC cracks to be vertical separations of the columnar yttria-stabilized zirconia (YSZ) top coat. Microscopic damage mechanisms in the form of plasticity (bending of YSZ columns) and TGO cracking were observed. Imperfections in the bond coat are associated with these vertical separations. Energy-dispersive element mapping of these imperfections indicated a composition of alumina and mixed Cr, Co, and Ni oxides.

I. Introduction

AIRFOILS in aeroturbines experience a wide range of thermomechanical fatigue (TMF) loadings that can adversely influence the durability of the system.[1–3] The loadings can be in phase or out of phase in both tension and compression. The range of responses is diverse. The intent of the present article is to examine one of these TMF modes to ascertain whether it presents a concern for the performance of the airfoil. The scenario envisages an airfoil subject to unidirectional inertial loads imposed in phase with the temperature. Namely, it addresses regions of the airfoil where the temperature increases in phase with the inertial loads so that the maximum load is imposed simultaneously with the maximum temperature. To be representative of an actual airfoil, the system to be investigated incorporates a bond coat and a thermal barrier coating (TBC). Prior in-phase TMF experiments have been conducted on hollow cylindrical (axisymmetric) and plate type (uniaxial) TBC specimens.[1–4]

In the design of the test, the goal has been to replicate as closely as possible the thermomechanical circumstances expected in the engine. Specifically, the test scheme imposes a thermal gradient through the TBC, while cycling the temperature in phase with a tensile load. The temperatures are chosen to be large enough to induce creep strains in the superalloy and the

bond coat, so that elongation occurs as the system cycles. Indeed, one objective of the study is to investigate the combined influence of creep and thermal strains on the response of a system with a brittle external coating.

During that portion of the thermal cycle when the interface between the TBC and bond coat is at its hottest, a thermally grown oxide (TGO) forms, comprised primarily of fine grained, polycrystalline α-Al_2O_3.[5,6] This material experiences creep at elevated temperatures. Absent an applied stress, the creep balances growth-induced strains, resulting in a small compressive (growth) stress as it thickens.[7,8] On cooling, substantial thermal expansion misfit between the α-Al_2O_3 and the substrate causes it to develop large in-plane compressive stress (of order 4 GPa) at ambient.[9,10] This stress can be measured by using a laser to excite fluorescence.[10–12] Moreover, because the TBC is transparent at certain laser wavelengths, the stress can be determined even when a TBC is present.[13–15] Probing this stress at various stages of TMF cycling provides an independent assessment of the state of the system.

In this investigation, TMF tests have been used to evaluate stress evolution in a TBC system. In service, TBCs are subjected to various loading modes, stress concentrations, temperature/load conditions etc. To simulate complicated degradation behavior in actual gas turbine blades, one idea is to use a holed specimen, because such a specimen allows well-controlled stress/strain distributions. When a through-holed specimen is used, the stress around the hole varies between $3\sigma_a$ and $-\sigma_a$, where σ_a is the remote applied stress. Although they are local stresses, rather than uniform stresses, the test is useful for the observation of the effects of superimposed applied cyclic stress and temperature on the degradation behavior of a TBC system. The test is also interesting because actual gas turbine blades have many through holes for air cooling, and there is concern that coating failure may initiate at holes. However, no work has been published on the effect of stress concentration on the degradation of TBC systems.

The article is organized as follows. Initially, and the material system and specimen preparation procedures are presented, and the thermomechanical test system and measurement methods are described. Then the experimental measurements and observations are presented and discussed. Finally, the experimental results are summarized and some concluding remarks are given.

II. Experimental Procedure

(1) TBC Specimen

Before application of the bond coat and TBC layers, the substrate superalloy Inconel 738LC (Mitsubishi Materials Corp., Tokyo, Japan) (16.0-Cr, 8.5-Co, 3.4-Al, 3.4-Ti, 1.7-Mo, 2.6-W, 1.75-Ta, 0.9-Nb, 0.11-C, balance-Ni, and comprised of a $\sim$3 mm thick

C. Levi—contributing editor

Manuscript No. 28700. Received September 30, 2010; approved March 4, 2011.
[†]Author to whom correspondence should be addressed. e-mail: kagawa@rcast.u-tokyo.ac.jp

Fig. 1. A typical polished transverse section of the EB–PVD TBC system. EB–PVD, electron-beam physical vapor deposition; TBC, thermal barrier coating.

plate), was machined into flat tensile test specimens. Both of the flat surfaces were then polished to allow good bonding between the bond coat and the substrate. The entire coated material system will hence be referred to as the "TBC system". The bond coat layer was 150 μm thick and was applied by low-pressure plasma spray. The chemical composition of the CoNiCrAlY bond coat (wt%) was: 32.0-Ni, 21.0-Cr, 8.0-Al, 0.5-Y, with the remainder Co. Following application of the bond coat, a 4 mol% Y_2O_3–ZrO_2 TBC layer, with a thickness of ∼200 μm, was applied via electron-beam physical vapor deposition (EB–PVD). The yttria-stabilized zirconia (YSZ) material was supplied by Japan Fine Ceramics Center (JFCC, Nagoya, Japan). Figure 1 shows a typical example of a polished transverse section of the as-deposited EB–PVD TBC system: the three different layers, i.e., the TBC, TGO, and bond coat, are clearly distinguished in the cross section. The TBC layer consists of columnar grains with a rough feather-like surface appearance: this surface appearance is typical of an EB–PVD coating, as reported elsewhere.[16]

A center hole, 1 mm in diameter, was introduced by an ultrasonic machining process. To avoid damage to the TBC layer, the entire specimen was embedded in an epoxy resin before machining. Once the hole had been machined, the embedded specimen was treated with CH_2Cl_2 solution (at 20°C) for 24 h to

Fig. 2. (a) Shape and dimensions of the thermomechanical fatigue test specimen, and (b) surface appearance of the specimen (thermal barrier coating surface).

Fig. 3. (a) Heating/cooling system used in the thermomechanical fatigue (TMF) test, and (b) temperature-time relations during the in-phase TMF test.

remove the epoxy resin. Subsequently, the specimen was heat treated for 30 min at 400°C to completely remove all traces of the epoxy. No additional coating was applied inside the hole, i.e., the as-machined surface was employed. The final shape and dimensions of the TMF specimen, as well as the surface appearance of the specimen (TBC side), are shown in Fig. 2. In this investigation, the applied stress at maximum temperature was 60 MPa, which is smaller than the yield stress of the substrate alloy (345 MPa at 980°C [17]). The estimated in-plane stress distribution in the specimen under the applied remote stress, σ_a, is presented in Appendix A (Fig. A1). As previously mentioned, the maximum stress concentration in the holed specimen varies between $3\sigma_a$ and $-\sigma_a$. That is, a stress distribution is present, making the observation of damage evolution under various loading conditions possible.

(2) TMF Test, Stress Measurement, and Observation

Cyclic, in-phase thermomechanical fatigue (TMF) tests were performed in ambient air using a specially designed TMF test system, the details of which are reported elsewhere.[18,19] The surface temperature of the TBC layer and substrate were measured using two R-type thermocouples. The heating/cooling arrangement is shown in Fig. 3(a), and the temperature-time relationship during the test is presented in Fig. 3(b). The temperature of the substrate was controlled by compressed cooling air blown from a T-shaped, flat-type nozzle, containing pinholes. Using this procedure, the surface of the substrate was kept at a constant 1000°C, while the maximum temperature at the TBC surface was held at ≈1150°C. Temperature variation using this procedure was ±5°C. Preliminary temperature measurement confirmed that a uniform surface temperature region exists in the ≈10 mm × 10 mm central region of the specimen. This indicates that a temperature gradient of $\Delta T \approx 150°C$ between the TBC surface and the substrate, was achieved using this heating/cooling system. The total time period of a TMF cycle was 940 s, with heating and cooling rates of 5°C/s and a maximum temperature dwell time of 600 s.

During the maximum temperature hold, a uni-axial tensile load of 1.8 kN was applied to the specimen, which corresponded to a maximum applied stress of $\sigma_a = 60$ MPa. Applied stress was calculated using: $\sigma_a = P_a/((h_{tbc}+h_{bc}+h_s)w)$, where P_a is the applied load, h is the thickness, w is the width of the gage section of the specimen, and the subscripts "tbc", "bc" and "s" refer to the TBC, bond coat and substrate, respectively. The specimen was attached to the test machine via a water-cooled mechanical grip. Axial strain in the specimen was measured by a contact type displacement gage (nominal strain resolution: 0.008%). After selected numbers of applied thermomechanical cycles ($N = 120, 240, 360$, and 480 cycles), the test was interrupted and the specimen removed from the test machine. The TBC surface of the tested specimen was then examined using scanning electron microscopy (SEM). After 480 cycles, the test was terminated. Stress in the thermally grown oxide (TGO) layer was measured using Cr^{3+} luminescence spectroscopy. The measurement was performed through the TBC layer, at a fully controlled temperature of 20°C, in ambient air using a microconfocal laser luminescence spectroscopy system (NRS-1000: Special Version, JASCO Corp., Tokyo, Japan). The same experimental equipment has already been used successfully for prior studies of TGO stress in the same TBC system.[20] The frequency shift of the R_2 peak is known to have a linear relationship with stress. The measured R_2 peak shift, Δv, and average stress in the TGO layer, $\bar{\sigma}_{tgo}$, has the following relationship[11]:

$$\bar{\sigma}_{tgo} = \frac{1}{\Pi}\Delta v = \frac{1}{3}(\sigma^x_{tgo} + \sigma^y_{tgo} + \sigma^z_{tgo}) \quad (1)$$

where $\bar{\sigma}_{tgo}$ is the average stress in the TGO layer, Π is the trace of the piezospectroscopic coefficient tensor, and has a value of 7.62 cm/GPa,[21] Δv is the frequency shift (cm^{-1}) of the Ruby line at 14430 cm^{-1} (R_2) Assuming that the entire TBC specimen is under plane stress conditions, i.e., $\sigma^z_{tgo} = 0$, the in-plane TGO stress is given by

$$\bar{\sigma}_{tgo} = \frac{1}{3}(\sigma^x_{tgo} + \sigma^y_{tgo}) = \frac{2}{3}\sigma^x_{tgo}\left(\equiv \frac{2}{3}\bar{\sigma}^{x,y}_{tgo}\right) \quad (2)$$

This relation is valid when luminescence spectroscopy measurements are performed through the TBC layer, and the laser incident direction is perpendicular to the TGO. Though the laser beam diameter is ≈ 15 μm, the spatial resolution is ~ 150 μm because of scattering through the TBC layer.[22]

Following luminescence spectroscopy, the specimen was sectioned both parallel to and perpendicular to the loading direction. The cross-sections were initially ground using SiC paper before polishing with diamond suspension. The specimens were then coated with gold and examined using SEM. Compositional information of the TGO and bond coat was obtained using backscattered electron imaging and energy-dispersive spectroscopy (EDS) element mapping.

III. Experimental Results and Discussion

(1) Deformation Behavior of the TBC System

Maximum and minimum tensile strains in the specimen (along the stress axis) evolving with the number of applied TMF cycles, are plotted in Fig. 4(a). The strain is defined as elongation measured by the gage, Δl_g, divided by the initial gage length, l_g, i.e, $\varepsilon_a = \Delta l_g/l_g$. The effect of local strains caused by the existence of the hole is neglected. The maximum strain (ε_{max}) is defined as the value measured at the end point of the maximum temperature hold ($T_{tbc} \approx 1150°C$) during each cycle, and the minimum strain (ε_{min}) is defined as the value at the minimum temperature ($T_{tbc} \approx 300°C$). Steady state creep behavior occurs up to about $N \approx 400$ with a creep rate $d\varepsilon/dN \approx 1.78 \times 10^{-3} N^{-1}$. The ratio of load applied to the substrate versus that applied to the entire specimen (P_s/P_a) is given by

$$P_s/P_a = E_s h_s/(E_{tbc} h_{tbc} + E_s h_s) \quad (3)$$

Fig. 4. (a) Maximum and minimum tensile strains vs. applied number of cycles, and (b) ratcheting behavior during the thermomechanical fatigue test.

where E is Young's modulus and h is the thickness, subscripts "s" and "tbc" refer to the substrate and TBC layer, respectively. Ratcheting behavior develops during the test (Fig. 4(b)) and the behavior becomes more apparent with increasing number of TMF cycles. The material properties derived from this estimation are presented in Table I The estimated P_s/P_a is ≈ 0.99. Therefore, the observed steady state creep behavior of the TBC system seems reasonable, because the substrate is much thicker than the TBC layer, and almost all of the applied load is supported by the superalloy substrate. An accelerated creep stage follows, with a creep rate $d\varepsilon/dN \approx 7.83 \times 10^{-3}/N$, in the range of $400 \leq N \leq 480$. The creep rate in the steady state stage coincides with that of similar specimens without holes.[19] However, the specimen in this TMF test did not fracture even though the total number of TMF cycles exceeded the number of cycles to failure observed in specimens without holes. The ultimate cycle in this TMF test seems to correspond to the onset of the final creep stage, when compared to creep curves of specimens without holes.

Table I. Mechanical and Physical Properties of Materials Used in the Investigation

Materials	Layer Thickness (μm)	Young's Modulus (GPa)	Coefficient of Thermal Expansion (10^{-6} K^{-1})
TBC	200	20	11
TGO	3	280[25]	8
Substrate	3000	200	15

TBC, thermal barrier coating; TGO, thermally grown oxide.

Fig. 5. (a) Appearance of the specimen surface near a hole after $N = 480$ cycles, and cross section markers (dashed lines) for cross sections shown in Fig. 6. (b) Micrograph showing inside of hole after $N = 480$ cycles. (c) Higher-magnification image indicated by dashed box in Fig. 5(b).

(2) Damage Evolution Observed at the TBC Surface

The appearance of the specimen surface near the hole after 480 cycles is shown in Fig. 5. The applied loading direction and assigned coordinates are also shown. Transverse cracks, which are shorter than 1 mm in length, and initiate from a maximum

Fig. 6. Scanning electron microscopy micrographs of polished cross sections: (a) parallel to loading direction (x–z section in Fig. 5(a)), (b) perpendicular to loading direction (y–z section in Fig. 5(a)).

tensile stress site ($3\sigma_a$), occur symmetrically with respect to the center of the loading axis. Formations of other transverse cracks, which are not directly connected to the free surface of the hole, are also observed inside the hole (Figs. 5(b) and (c)). Conversely, no visible damage formation in compressive stress sites ($-\sigma_a$) is observed. A notable characteristic of these cracks is that the crack mouths are open even when the applied load is completely removed. Note also that the spheroidal material (diameter $100 \sim 200\ \mu m$) observed on the TBC surface in Fig. 5 are actually the tops of conical defects formed in the YSZ during the EB–PVD process and are not related to the cracking behavior of the TBC layer. Cracks in the TBC layer do not extend the entire width of the specimen. The spacing of the transverse cracks is about $1 \sim 1.3$ mm, except in the case of cracks formed near the edge of the hole. This suggests that tensile stress concentration during maximum temperature hold, $3\sigma_a$, is responsible for the formation of the transverse cracks, and compressive stress is effective at arresting the cracks.

(3) Observation of Microscopic Damage Evolution

Polished cross sections, in both x–z and y–z orientations, are shown in Figs. 6(a) and (b), respectively. Parallel to the loading

Fig. 7. Higher-magnification images showing the transverse crack nucleated near the edge of the hole.

(a)
Area A

(b)
Area B

50 μm

Fig. 8. Creep voids in the bond coat layer at areas A and B in Fig. 6.

10 μm

Fig. 9. Scanning electron microscopy micrograph of vertical separation in the thermal barrier coating layer and imperfection in the thermally grown oxide layer.

direction (Fig. 6(a), x–z section), cracks or vertical separations in the TBC layer are observed at points that coincide with a stress of $3\sigma_a$. These TBC cracks generally stop at the TGO interface. A detailed image of the widest separation is shown in Fig. 7. The crack tip is blunted in the bond coat layer and an irregularly shaped imperfection extends into the bond coat. Within the bond coat below the crack, the number of voids has increased, compared with that observed in the as-deposited state, or in an undamaged region (Fig. 8). This is expected, because the voids are formed via bond coat creep in the regions adjacent to the TBC crack, where localized plastic deformation will occur. Within the voids, new oxide formation is clearly observed in the polished section (Fig. 7(b)), indicating that the voids are connected to the central breakthrough feature. Perpendicular to the loading direction (Fig. 6(b), y–z section), the specimen shows a slight convex appearance with the apex at the maximum compressive stress, $-\sigma_a$. The vertical separations in the TBC are also apparent in this section, but have a different appearance. This is due to the fact that the TBC cracks are not straight, but somewhat wavy.

Another vertical separation in the TBC layer is shown in Fig. 9. An imperfection in the TGO layer is clearly visible at the root of the crack. Microcracking, both parallel and perpendicular to the TGO interface is associated with the imperfection. Bending of the TBC columns to the right of the separation is also visible. This implies that the YSZ columns are undergoing plastic deformation during the maximum temperature phase of the TMF cycle. No such bending of the TBC columns is asso-

ciated with the TBC crack in Fig. 7. The reason for this not entirely clear, but it should be noted that the crack shown in Fig. 7 lies at the bottom of the concave undulation of the TBC, where bending stresses may be at a minimum. In any case, it would appear that two potentially competing microdamage mechanisms are operative within the TBC system; YSZ plasticity and TGO cracking. Further work needs to be carried out to assess the relative effects of these mechanisms on the overall integrity and service life of the TBC system.

Figure 10 shows an EDS element map taken in the region of the TGO imperfection shown in Fig. 9. The map indicates that walls of the imperfection and nearby void are covered by a thin Al_2O_3 layer. The presence of Cr, Co, and Ni may indicate that mixed Cr, Co, and Ni oxides are also associated with the entrained void, although this is not completely clear.

Another interesting observation is the difference in the TGO morphology between the two sections. Parallel to the loading direction, the TGO layer is flat (Fig. 11(a)). However, perpendicular to loading direction, the TGO shows clear undulation behavior (Fig. 11(b)). The average thickness of the TGO layer is ~ 4 μm. The undulation amplitude is approximately $2\delta \approx 22.7$ μm in the y–z section, while that in the x–z section is $2\delta \approx 0$. The large undulation amplitude in the y–z section would be associated with the compressive stress, σ_y and is attributed to cyclic plasticity caused by bond coat creep in compression. Meanwhile, the flat TGO in the x–z section is associated with elongation of the bond coat. Similar anisotropic TGO undulation behavior has also been observed in previous studies,[23,24] including under compressive thermomechanical testing on a (Ni,Pt)-Al coated superalloy.[23] The present result indicates that the anisotropic TGO undulation behavior occurs despite the presence of the YSZ layer, which suggests that such behavior probably occurs in actual TBC systems used in gas turbine engines.

(4) Change of TGO Stress Distribution with Number of TMF Cycles

Visible damage evolution and associated in-plane TGO stress are shown in Fig. 12. The surface appearance shows a sequence of transverse cracking events. After 360 cycles, transverse cracks appear at the edge of the hole. Further initiation and growth of transverse cracks is also observed with increasing number of applied cycles. Generally, the TGO compressive stress increases with number of cycles, N. However, detailed observation suggests the TGO stress change depends on location, i.e., the stress change depends on the remote applied stress. Far from the hole, the TGO compressive stress continuously increases with increasing number of cycles. The TGO stress at sites of maximum compressive stress, $-\sigma_a$, is slightly lower than that far from the

Fig. 10. Energy-dispersive spectroscopy element map taken in the region of the thermally grown oxide imperfection shown in Fig. 9.

hole (σ_a). The stress at $3\sigma_a$ shows nearly the same variation as that at σ_a. However, after 360 cycles, the TGO stress in the maximum tensile stress region clearly decreases, and this tendency continues with increasing number of TMF cycles.

Plots of average in-plane compressive stress in the TGO layer far from the hole vs. applied number of TMF cycles, N, are presented in Fig. 13, with the total time at maximum temperature also shown for comparison purposes. The result shows gradually increasing compressive stress with increasing number of cycles. Typically, TGO stress is estimated assuming that all the stress originates from thermal expansion mismatch. The estimated thermal stress is given by:

$$\sigma_{tgo}^T = \frac{E_{tgo}^* \Delta T [E_{tbc}^* h_{tbc}(\alpha_{tbc}^* - \alpha_{tgo}^*) + E_s^* h_s(\alpha_s^* - \alpha_{tgo}^*)]}{E_{tbc}^* h_{tbc} + E_{tgo}^* h_{tgo} + E_s^* h_s} \quad (4)$$

where for plane strain conditions, $E^* = E/(1 - \nu^2)$, $\nu^* = \nu/(1 - \nu)$, and $\alpha^* = (1 + \nu)\alpha$, E is Young's modulus, ν is Poisson's ratio, α is the thermal expansion coefficient, h is thickness, and the subscripts represent materials. At the beginning of the TMF test, the measured compressive stress is 3 GPa and the estimated value using Eq. (4) is 2.8 GPa, so they are in good agreement. During TMF cycling, sintering of the TBC layer and growth of TGO layer occur, and these factors affect the TGO stress. Considering the contribution of individual materials properties, the increase of Young's modulus of the TBC layer and increase in the thickness of the TGO layer, lead to the stress decrease. In contrast, an increase of the Young's modulus of the TGO causes an increased in-plane compressive stress. The increase of Young's modulus of the TGO with increasing thermal exposure is reported in,[25] and is attributed to the closure of microcracks. Assuming a 10% increase of Young's modulus, the stress increase is also ≈10%. The estimation only incorporates linear elastic considerations; creep deformation behavior, temperature gradient, ratcheting effect, growth strain of the TGO, etc. are not considered. The increase of the Young's modulus of the TGO is only one possibility; the reason for the increase of compressive stress with increasing number of TMF cycles is still an open problem.

IV. Summary and Concluding Remarks

In-phase TMF tests on a center-holed plate type specimen with an applied TBC system have been conducted. Damage evolution in the form of cracks is observed in the TBC adjacent to the hole. These cracks run perpendicular to the direction of the applied load. The cracking initiates at the edge of the hole exhibiting the highest tensile stress. This suggests that the cracking behavior is strongly correlated with tensile stress at maximum temperature. Changes in the TGO stress state with applied number of TMF cycles have been measured by luminescence spectroscopy. The results indicate that the compressive residual stress in the TGO increases with increasing number of cycles. The Increase in Young's modulus of the TGO layer, possibly as a result of the closure of microcracks, is believed to be the explanation for the increasing TGO stress. Anisotropic TGO growth behavior is also observed. Parallel to the loading direction, the TGO is flat; however, perpendicular to loading direction, the TGO exhibits marked undulations. Employing holed specimens in TMF tests is an effective way to observe the effect of applied stress on damage evolution in a TBC system, because the specimen has a stress distribution of $-\sigma_a$–$3\sigma_a$, for an applied tensile stress, σ_a.

Detailed SEM analysis of the cross-sectioned specimen has determined that the cracks in the TBC are vertical separations in the YSZ columnar structure, that extended to the TGO interface. Imperfections in the TGO and increased incidence of bond coat voids occur at the roots of these separations. These voids

Fig. 11. Differences in the thermally grown oxide (TGO) morphology in the two orthogonal directions, (a) *x–z* and (b) *y–z*.

Fig. 3. Hysteresis loops for bare René N5 at start of cycling and at 50% of total SPLCF life.

stress approximately equal to the magnitude of the stress relaxation. The tensile stresses at the end of the cycle continue to increase until approximately 20% of the total life, where they stabilize. At 50% of life for $\Delta\varepsilon_t = 0.35\%$ at 1093°C, a maximum tensile stress of approximately 190 MPa is reached upon returning to zero strain, Fig. 3. Hysteresis loops are similar for the vapor phase aluminide-coated sample, however, the total cycles to failure increased to $N_f = 10\,200$. Thus the rate of damage accumulation is apparently higher in the uncoated sample.

The dependence of SPLCF life on total strain range for Pt-aluminide-coated samples at 982°C is shown in Fig. 4. The SPLCF life increases by a factor of approximately $10\times$ as the strain range is reduced from 0.96% to 0.56%. Under the conditions studied, the compressive hold can degrade the fatigue life (with no compressive hold) by a factor of up to $100\times$, depending on the temperature and strain range.

SPLCF cycling of samples to failure as well as experiments interrupted at various fractions of the average failure life have been conducted at 1093° and 982°C. At 1093°C a series of bare and vapor phase aluminide-coated samples were interrupted at various fractions of total fatigue life. Vapor phase aluminide-coated samples were removed from SPLCF tests after 2000, 4000, 6000, and 8000 cycles (approximately 20%, 40%, 60% and 80% of life, respectively) at 1093°C and $\Delta\varepsilon_t = 0.35\%$ and sectioned longitudinally. While the details of these sectioning studies have been reported elsewhere,[6] it was observed that cracks do not penetrate beyond the interdiffusion zone until

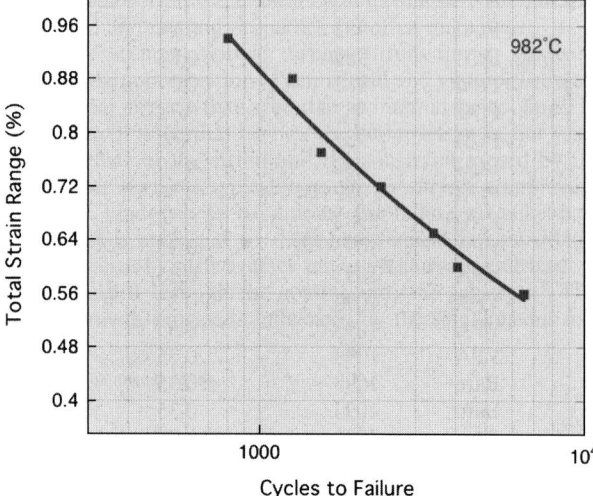

Fig. 4. SPLCF failure data for Pt-aluminide coated René N5 at 982°C cycled with 120-s compressive holds.

approximately 80% of the life is consumed by cycling. Similarly, testing of a Pt aluminide-coated sample at 982°C with $\Delta\varepsilon_t = 0.56\%$ demonstrates that at failure many cracks have not progressed beyond the interdiffusion zone, Fig. 5. Among the 75 cracks examined along the length of the gage section after failure, only three cracks extended to depths of 150–250 μm, not including the one longer crack that ultimately caused failure of the sample, Fig. 5.

The experiments at two temperatures (982° and 1093°C) with two different coatings (Pt aluminide and vapor phase aluminide) collectively reveal four successive stages of the failure process (Figs. 6 and 7): crack extension from the surface through the (I) bond coat (BC), (II) interdiffusion zone (IDZ), (III) superalloy substrate, and (IV) long crack growth. In the early stages of cycling, many cracks form in the bond coat along the length of the gage section. These cracks gradually progress through the bond coat and interdiffusion zone and into the superalloy substrate. The cracks are filled with an alumina layer a few micrometers thick, Fig. 8, that is relatively uniform in structure and thickness along the surface of the crack. After growth into the superalloy to a depth of a few hundred micrometers, Stage IV is approached and the cracks grow at an inclined angle with respect to the axis of the applied compressive stress, Fig. 7. In many cases coarse striations are visible on the fracture surface in Stage IV, indicating that rapid mixed-mode crack growth occurs in this final stage.

To experimentally assess the role of coatings, interrupted SPLCF cycling tests were also conducted on bare René N5 samples. SPLCF experiments were interrupted at 1000, 2000, 4000, 6000, and 8000 cycles at 1093°C with $\Delta\varepsilon_t = 0.35\%$. The average and maximum crack depths are given in Table II. Similar to the coated samples, the cracks do not extend to depths beyond 100 μm until very late in life.

IV. Finite Element Modeling of the SPLCF Process

A series of ABAQUS-based finite element models were developed to elucidate the properties of the substrate, interdiffusion zone, β-coating and oxide layer that most strongly influence the failure process. The details of a finite element mesh that contains an oxide layer of thickness h on the superalloy surface and within a shallow crack of depth a are shown in Fig. 9. Initial analyses of the crack growth process for an oxide penetrating into the superalloy in this configuration have been presented elsewhere.[7] These analyses are extended here to consider coatings and the conditions under which damage development is strongly affected by their presence. To study the role of the β-coating and interdiffusion zone, additional meshes were developed with oxide-containing cracks within the successive layers, as shown schematically in Fig. 10.

High-temperature properties of the substrate have been measured[29] and these are considered in the model. Two versions of the model have previously been developed[7] for the oxide on the superalloy: one where creep properties are explicitly considered and a second where an equivalent plasticity analysis is implemented for numerical efficiency. Steady-state creep can be adequately represented by power law behavior of the form: $\dot{\varepsilon} = \dot{\varepsilon}_o \left(\frac{\sigma}{\sigma_o}\right)^n$, where σ is the local Mises stress, σ_o is a reference stress, $\dot{\varepsilon}_o$ a reference strain rate, and n the creep exponent. In the present calculations $\sigma_o = 100$ MPa and $n = 10$. The high-temperature properties of the Pt aluminide bond coat measured in samples removed directly from coatings have been incorporated into the model.[30] While the properties of the interdiffusion zone have not been directly measured, the properties of alloys with compositions similar to those observed in the interdiffusion zone indicate that the creep rates are likely to be intermediate to those of the weak β-coating and the precipitation-strengthened superalloy.[31] To capture the lateral straining in the TGO, in the model, $\dot{\varepsilon}_{growth}$ is imposed during those stages of the strain cycle when the crack is open. It is imposed at a uniform rate governed by the TGO thickness, h. This strain-rate causes creep of the

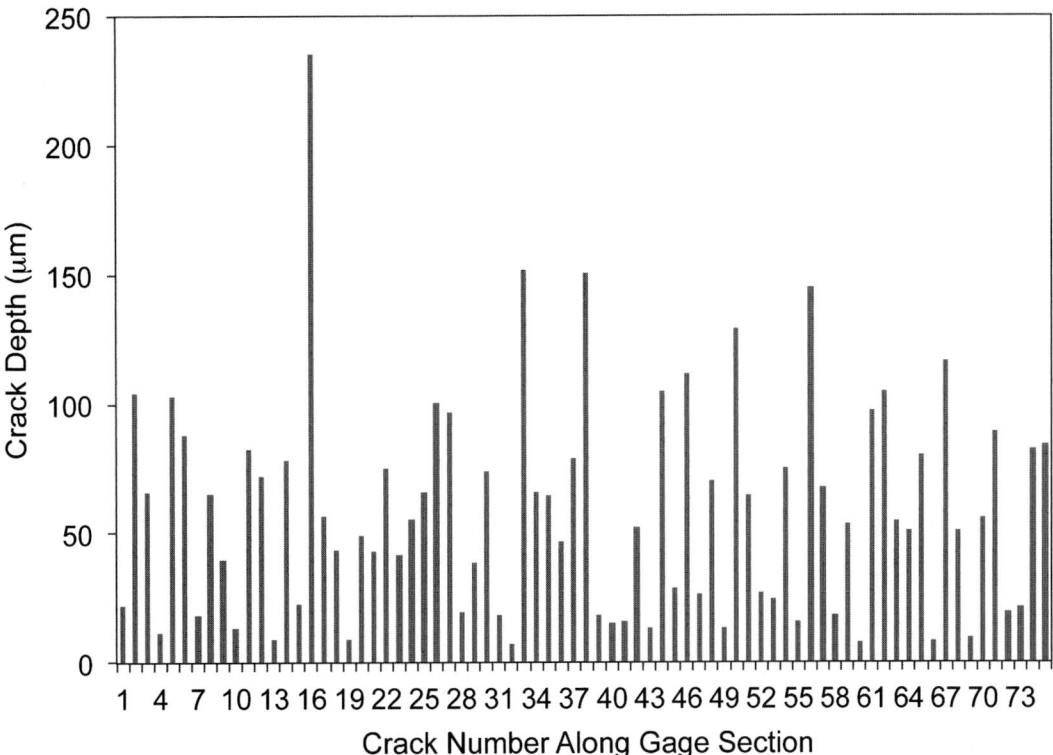

Fig. 5. Crack lengths along the gage section at failure following cycling of a Pt aluminide-coated sample at 982°C at $\Delta\varepsilon_t = 0.56\%$.

TGO,[27] replicated by ensuring that the Mises stress does not exceed the growth stress, $\sigma_{eq} \leq \sigma_{growth}$. The ensuing response of the TGO is governed by the creep deformation of the surrounding material (bond coat, IDZ, or substrate). Namely, creep deformation occurring around the tip accommodates elongation of the TGO along the "y" direction (Fig. 9), leading to a crack extension per cycle. The five parameters affecting the fatigue crack growth rate are thus: the TGO growth stress, σ_{growth}, the lateral strain-rate experienced by the TGO as it grows, $\dot{\varepsilon}_{growth}$, the oxide thickness, h, the creep characteristics of the surrounding material, $\dot{\varepsilon}_{sub} = \dot{\varepsilon}_o \left(\frac{\sigma}{\sigma_o}\right)^n$, and the crack length, a. The oxide thickness was assumed to be $h = 3$ μm in all analyses presented here.

The principal features of the damage growth process can also be elucidated by using a plasticity model for the substrate and bond coat (with power law hardening) and by invoking the reference stress method to convert to power law creep.[7,18] Results presented here are primarily from plasticity analyses. The sequential stages of damage development are considered for bare and coated materials and the conditions where the degradation is dominated by the superalloy, bond coat, and IDZ properties, respectively, are highlighted.

(1) Superalloy Dominated Growth

Initial calculations considered damage growth for a TGO formed on bare superalloy, to establish a baseline, since the properties of the substrate are well established. Figure 11 considers the early stages of the TGO penetration ($a_o = 20$ μm) into the superalloy substrate as a function of the strength of the substrate. For a moderate applied compressive strain of 0.4% and a superalloy strength of 100 MPa (comparable to René N5 at 1093°C) the oxide elongates at a rate of approximately 6.8 nm/cycle, Fig. 11(b). Across the range of conditions examined, the TGO penetration rate scaled linearly with the oxide thickness, so penetration rates are normalized by the oxide thickness, Fig. 11(a). As the strength of the superalloy approaches approximately 2/3 that of the TGO, the resistance to deformation along

Fig. 6. The first three stages of crack progression with cracking in the Pt aluminide bond coat (Stage I), interdiffusion zone (Stage II), and superalloy substrate (Stage III). René N5 tested at 982°C.

Fig. 7. Longitudinal section through an SPLCF sample showing the four stages of failure, with an inclined crack causing final failure in Stage IV.

Fig. 8. SEM secondary electron (SE) image and corresponding X-ray maps showing alumina in the crack tip after SPLCF testing of René N5 with Pt aluminide coating at 982°C.

the surface of the oxide-lined crack face and at the crack tip rises to a sufficiently high level to almost completely suppress crack growth. For an applied compressive strain of 0.35% and a substrate strength of 100 MPa, the TGO penetration rate is 6 nm/cycle, Fig. 11(a) and rises to 12 nm/cycle for 50 MPa strength. Experimentally, for an applied compressive strain of 0.35%, uncoated René N5 samples were interrupted after 2000 and 4000 cycles, Table I. The maximum crack depth was measured as 36 μm at 2000 cycles and 19 μm at 4000 cycles (two distinct samples), corresponding to a crack growth rates in the range of 5 nm/cycle–18 nm/cycle, in good agreement with the model. A limited number of high-temperature, *in-situ* experimental investigations on growth strains in the TGO formed on aluminide coatings have been conducted in the temperature range of

$1000°–1100°C$,[26,32] however, no measurements of growth strains are available for the TGO on superalloys. Bond coats experience compressive growth strains varying from -0.01% to -0.10%, so growth strains over this range were used in the modeling for both the bond coat and the superalloy. Figure 12 shows the influence of oxide growth strains on the rate of penetration of the TGO into the substrate. With no growth strains in the oxide, there is still a tendency for the oxide to push into the substrate due to transverse compression of the oxide during the hold period. Growth strains as low as 0.03%/cycle accelerate the rate of crack growth by more than a factor of $5\times$ as the cracks grow beyond 30–40 μm in length. The influence of the growth strains on the damage development rate, however, are less pronounced at $a = 20$ μm and below. Comparing again with experimental results, for the 20 μm crack length in Fig. 12, the predicted growth rate for growth strains of 0.03%/cycle is 12 nm/cycle; again this is within the range of rates observed experimentally. These results suggest that the model is capturing some of the essential details of the crack growth process in the early stages.

It is worth considering more deeply penetrating oxides, since cracks that pass through the bond coat and interdiffusion zone will subsequently be embedded in the superalloy substrate. Figure 13 shows the TGO penetration rate normalized by the oxide thickness into the superalloy for an initial crack depth of 200 μm, which is well beyond the combined 100 μm thickness of the bond coat and IDZ. The predicted crack growth rate in Fig. 13 for $\Delta\varepsilon_t = 0.35\%$ is approximately 25 nm/cycle, assuming

Table II.   Crack Length Data for Bare René N5 Sample Cycled at 1093°C with $\Delta\varepsilon_t = 0.35\%$

Cycles	Average depth (μm)	Maximum depth (μm)
1000	4.1	21.2
2000	3.8	36.0
4000	4.1	19.0
6000	7.2	49.7
6755	5.9	36.7
8000	7.3	209.7

Fig. 9. Details of finite element mesh with oxide on surface and in crack tip.

a growth strain of 0.1%/cycle and bond coat and IDZ strengths of 50 MPa. In the interrupted tests on uncoated samples, a maximum crack depth of 200 μm was reached at approximately 80% of life, Table I. At this point the crack growth rate was ≈150 nm/cycle. Earlier experimental measurements at 80% of life with a vapor phase aluminide-coated sample indicate a crack growth rate of about 40 nm/cycle at 80% life,[6] so again the model predicts growth rates within a factor of 5. It should be noted that the crack growth rate increases rapidly in the later stages of life, rising from 40 nm/cycle at 80% life to 200 nm/cycle at 95% of life, so comparison of experiments with the model is more challenging in the later stages of damage growth. Ultimately, as the crack penetrates deeply into the superalloy substrate, the rate of growth depends on the difference in the substrate and TGO strengths, Fig. 13(b).

The formation of alumina along the crack faces by depletion of Al from the superalloy substrate can lead to a layer of soft γ-Ni adjacent to the oxide-filled crack. For this reason simulations were also conducted with soft layers surrounding the crack tip with thicknesses in the range of 1–3 μm with growth strains ranging 0.02%–0.20%. Decreasing the strength of the soft layer from 100 MPa (the same as the superalloy strength) to 10 MPa accelerated the crack growth rate by an additional factor of 2–3 ×.

(2) Role of the Bond Coat and IDZ

For the purposes of modeling the influence of coating and IDZ properties, the β-NiAl and interdiffusion zone are each considered to be 50 μm in thickness with a 3-μm-thick TGO. The oxide penetration rate in each of the three stages has been addressed separately by generating a series of meshes for varying crack depths. The properties of the β-NiAl coating, IDZ, and superalloy substrate are varied along with oxide growth strains over bounds that might reasonably be expected in present and future systems, with bond coat and IDZ strengths varied parametrically from 10 to 100 MPa.

Figure 14 shows the influence of the bond coat strength on the TGO crack tip lengthening rate for Stage I growth with a crack depth of 20 μm and a TGO thickness of 3 μm. At this early stage of the process where the cracks are shallow and the bond coat is softer than the substrate, the lengthening of the oxide causes not only penetration into the depth of the β-NiAl, but also a slight upward displacement of the bond coat surface of the order of 0.1 μm in the "y" direction, (direction indicated in Fig. 9). Determination of the overall crack growth rate in this early stage is complicated by the uplift as well as the included angle of the crack. Nevertheless, stronger bond coats do inhibit the rate of TGO lengthening as they increase in strength from 20 to 100 MPa. At this early stage in life, interdiffusion zone strengths have very limited influence on the overall response, Fig. 14, since the crack has not yet reached this zone. The overall crack growth rate within this stage for a bond coat strength close to β-NiAl is calculated to be about 4.1 nm/cycle, which is in reasonable agreement with the experimentally observed growth rate of 9 nm/cycle.[6] It is important to note that about half of the overall life is spent in Stage I, so strengthening the bond coat is likely to have an overall beneficial effect on SPLCF life.

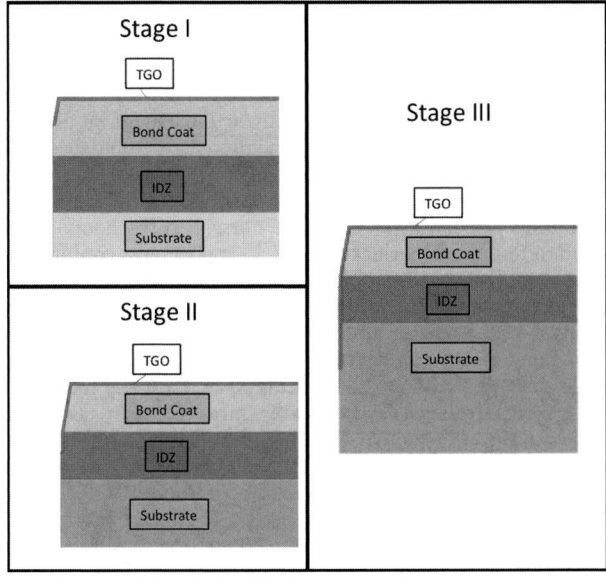

Fig. 10. Schematic of finite element models for the three stages of crack growth.

Fig. 16. Influence of fatigue cycling on crack growth through the interdiffusion zone in Stage II.

As the crack is well into Stage III growth, the crack growth rate, da/dN, is dependent on the difference in the TGO-substrate strength such that, Fig. 13(b):

$$\frac{da}{dN} = \lambda h \varepsilon_{growth} \left(\frac{\sigma_y^{TGO} - \sigma_y^{Sub}}{\sigma_y^{Sub}} \right)$$

where σ_y^{TGO} and σ_y^{Sub} are the yield strengths of the TGO and substrate at the test temperature, h is the thickness of the oxide, ε_{growth} is the oxide growth strain per cycle in the superalloy and a constant,[18] $\lambda \approx 4.5$ for the René N5 substrate. The crack is expected to grow at a steady rate until the threshold for macroscopic fatigue crack propagation is exceeded in the tension portion of the cycle, thereby entering Stage IV. For René N5 at $\Delta\varepsilon_t = 0.35\%$, the maximum tensile stress is 190 MPa, Fig. 3. Considering a typical long crack threshold of $\Delta K_{th} = 7$ MPa·m$^{1/2}$, this transition would occur for a crack depth of 430 μm, in rough agreement with the crack angle transition observed in Fig. 7. Beyond this, rapid crack growth is expected, independent of coating properties. However, since this represents only a small fraction of overall SPLCF life, strategies for

Fig. 17. Crack growth rates converge to a common rate after penetrating 250–300 μm into the substrate.

improving SPLCF resistance should focus on slowing the rate of damage growth during Stages I–III.

VI. Conclusions

1. Failure due to fatigue cycling with compressive holds in single crystal samples with or without aluminide coatings occurs due to oxide-assisted crack growth from sample surfaces at 982° and 1093°C.

2. Oxide-filled cracks do not penetrate the bond coat and interdiffusion zone until approximately 80% of the cyclic life.

3. A model that accounts for relaxation of oxide growth stresses as the primary driving force for extension of the oxide through the coating and into the substrate provides reasonable predictions of the rate of crack growth through the bond coat, interdiffusion zone, and superalloy single crystal substrate.

4. The model predicts that cyclic life can be enhanced by strengthening of the bond coat and interdiffusion zone or by reduction of oxide growth stresses.

Acknowledgments

This paper is dedicated to the memory of Tony Evans, whose original insights to this problem motivated this research. The authors also acknowledge the support of the General Electric Company and useful discussions with D. Konitzer, J. Rigney, B. Hazel, and M. F. X. Gigliotti.

References

[1]A. G. Evans, D. R. Mumm, J. W. Hutchinson, G. H. Meier, and F. S. Pettit, "Mechanisms Controlling the Durability of Thermal Barrier Coatings," *Prog. Mater. Sci.*, **46**, 505–53 (2001).

[2]T. M. Pollock and S. Tin, "Nickel-Based Superalloys for Advanced Turbine Engines: Chemistry, Microstructure and Properties," *AIAA J.: Propulsion Power*, **22** [2] 361–74 (2006).

[3]A. G. Evans, D. R. Clarke, and C. G. Levi, "The Influence of Oxides on the Performance of Advanced Gas Turbines," *J. Eur. Ceram. Soc.*, **28**, 1405–919 (2008).

[4]Y. L. Bihan, P.-Y. Joubert, and D. Placko, "Wall Thickness Evaluation of Single-Crystal Hollow Blades by Eddy Current Sensor," *NDT&E Int.*, **34**, 363–8 (2001).

[5]R. C. Reed, *The Superalloys: Fundamentals and Applications*. Cambridge University Press, Cambridge, 2006.

[6]A. Suzuki, M. F. X. Gigliotti, B. T. Hazel, D. G. Konitzer, and T. M. Pollock, "Crack Progression During Sustained Peak Low Cycle Fatigue in René N5," *Metall. Mater. Trans.*, **41A**, 948–56 (2010).

[7]A. G. Evans, M. Y. He, A. Suzuki, M. Gigliotti, B. Hazel, and T. M. Pollock, "The Mechanism Governing Sustained Peak Low Cycle Fatigue of Coated Superalloys," *Acta Mater.*, **57**, 2969–83 (2009).

[8]T. E. Strangman, "Thermal-Mechanical Fatigue Life Model for Coated Superalloy Turbine Components"; pp. 795–804 in *Superalloys 1992*, Edited by S. D. Antolovich., et al TMS, Warrendale, PA, 1992.

[9]J. W. Holmes and F. A. McClintock, "The Chemical and Mechanical Processes of Thermal Fatigue Degradation of an Aluminide Coating," *Metall. Trans.*, **21A**, 1209–22 (1990).

[10]P. Moretto and J. Bressers, "Thermomechanical Fatigue Degradation of a Nickel-Aluminide Coating on a Single Crystal Superalloy," *J. Mater. Sci.*, **31**, 4817–29 (1996).

[11]T. C. Totemeier and J. E. King, "Isothermal Fatigue of an Aluminide-Coated Single-Crystal Superalloy: Part II. Effects of Brittle Precracking," *Metall. Mater. Trans.*, **27A**, 353–61 (1996).

[12]J. S. Crompton and J. W. Martin, "Crack Growth in a Single Crystal Superalloy at Elevated Temperature," *Metall. Trans.*, **15A**, 1711–9 (1984).

[13]H. Zhou, H. Harada, Y. Ro, and I. Okada, "Investigations on the Thermo-Mechanical Fatigue of Two Ni-Based Single Crystal Superalloys," *Mater. Sci. Eng.*, **A**, 161–7 (2005).

[14]N. Isobe and S. Sakurai, "Compressive Strain Hold Effect on High Temperature Low-Cycle Fatigue Crack Growth in Superalloys," *Mater. Sci. Res. Int.*, **9**, 29–33 (2003).

[15]R. Nutzl, E. Affeldt, and M. Goken, "Damage Evolution During Thermomechanical Fatigue of a Coated Monocrystalline Nickel-Base Superalloy," *Intl. J. Fatigue*, **30**, 314–7 (2008).

[16]Y. H. Zhang, D. M. Knowles, and P. J. Withers, "Micromechanics of Failure of Aluminide Coated Single Crystal Ni Superalloy Under Thermomechanical Fatigue," *Scr. Mater.*, **37**, 815–20 (1997).

[17]E. Fluery and L. Remy, "Behavior of Nickel-Base Superalloy Single Crystals Under Thermomechanical Fatigue," *Metall. Mater. Trans.*, **25A**, 99–109 (1994).

[18]M. Y. He and A. G. Evans, "A Model for Oxidation-Assisted Low Cycle Fatigue of Superalloys," *Acta Mater.*, **58**, 583–91 (2010).

[19]J. A. Nychka and D. R. Clarke, "Quantification of Aluminum Outward Diffusion During Oxidation of FeCrAlY Alloys," *Oxidation Metals*, **63** [5–6] 325–52 (2005).

[20]V. K. Tolpygo and D. R. Clarke, "Microstructural Evidence for Counter-Diffusion of Aluminum and Oxygen during the Growth of Alumina Scales," *Mater. High Temp.*, **20**, 261–71 (2003).

[21]V. K. Tolpygo and D. R. Clarke, "Oxidation-Induced Failure of BPVD Thermal Barrier Coatings," *Surf. Coat. Tech.*, **146–147**, 124–31 (2001).

[22]D. S. Balint and J. W. Hutchinson, "An Analytical Model of Rumpling in Thermal Barrier Coatings," *J. Mech. Phys. Solids*, **53**, 949–73 (2005).

[23]A. W. Davis and A. G. Evans, "Some Effects of Imperfection Geometry on the Cyclic Distortion of Thermally Grown Oxides," *Oxidation Metals*, 1573–4889 (2006).

[24]A. M. Karlsson, J. W. Hutchinson, and A. G. Evans, "A Fundamental Model of Cyclic Instabilities in Thermal Barrier Systems," *J. Mech. Phys. Solids*, **50**, 1565–89 (2002).

[25]A. H. Heuer, A. Reddy, D. B. Hovis, B. Veal, A. Paulikas, A. Vlad, and M. Rühle, "The Effect of Surface Orientation on Oxidation-Induced Growth Strains in a Model NiCrAlY Bond Coat Alloy," *Scr. Mater.*, **54**, 1907–12 (2006).

[26]D. Hovis, L. Hu, A. Reddy, A. H. Heuer, A. P. Paulikas, and B. W. Veal, "In-Situ Studies of TGO Growth Stresses and the Martensitic Transformation in the B2 Phase in Commercial Pt-Modified NiAl and NiCoCrAlY Bond Coat Alloys," *Intl. J. Mater. Res.*, **98**, 1209–13 (2007).

[27]B. W. Veal, A. P. Paulikas, B. Gleeson, and P. Y. Hou, "Creep in α-Al_2O_3 Thermally Grown on β-NiAl and NiAlPt Alloys," *Surf. Coat. Tech.*, **202**, 608–12 (2007).

[28]B. Tryon, K. S. Murphy, C. G. Levi, J. Yang, and T. M. Pollock, "Hybrid Intermetallic Ru/Pt-modified Bond Coatings for Thermal Barrier Systems," *Surf. Coat. Tech.*, **202**, 349–61 (2007).

[29]T. M. Pollock and R. D. Field, "Dislocations and High Temperature Plastic Deformation of Superalloy Single Crystals"; pp. 547–618 in *Dislocations in Solids*, Vol 11, Edited by F. R. N. Nabarro, and M. S. Duesbery. Elsevier, Amsterdam, 2002.

[30]D. Pan, M. W. Chen, P. K. Wright, and K. J. Hemker, "Evolution of a Diffusion Aluminide Bond Coat for Thermal Barrier Coatings During Thermal Cycling," *Acta Mater.*, **51**, 2205–17 (2003).

[31]J. S. Van Sluytman, A. Suzuki, R. Helmik, A. Bolcavage, and T. M. Pollock, "Gamma Prime Morphology and Creep Properties of Nickel-Base Superalloys with Platinum Group Metal Additions"; pp. 499–507 in *Superalloys 2008*, Edited by R. C. Reed, K. A. Green, P. Caron, T. P. Gabb, M. G. Fahrmann, E. S. Huron, and S. R. Woodard TMS, Warrendale, PA, 2008.

[32]P. Y. Hou, A. P. Paulikas, and B. W. Veal, "Growth Strains in Thermally Grown Al_2O_3 Scales Studied Using Synchrotron Radiation," *JOM*, **61**, 51–5 (2009). □

Fig. 3. $W_{sep-All}$ from first-principles calculations versus $W_{sep-All}$ predicted by Eq. (6) for the stoichiometric metal/alumina interfaces. All the data are listed in Table IV.

in which the empirical constant c was estimated to be 0.22, N is Avogadro's constant, and V_{Me} is the molar volume of the metal Me. $\Delta\bar{H}^{\infty}_{O(Me)}$ and $\Delta\bar{H}^{\infty}_{Al(Me)}$ are the enthalpies of mixing at infinite dilution of oxygen and aluminum atoms in the bulk metal Me, respectively.

Li[55] regarded the $W_{sep-All}$ as a function of the electron density parameter n_{ws}, although no straightforward expression was given. From Table IV, it is observed that most of the results obtained here and Chatain and colleagues' model predictions agree well with experimental data, while there are several large discrepancies for Li's model, such as the Pd-, Pt-, Cu-, Au-, Ag-, and Al-related systems. Considering the mixed ionic-covalent and metallic character of chemical bonding at the $M/(Al_2O_3)_{All}$ interfaces, the electron density parameter n_{ws} is not sufficient to describe the adhesion of the stoichiometric interface. The contribution from the metal–oxygen ionic–covalent bonding has to be considered, even though it is not as large.

(4) All Stoichiometries of M/α-Al₂O₃ Interfaces

Combining the results for the interfaces of different terminations, a simple, semiempirical relationship ensues for the work

of separation, applicable to all the metals and interfacial stoichiometries:

$$W_{sep} = a n_{ws} + b|EOF| + c \tag{8}$$

The nature of the chemical bonding of metal/α-Al₂O₃ interfaces underlying Eq. (8) is now clear. The coefficients a–c are different for the different stoichiometries. However, the above comparisons with results of experiments and first-principles computations demonstrate clearly that there does exist a group of (a–c) parameters to describe the adhesion of metals with α-Al₂O₃ for stoichiometric, aluminum-rich, and oxygen-rich interfacial stoichiometries. It is reasonable to expect that the model has some applicability to other interfaces between metals and oxides.

IV. Conclusions

Through analyses of interfacial stoichiometry, atomic relaxation, electronic structure, chemical bonding, and adhesion between α-Al₂O₃ and a series of metals including Al, Ni, Cu, Au, Ag, Rh, Ir, Pd, Pt, Nb, and β-NiAl, one can conclude the following. Chemical bonding is found to be metallic for the aluminum-rich interfacial stoichiometry, primarily ionic with some covalent contributions for the oxygen-rich stoichiometry, and of mixed ionic-covalent and metallic for the stoichiometric interfaces. Simple empirical relationships were identified. It was found that the metallic component of bonding scales with $(B/V)^{1/2}$, where B is the bulk modulus and V is molar volume of the metal. Similarly, it was found that the ionic-covalent component of bonding is equal to a constant plus a term that scales with the enthalpy of oxide formation $|EOF|$ of the metal. A linear combination of a constant, $(B/V)^{1/2}$, and the enthalpy of oxide formation, both of the latter of the metal, provides a unified relationship to evaluate the works of separation for all the stoichiometries: aluminum-rich, oxygen-rich, and stoichiometric metal/α-Al₂O₃ interfaces. This relationship is consistent with the results of first-principles calculations and experimental measurements for all three potential stoichiometries of the eleven metal/α-Al₂O₃ interfaces. These analytic functions of a constant, $(B/V)^{1/2}$, and $|EOF|$ suggest also that Al-metal–O interactions are important contributors to stoichiometric effects in bonding at these interfaces. In general, higher surface energies and lower interfacial energies lead to the higher works of separation found for O-rich and Al-rich interfaces.

Acknowledgment

The authors would like to acknowledge Prof. Dominique Chatain for providing the useful information of his model and experimental data. H. Li and W. Zhang thank the SCCAS for time on the Shengteng7000 supercomputer. J. R. Smith is supported for this work by the U.S. Office of Naval Research under Grant N00014-08-1-1164.

Fig. 4. $W_{sep-All}$ by experimental measurements versus $W_{sep-All}$ predicted by Eq. (6) for the stoichiometric metal/alumina interfaces. The slope of solid line is unity. All the data are listed in Table IV, and the short vertical dotted line designates the range of experimental data[17] for the $W_{sep-All}$ of $Mn/(Al_2O_3)_{All}$. *Experimental data correspond to the work of adhesion (W_{ad}) for the $M/(Al_2O_3)_{All}$ interfaces as discussed in the paragraph below Eq. (2).*

References

[1]A. Rabiei and A. G. Evans, "Failure Mechanisms Associated with the Thermally Grown Oxide in Plasma-Sprayed Thermal Barrier Coatings," *Acta Mater.*, **48**, 3963–76 (2000).

[2]N. P. Padture, M. Gell, and E. H. Jordan, "Materials Science-Thermal Barrier Coatings for Gas-Turbine Engine Apptications," *Science*, **296**, 280–4 (2002).

[3]D. W. Goodman, "Model Studies in Catalysis Using Surface Science Probes," *Chem. Rev.*, **95**, 523–36 (1995).

[4]R. C. Santana, S. Jongpatiwut, W. E. Alvarez, and D. E. Resasco, "Gas-phase Kinetic Studies of Tetralin Hydrogenation on Pt/Alumina," *Ind. Eng. Chem. Res.*, **44**, 7928–34 (2005).

[5]F. Ahmed, M. K. Alam, A. Suzuki, M. Koyama, H. Tsuboi, N. Hatakeyama, A. Endou, H. Takaba, C. A. Del Carpio, M. Kubo, and A. Miyamoto, "Dynamics of Hydrogen Spillover on Pt/Gamma-Al₂O₃ Catalyst Surface: A Quantum Chemical Molecular Dynamics Study," *J. Phys. Chem. C*, **113**, 15676–83 (2009).

[6]C. C. Young, J. G. Duh, and C. S. Huang, "Improved Characteristics of Electroless Cu Deposition on Pt-Ag Metallized Al₂O₃ Substrates in Microelectronics Packaging," *Surf. Coat. Technol.*, **145**, 215–2 (2001).

[7]H. J. Grabke, D. Wiemer, and H. Viefhaus, "Segregation of Sulfur During Growth of Oxide Scales," *Appl. Surf. Sci.*, **47**, 243–50 (1991).

[8]H. J. Grabke, G. Kurbatov, and H. J. Schmutzler, "Segregation Beneath Oxide Scales," *Oxid. Met.*, **43**, 97–114 (1995).

[9]J. A. Haynes, B. A. Pint, K. L. More, Y. Zhang, and I. G. Wright, "Influence of Sulfur, Platinum, and Hafnium on the Oxidation Behavior of CVD NiAl Bond Coatings," *Oxid. Met.*, **58**, 513–44 (2002).

[10]P. Y. Hou, "Segregation Phenomena at Thermally Grown Al₂O₃/Alloy Interfaces," *Annu. Rev. Mater. Res.*, **38**, 275–98 (2008).

[11]F. Ernst, "Metal–Oxide Interfaces," *Mater. Sci. Eng.*, **R 14**, 97–156 (1995).

[12]M. W. Finnis, "The Theory of Metal–Ceramic Interfaces," *J. Phys.: Condens. Matter*, **8**, 5811–36 (1996).

[13]J. M. Howe, "Bonding, Structure, and Properties of Metal–Ceramic Interfaces .1. Chemical Bonding, Chemical-Reaction, and Interfacial Structure," *Int. Mater. Rev.*, **38**, 233–56 (1993).

[14]E. Saiz, A. P. Tomsia, and R. M. Cannon, "Wetting and Work of Adhesion in Oxide/Metal Systems"; pp. 65–82 in *Ceramic Microstructure: Control at the Atomic Level*, Edited by A.P Tomsia, and A. M. Glaeser. Plenum Press, New York, 1998.

[15]E. Saiz, R. M. Cannon, and A. P. Tomsia, "Energetics and Atomic Transport at Liquid Metal/Al₂O₃ Interfaces," *Acta Mater.*, **47**, 4209–20 (1999).

[16]E. Saiz, R. M. Cannon, and A. P. Tomsia, "High-Temperature Wetting and the Work of Adhesion in Metal/Oxide Systems," *Annu. Rev. Mater. Res.*, **38**, 197–226 (2008).

[17]D. Chatain, I. Rivollet, and N. Eustathopoulos, "Thermodynamic Adhesion in Non-Reactive Liquid Metal–Alumina Systems," *J. Chim. Phys.*, **83**, 561–7 (1986).

[18]D. Chatain, I. Rivollet, and N. Eustathopoulos, "Estimation of The Thermodynamic Adhesion and the Contact-Angle in the Nonreactive metal-Ionocovalent Oxide Systems," *J. Chim. Phys.*, **84**, 201–3 (1987).

[19]D. Chatain, L. Coudurier, and N. Eustathopoulos, "Wetting and Interfacial Bonding in Ionocovalent Oxide-liquid Metal Systems," *Rev. Phys. Appl.*, **23**, 1055–64 (1988).

[20]M. Degraef, B. J. Dalgleish, M. R. Turner, and A. G. Evans, "Interfaces Between Alumina and Platinum: Structure, Bonding and Fracture Resistance," *Acta Metal. Mater.*, **40**, S333–S344 (1992).

[21]M. Rühle, "Structure and Composition of Metal/Ceramic Interfaces," *J. Eur. Ceram. Soc.*, **16**, 353–65 (1996).

[22]N. Eustathopoulos and B. Drevet, "Determination of the Nature of Metal-Oxide Interfacial Interactions from Sessile Drop Data," *Mater. Sci. Eng. A*, **249**, 176–83 (1998).

[23]N. Eustathopoulos, B. Drevet, and M. L. Muolo, "The Oxygen-Wetting Transition in Metal/Oxide Systems," *Mater. Sci. Eng. A*, **300**, 34–40 (2001).

[24]N. Eustathopoulos, "Progress in Understanding and Modeling Reactive Wetting of Metals on Ceramics," *Curr. Opin. Solid State Mater. Sci.*, **9**, 152–60 (2005).

[25]A. Bogicevic and D. R. Jennison, "Variations in the Nature of Metal Adsorption on Ultrathin Al₂O₃ Films," *Phys. Rev. Lett.*, **82**, 4050–3 (1999).

[26]C. Verdozzi, D. R. Jennison, P. A. Schultz, and M. P. Sears, "Sapphire(0001) Surface, Clean and with D-Metal Overlayers," *Phys. Rev. Lett.*, **82**, 799–802 (1999).

[27]J. R. Smith and W. Zhang, "Stoichiometric Interfaces of Al and Ag with Al₂O₃," *Acta Mater.*, **48**, 4395–403 (2000).

[28]Z. Łodziana and J. K. Norskov, "Adsorption of Cu and Pd on Alpha-Al₂O₃(0001) Surfaces with Different Stoichiometries," *J. Chem. Phys.*, **115**, 11261–7 (2001).

[29]D. J. Siegel, L. G. Hector, and J. B. Adams, "Adhesion, Atomic Structure, and Bonding at the Al(111)/Alpha-Al₂O₃(0001) Interface: A First Principles Study," *Phys. Rev. B*, **65**, 085415, 19pp (2002).

[30]I. G. Batirev and L. Kleinman, "In-plane Relaxation of Cu(111) and Al(111)/ Alpha-Al₂O₃ (0001) Interfaces," *Phys. Rev. B*, **64**, 033410, 4pp (2001).

[31]I. G. Batirev, A. Alavi, M. W. Finnis, and T. Deutsch, "First-Principles Calculations of the Ideal Cleavage Energy of Bulk Niobium(111)/Alpha-Alumina(0001) Interfaces," *Phys. Rev. Lett.*, **82**, 1510–3 (1999).

[32]I. G. Batyrev, A. Alavi, and M. W. Finnis, "Equilibrium and Adhesion of Nb/ Sapphire: The Effect of Oxygen Partial Pressure," *Phys. Rev. B*, **62**, 4698–706 (2000).

[33]W. Zhang and J. R. Smith, "Nonstoichiometric Interfaces and Al₂O₃ Adhesion with Al and Ag," *Phys. Rev. Lett.*, **85**, 3225–8 (2000).

[34]W. Zhang and J. R. Smith, "Stoichiometry and Adhesion of Nb/Al₂O₃," *Phys. Rev B*, **61**, 16883–9 (2000).

[35]W. Zhang, J. R. Smith, and A. G. Evans, "The Connection between Ab Initio Calculations and Interface Adhesion Measurements on Metal/Oxide Systems: Ni/ Al₂O₃ and Cu/Al₂O₃," *Acta Mater.*, **50**, 3803–16 (2002).

[36]J. W. Feng, W. Q. Zhang, and W. Jiang, "Ab Initio Study of Ag/Al₂O₃ and Au/Al₂O₃ Interfaces," *Phys. Rev. B*, **72** [1–11] 115423 (2005).

[37]W. Zhang, J. R. Smith, and X. G. Wang, "Thermodynamics from Ab Initio Computations," *Phys. Rev. B*, **70**, 024103, 8pp (2004).

[38]J. Bruley, R. Brydson, H. Mülleejans, J. Mayer, G. Gutekunst, W. Mader, D. Knauss, and M. Rühle, "Investigations of the Chemistry and Bonding at Niobiumsapphire Interfaces," *J. Mater. Res.*, **9**, 2574–83 (1994).

[39]G. Dehm, M. Ruhle, G. Ding, and R. Raj, "Growth and Structure of Copper Thin-Films Deposited on (0001) Sapphire by Molecular-Beam Epitaxy," *Philos. Mag. B*, **71**, 1111–24 (1995).

[40]C. Scheu, G. Dehm, M. Rühle, and R. Brydson, "Electron-Energy-Loss Spectroscopy Studies of Cu-Alpha-Al₂O₃ Interfaces Grown by Molecular Beam Epitaxy," *Philos. Mag.A*, **78**, 439–65 (1998).

[41]J. P. Perdew, K. Burke, and M. Ernzerhof, "Generalized Gradient Approximation Made Simple," *Phys. Rev. Lett.*, **77**, 3865–8 (1996).

[42]G. Kresse and J. Hafner, "Ab Initio Molecular Dynamics for Liquid Metals," *Phys. Rev. B*, **47**, 558–61 (1993).

[43]G. Kresse and J. Furthmuller, "Efficient Iterative Schemes for Ab Initio Total-Energy Calculations using a Plane-Wave Basis Set," *Phys. Rev. B*, **54**, 11169–86 (1996).

[44]P. E. Blöchl, "Projector Augmented-Wave Method," *Phys. Rev. B*, **50**, 17953–79 (1994).

[45]G. Kresse and D. Joubert, "From Ultrasoft Pseudopotentials to the Projector Augmented-Wave Method," *Phys. Rev. B*, **59**, 1758–75 (1999).

[46]H. T. Li, J. W. Feng, W. Q. Zhang, W. Jiang, H. Gu, and J. R. Smith, "Ab Initio Thermodynamic Study of the Structure and Chemical Bonding of a Beta-Ni₁₋ₓAlₓ/Alpha-Al₂O₃ Interface," *Phys. Rev. B*, **80**, 205422, 12pp (2009).

[47]I. Batyrev, A. Alavi, and M. W. Finnis, "Ab Initio Calculations on the Al₂O₃ (0001) Surface," *Faraday Discuss.*, **114**, 33–43 (1999).

[48]X. G. Wang, A. Chaka, and M. Scheffler, "Effect of the Environment on Alpha-Al₂O₃ (0001) Surface Structures," *Phys. Rev. Lett.*, **84**, 3650–3 (2000).

[49]F. R. D. Boer, R. Boom, W. C. M. Mattens, A. R. Miedema, and A. K. Niessen, *Cohesion in Metals*, North-Holland, New York, 1988.

[50]A. K. Niessen and A.R Miedema, "The 'Macroscopic Atom' Model: An Easy Tool to Predict Thermodynamic Quantities"; pp. 29–54 in *Thermochemistry of Alloys*, edited by H. Brodowsky, and H. J. Schaller. Kluwer Academic Publishers, Boston, 1987.

[51]A. R. Miedema, "Surface Energies of Solid Metals," *Z. Metallkd.*, **69**, 287–92 (1978).

[52]R. S. Mulliken, "Electronic Population Analysis on Lcao-Mo Molecular Wave Functions .1," *J. Chem. Phys.*, **23**, 1833–40 (1955).

[53]O. Kubaschewski, C. B. Alcock, and P. J. Spencer, *Materials Thermochemistry*. Pergamon Press, Oxford, 1993.

[54]D. R. Lide, *CRC Handbook of Chemistry and Physics*. CRC Press, Boca Raton, 1997.

[55]J. G. Li, "Wetting and Interfacial Bonding of Metals with Ionocovalent Oxides," *J. Am. Ceram. Soc.*, **75**, 3118–26 (1992).

[56]D. M. Lipkin, J. N. Israelachvili, and D. R. Clarke, "Estimating the Metal-Ceramic Van der Waals Adhesion Energy," *Philos. Mag. A*, **76**, 715–28 (1997).

[57]C. Kittel, *Introduction to Solid State Physics*. Wiley, New York, 1971.

[58]J. E. McDonald and J. G. Eberhart, "Adhesion in Aluminum Oxide-Metal Systems," *Trans. Metall. Soc., AIME*, **233**, 512–7 (1965).

[59]Y. V. Naidich, "The Wettability of Solids by Liquid Metals," *Prog. Surf. Membr. Sci.*, **14**, 353–484 (1981).

[60]A. M. Stoneham and P. W. Tasker, "Metal Non-Metal and Other Interfaces—The Role of Image Interactions," *J. Phys. C: Solid State Phys.*, **18**, L543–L548 (1985).

[61]M. W. Finnis, "Metal Ceramic Cohesion and the Image Interaction," *Acta Metall. Mater.*, **40**, S25–37 (1992).   ☐

Fig. 2. Time sequences of axial tensile stress for (a) normal impact $\phi = 0°$ and (b) inclined impact at $\phi = 45°$. The white zone denotes compression.

Then, define the normalized maximum tensile stress $\bar{\sigma}_{22}$ as

$$\bar{\sigma}_{22} = \frac{(\sigma_{22})_{max}}{V_0 \sqrt{E_{TBC} \rho_{TBC}}} \tag{1}$$

The denominator of Eq. (1) is a reference tensile stress: it is the uniaxial tensile stress in an elastic rod (made from TBC material) when subjected to a step-wise jump in the axial velocity of V_0 at one end.[13] An assessment is now made of the value of $\bar{\sigma}_{22}$ that leads to cracking from the edge of the columns. Then, in the remainder of this section, we shall explore the sensitivity of $\bar{\sigma}_{22}$ to the size and initial velocity of the impacting particle for both the plane strain and axisymmetric problems.

(A) Assessment of Crack Initiation Under the Transient Stress Field: It is instructive to assess whether the stress level $\bar{\sigma}_{22}$ from the elastodynamic simulations is sufficiently high to lead to tensile failure from the edge of the feathery TBC columns. Here, we perform a preliminary fracture mechanics assessment and calculate the maximum mode I stress intensity factor for incipient flaws at the edge of each TBC column. The role of material inertia is ignored in the calculation of the mode I stress intensity factor from the stress state $\bar{\sigma}_{22}$: it is appreciated that this is only an approximation, but our intent is to provide an overall assessment of whether the elastodynamic stress field is sufficiently intense to lead to erosion of the coating.

Microstructural observations of TBC coatings suggest that the feathery edges of the EB-PVD columns possess crack-like flaws of length a on the order of 1 μm along the sides of each column. Further, it is observed from the finite element simulations of the present study that the transient tensile stress distribution occurs to a depth on the order of 1 μm from the free edge; this stress state is approximated by a triangular waveform from a maximum value at the edge of the column to zero at the crack tip. Now, the stress intensity factor K_I for an edge crack of length a in a semiinfinite plate,[¶] subjected to a linear distribution of traction from σ at the

crack mouth to zero at the crack tip, is given by $K_I = 1.208$ $(1 - 2/\pi)\sigma\sqrt{\pi a}$.[14] Assume a representative value of fracture toughness of the TBC layer to be $K_{IC} = 1$ MPa $\sqrt{m}$ and, with $a = 1$ μm, this yields a critical value of stress $\sigma_C = 1.65$ GPa. For an impact velocity $V_0 = 300$ ms^{-1}, along with $E_{TBC} = 140$ GPa and $\rho_{TBC} = 5900$ kg/m^3, relation (1) implies that the critical value of $\bar{\sigma}_{22}$ is given by $\bar{\sigma}_{22} = \bar{\sigma}_C = 0.19$. We shall show below that the level of peak axial stress $\bar{\sigma}_{22}$ in many of the elastodynamic simulations exceeds this threshold value and consequently crack growth from the edge flaws is anticipated. It is recognized that the loading duration by wave propagation is on the order of nanoseconds. However, the loading period is much greater than the time period for atomic vibration, and is thereby sufficiently long to allow for the possibility of cleavage fracture.

(4) The Effect of Particle Size and Initial Impact Velocity on the Transient Tensile Stress

The FE simulations reveal that the maximum transient tensile stress occurs at the edge of the column and at a depth of 0.2–2.2 d. The nondimensional peak stress $\bar{\sigma}_{22}$ is evaluated for selected values of V_0/c and is plotted as a function of d/R in Fig. 3(a) for the plane strain problem, and in Fig. 3(b) for the axisymmetric case. For both problems, $\bar{\sigma}_{22}$ decreases somewhat with increasing d/R. The mild but nonmonotonic dependence of $\bar{\sigma}_{22}$ upon V_0/c is associated with the details of contact evolution between columns. In broad terms, the plane strain model adequately mimics the elastodynamic response of the axisymmetric model but predicts somewhat higher stress levels. For both types of models, the impacting particle has retarded to a minor degree (5%–20% depending on the value of d/R) over the time interval for the attainment of maximum tensile stress. An analytical model for the deceleration of the particle has been given by Fleck and Zisis,[10] including validation by finite element simulation. They show that the contact pressure exerted by the columnar TBC on the particle scales with the instantaneous particle velocity V according to

$$p = V \sqrt{E_{TBC} \rho_{TBC}} \tag{2}$$

[¶]It is appreciated that a more realistic estimate of K_I can be obtained by considering a single edge-notched specimen rather than an edge crack in a semi-infinite plate. The K_I value will be increased by about 10% for the finite width plate.

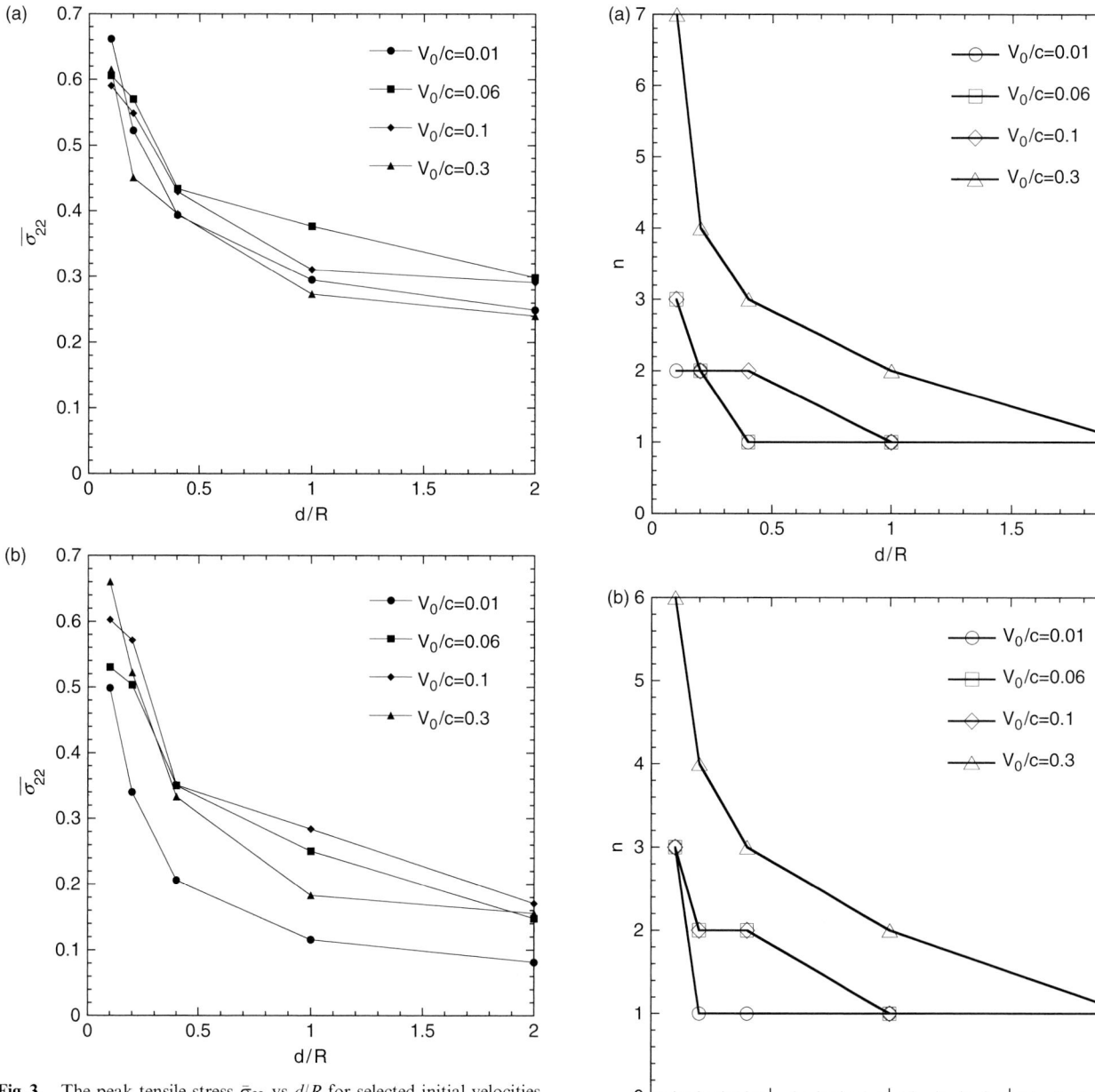

Fig. 3. The peak tensile stress $\bar{\sigma}_{22}$ vs d/R for selected initial velocities. (a) plane strain case, (b) axisymmetric case.

Fig. 4. The column number n for which the tensile axial stress has a global and temporal maximum value versus d/R. (a) plane strain and (b) axisymmetric case.

The column number n for which the maximum tensile stress occurs, counted from the initial contact point, is plotted against the normalized particle size d/R in Fig. 4(a) for the plane strain model and in Fig. 4(b) for the axisymmetric case. Both models show similar trends: the column that suffers the highest transient stress is located further from the center line with increasing V_0/c and decreasing d/R. The surprising result that the peak stress is attained in a column several diameters away from the impacted column is a consequence of the highly nonlinear contact interactions along with the ensuing complex pattern of wave motion.

III. Oblique Impact of the Columnar TBC Layer

It is of interest to determine the sensitivity of axial stress to the obliquity of impact and to the impact site at the top of the columns. We limit attention to the 2D plane strain case, and assume that the cylindrical particle has an incoming velocity V_0 at an angle ϕ to the normal, and impacts a column at a distance s from its midplane, as defined in Fig. 1(a). The same FE mesh, boundary conditions and time-stepping conditions were used for oblique impact as for normal impact. Again, the maximum value of transient tensile stress along the axis of the

columns was determined over the first 50 ns required for stress wave propagation to the bottom of the TBC columns.

The impact problem now involves a larger number of geometric parameters. The sensitivity of elastodynamic response to the intercolumnar gap size g and height of column H_{TBC} has been addressed previously[10] for the case of normal impact. Accordingly, in the remainder of this study, the values for (g, H_{TBC}) are fixed at $g = 0.01$ d and $H_{TBC} = 20$ d. We also limit attention to a circular cylinder of radius $R = 2.5$ d and of density $\rho_p = 0.34\rho_{TBC}$. The particle impinges the underlying TBC at an initial velocity $V_0 = 0.062\sqrt{E_{TBC}/\rho_{TBC}}$ and, for the case of a TBC made from YSZ, this corresponds to an impact velocity of the tip of the turbine blades, $V_0 = 300$ ms^{-1}.

The sensitivity of the maximum axial stress to the impact angle ϕ for three selected values of impact location $s = -d/4$, 0, and $d/4$ is given in Fig. 5 for the frictionless case $\mu_s = \mu_c = 0$. The peak stress declines by a factor of about four when the angle of obliquity ϕ is increased from $0°$ to $75°$, and increases slightly

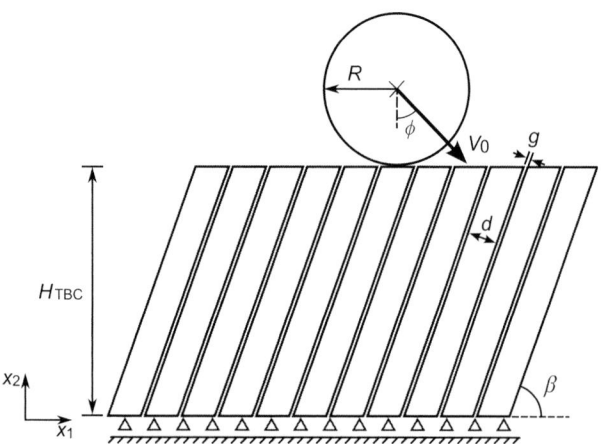

Fig. 10. Sketch of the plane strain model for the impact of inclined columns.

IV. The Elastodynamic Response of Inclined TBC Columns

Inclined TBC columns provide greater thermal insulation than vertical columns, but have a reduced erosion resistance. In this section, the elastodynamic response of inclined columns is compared with that of vertical columns, first for normal impact and then for inclined impact by circular cylindrical particles.

(1)　Normal Impact of Inclined Columns

The impact response has been explored for normal impact ($\phi = 0°$) of an array of inclined columns, as defined by the angle β in Fig. 10. Otherwise, the geometry and material parameters are the same as those used for the vertical columns as defined above. The rigid particle is a circular cylinder of radius $R = 2.5\ d$ and density $\rho_p = 0.34\rho_{TBC}$, and it impacts the underlying columnar coating perpendicularly at an initial velocity $V_0 = 0.062\sqrt{E_{TBC}/\rho_{TBC}}$ (corresponding to 300 ms^{-1} in a practical system). For the present calculations, all the contacts are assumed to be frictionless.

The commercial finite element software ABAQUS/Explicit was again used, with the implementation details and boundary conditions as described in previous sections of this paper. A typical elastodynamic response for columns inclined at $\beta = 60°$ is shown in Fig. 11: contours of the tensile stress σ_a coaxial with the columns are shown after 2 ns and after 8 ns. It is instructive

to compare the response with that shown in Fig. 2 for normal impact ($\phi = 0°$) of vertical columns. For the case of inclined columns ($\beta = 60°$), the peak axial stress (of 10.2 GPa) occurs shortly after impact at $t = 2$ ns, along the edge of the first column and at a depth of 0.15 d. Stress wave propagation follows into neighboring inclined columns via contact from one column to the next, and a complex distribution of dispersive bending waves is established in the array of columns. After $t = 8$ ns, the peak axial stress is of magnitude 3.9 GPa at the outer edge of the second column. This resembles the peak state of stress in the vertical array of columns that occurred after $t = 8$ ns; recall Fig. 2.

The sensitivity of peak axial stress to the inclination β of inclined columns has also been determined by a series of FE simulations for normal impact. The nondimensional maximum tensile stress $\bar{\sigma}_a$ in the axial direction is related to the temporal and spatial maximum tensile stress $(\sigma_a)_{max}$ within the columns according to

$$\bar{\sigma}_a = \frac{(\sigma_a)_{max}}{V_0\sqrt{E_{TBC}\rho_{TBC}}} \tag{3}$$

Fig. 11. Typical response of inclined columns, inclined at $\beta = 60°$, showing the axial tensile stress following normal impact. The white color denotes a state of compression.

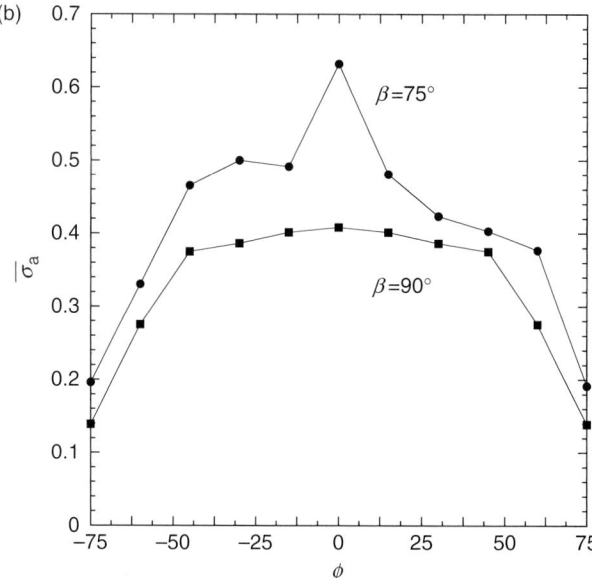

Fig. 12. (a) Peak axial stress versus inclination of the columns β for normal impact $\phi = 0°$ and oblique impact $\phi = 75°$; (b) peak axial stress versus impact angle ϕ for $\beta = 75°$ and $90°$.

The dependence of $\bar{\sigma}_a$ on β is plotted in Fig. 12(a) for normal impact. The tensile stress $\bar{\sigma}_a$ equals 1.2 for β in the range $30°$–$60°$, and then decreases to 0.43 as β is increased to the limit of vertical columns, $\beta = 90°$. The maximum transient tensile stress $(\sigma_a)_{max}$ occurs within 5 ns of impact and at $< 0.2\, d$ from the top surface of the columns. The sensitivity of $\bar{\sigma}_a$ to β is consistent with the observations of Nicholls[5]: vertical columns are more erosion resistant than inclined columns under normal impact.

(2) Oblique Impact of Inclined Columns

The dependence of $\bar{\sigma}_a$ on β is plotted for a particular impact angle $\phi = 75°$ in Fig. 12(a). The tensile stress $\bar{\sigma}_a$ gradually decreases from 0.62 to 0.14 when β is increased from $30°$ to $60°$. The stress levels for oblique impact are significantly below those for normal impact $\phi = 0°$. However, the location of peak stresses and the time to attain it are comparable for the two values of ϕ.

Recall from Fig. 5 that the peak axial stress within frictionless, vertical columns decreases with increasing angle of obliquity of impact ϕ. Does this conclusion still hold for the case of inclined columns? To investigate this, a limited number of impact simulations have been performed on an array of frictionless columns inclined at $\beta = 75°$. Otherwise, the finite element model is the same as that described in the previous section. The observed dependence of $\bar{\sigma}_a$ on ϕ is plotted is Fig. 12(b) for $\beta = 75°$ and for $\beta = 90°$ (from Fig. 5). For both values of β, the highest axial stresses are generated by normal impact, $\phi = 0°$. For $\beta = 75°$, the normalized tensile stress $\bar{\sigma}_a$ declines from 0.63 to 0.20 as the magnitude of ϕ increases from $0°$ to $75°$. Again, we conclude that vertical columns ($\beta = 90°$) are more erosion resistant than inclined columns, regardless of the impact angle.

V. Concluding Remarks

The explicit finite element method has been used to explore the elastodynamic response of the EB-PVD columnar structure to normal and oblique impact by a foreign particle. A complex transient stress pattern emerges from the contact, and contact between the columns can lead to the most intense tensile stresses occurring in columns adjacent to the impacted column. The level of peak axial stress is sufficiently high to induce cracking from the edge of the column and thereby erosion. It is found that the transient tensile stress is sensitive to the size, shape, and approach direction of the incoming particle and to the details of the TBC geometry. Moreover, the peak tensile stress is sensitive to the level of friction between columns and between the particle and the top of TBC coating. Additionally, the maximum tensile stress significantly depends on the inclination of the columns: vertical columns have a greater erosion resistance than inclined columns for the elastodynamic mechanisms. Consequently, the erosion of TBC coatings is highly stochastic in nature.

Acknowledgments

This research was initiated when Prof. Evans was on sabbatical from University of California, Santa Barbara, at Cambridge University. The two other authors are indebted to him for his contributions to this article, and for his wise counsel and friendship over the years.

References

[1]A. G. Evans, D. R. Mumm, J. W. Hutchinson, G. H. Meier, and F. S. Pettit, "Mechanisms Controlling the Durability of Thermal Barrier Coatings," *Prog. Mater. Sci.*, **46** [5] 505–53 (2001).

[2]D. D. Hass, A. J. Slifka, and H. N. G. Wadley, "Low Thermal Conductivity Vapour Deposited Zirconia Microstructures," *Acta Mater.*, **49** [6] 973–8 (2001).

[3]S. Gu, T. J. Lu, D. D. Hass, and H. N. G. Wadley, "Thermal Conductivity of Zirconia Coatings with Zig-Zag Pore Microstructures," *Acta Mater.*, **49** [13] 2539–47 (2001).

[4]J. R. Nicholls, M. J. Deakin, and D. S. Rickerby, "A Comparison Between the Erosion Behaviour of Thermal Spray and Electron Beam Vapour Deposition Thermal Barrier Coatings," *Wear*, **233**, 352–61 (1999).

[5]R. G. Wellman, M. J. Deakin, and J. R. Nicholls, "The Effect of TBC Morphology on the Erosion Rate of EB-PVD TBCs," *Wear*, **258** [1–4] 349–56 (2005).

[6]R. G. Wellman, M. J. Deakin, and J. R. Nicholls, "The Effect of TBC Morphology and Aging on the Erosion Rate of EB-PVD TBCs," *Tribol. Int.*, **38** [9] 798–804 (2005).

[7]A. G. Evans, N. A. Fleck, S. Faulhaber, N. Vermaak, M. Maloney, and R. Darolia, "Scaling Laws Governing the Erosion and Impact Resistance of Thermal Barrier Coatings," *Wear*, **260** [7–8] 886–94 (2006).

[8]X. Chen, M. Y. He, I. Spitsberg, N. A. Fleck, J. W. Hutchinson, and A. G. Evans, "Mechanisms Governing the High Temperature Erosion of Thermal Barrier Coatings," *Wear*, **256** [7–8] 735–46 (2004).

[9]Th. Zisis and N. A. Fleck, "Mechanisms of Elastodynamic Erosion of Electron-Beam Thermal Barrier Coatings," *Int. J. Mater. Res.*, **98** [12] 1196–202 (2007).

[10]N. A. Fleck and Th. Zisis, "The Erosion of EB-PVD Thermal Barrier Coatings: The Competition Between Mechanisms," *Wear*, **268** [11–12] 1214–2 (2010).

[11]R. G. Wellman and J. R. Nicholls, "Some Observations on Erosion Mechanisms of EB PVD TBCs," *Wear*, **242** [1–2] 89–96 (2000).

[12]Hibbitt, Karlsson and Sorensen Inc. *ABAQUS version 6.5 User's Manual.* Hibbitt, Karlsson and Sorensen Inc., Pawtucket, 2005.

[13]K. L. Johnson. *Contact Mechanics.* Cambridge Press, Cambridge, U.K., 1985.

[14]H. Tada, P. Paris, and G. Irwin, *The Stress Analysis of Cracks Handbook.* Del Research Corporation, St. Louis, 1973. □

J. Am. Ceram. Soc., **94** [S1] S178–S185 (2011)
DOI: 10.1111/j.1551-2916.2011.04448.x
© 2011 The American Ceramic Society

Journal

Chemical and Mechanical Consequences of Environmental Barrier Coating Exposure to Calcium–Magnesium–Aluminosilicate

Bryan J. Harder,[‡,∥] Joaquin Ramírez-Rico,[‡,††] Jonathan D. Almer,[§] Kang N. Lee,[¶] and Katherine T. Faber[†,‡]

[‡]Department of Materials Science and Engineering, Northwestern University, Evanston, Illinois 60208

[§]Argonne National Laboratory, Advanced Photon Source, Argonne, Illinois 60439

[¶]Rolls-Royce Corporation, Materials, Processes and Repair Technologies, Indianapolis, Indiana 46206

The success of Si-based ceramics as high-temperature structural materials for gas turbine applications relies on the use of environmental barrier coatings (EBCs) with low silica activity, such as $Ba_{1-x}Sr_xAl_2Si_2O_8$ (BSAS), which protect the underlying components from oxidation and corrosion in combustion environments containing water vapor. One of the current challenges concerning EBC lifetime is the effect of sandy deposits of calcium–magnesium–aluminosilicate (CMAS) glass that melt during engine operation and react with the EBC, changing both its composition and stress state. In this work, we study the effect of CMAS exposure at 1300°C on the residual stress state and composition in BSAS–mullite–Si–SiC multilayers. Residual stresses were measured in BSAS multilayers exposed to CMAS for different times using high-energy X-ray diffraction. Their microstructure was studied using a combination of scanning electron microscopy and transmission electron microscopy techniques. Our results show that CMAS dissolves the BSAS topcoat preferentially through the grain boundaries, dislodging the grains and changing the residual stress state in the topcoat to a nonuniform and increasingly compressive stress state with increasing exposure time. The presence of CMAS accelerates the hexacelsian-to-celsian phase transformation kinetics in BSAS, which reacts with the glass by a solution–reprecipitation mechanism. Precipitates have crystallographic structures consistent with Ca-doped celsian and Ba-doped anorthite.

I. Introduction

SILICON-BASED ceramics, such as SiC or Si_3N_4, are proposed candidates for high-temperature structural applications such as gas turbine components due to their high thermomechanical stability and low density.[1,2] One of the main drawbacks of these materials is their low oxidation resistance in combustion environments, as the protective SiO_2 that forms on the surface of Si-based ceramics reacts with water vapor at high temperatures to form a volatile silicon hydroxide, leading to surface recession and limiting component lifetime.[3–7] For this reason, environmental barrier coatings (EBCs) were developed as means of protecting Si-based ceramics from corrosion and recession in combustion environments.[2,8–16]

EBCs based on $Ba_{1-x}Sr_xAl_2Si_2O_8$ (BSAS) constitute the baseline performance of current EBCs, although other compositions such as rare-earth silicates are currently being explored.[2,11,17] BSAS EBCs are air plasma sprayed at high temperature ($\approx 1200°C$) on SiC/SiC substrates as the topcoat of multilayer coating systems. Typically, a bond coat of silicon ($\approx 125\,\mu m$) is deposited on the ceramic substrate to enhance adhesion of the subsequent layers, as well as to provide additional oxidation resistance.[2] A thicker ($\approx 200\,\mu m$) intermediate layer consisting of mullite ($3Al_2O_3 \cdot 2SiO_2$) or mullite plus $SrAl_2Si_2O_8$ (SAS) is then deposited and serves both to prevent the topcoat from reaction with the silica scale on the substrate and to better match the thermoelastic properties of the substrate with those of the topcoat to minimize residual stresses. Finally, a $\approx 200\,\mu m$ layer of BSAS ($x \leq 0.25$), constituting the EBC, is deposited.

In room-temperature plasma-sprayed systems, BSAS is deposited as an amorphous phase and crystallizes in the metastable hexacelsian phase (hexagonal) when heated. Heating the substrate to 1200°C during the plasma-spray process is sufficient to deposit BSAS in the hexacelsian phase. This hexacelsian phase then undergoes a phase transformation at temperatures over 1200°C into the stable celsian phase (monoclinic).[18–23] This transformation is accompanied by a 0.5% volume reduction and drastically changes the elastic modulus and coefficient of thermal expansion (CTE) of the topcoat (see Table I). Harder and Faber[24] showed that the transformation rate of plasma-sprayed BSAS was faster as a coating than when crushed into a powder due to the presence of unmelted celsian particles from the plasma-spray powder feed. The unmelted particles act as seeds, and thereby effectively reduce the activation energy of the hexacelsian-to-celsian transformation of freestanding plasma-sprayed BSAS bars compared with BSAS powders obtained by crushing and grinding plasma-sprayed material.

The CTEs and compliances of the different layers within the EBC system must be matched to minimize residual stresses that arise upon thermal cycling of the multilayer. The celsian/hexacelsian phase fraction in BSAS was shown to have a high impact on the residual stresses due to CTE differences.[9] The high CTE of the hexacelsian phase induced large tensile stresses in the topcoat that lead to through-thickness cracking upon cooling. The stress drop due to the formation of these cracks was detected *in situ* using a high-energy diffraction technique,[10] which showed that when the topcoat was completely transformed to the celsian phase by a static heat treatment, the resulting compressive residual stress inhibited crack formation.

A challenge for EBC operation is the deleterious effect of reactive glasses that deposit onto hot surfaces within the engine, which was identified by Evans and colleagues as one of the *extrinsic* factors governing thermal barrier coating (TBC) lifetimes.[25,26] These glasses can form from dust and sand deposits in engines operated at temperatures over 1200°C[27–31] and can infiltrate pores and small cracks to subsequently react with the coating, limiting its lifespan by causing premature failure. The composition of deposits on turbine blades coated with a

C. Levi—contributing editor

Manuscript No. 28705. Received September 30, 2010; approved January 14, 2011.
This work was supported by the Department of Energy, Office of Basic Energy Science, under contract number DE-AC02-06CH11357.
[†]Author to whom correspondence should be addressed. e-mail: k-faber@northwestern.edu
[∥]Present address: NASA Glenn Research Center, Cleveland, Ohio.
[††]Present address: University of Seville, Spain.

Table I. Thicknesses, Elastic Coefficients for the Crystallographic Planes Measured by Transmission Diffraction and Coefficient of Thermal Expansions (CTEs) of the Multilayer System[9]

Phase	Thickness (μm)	hkl	E_{hkl} (GPa)	ν_{hkl}	α ($10^{-6}\,^{\circ}C^{-1}$)
BSAS (Hexacelsian)	200	(110)	129	0.35	8.37
BSAS (Celsian)	200	(112)	81	0.30	4.28
Mullite	200	(220)	145	0.16	5.50
SAS	200	(041)	97	0.37	4.28
Silicon	125	(113)	187	0.15	4.44
Silicon carbide	≈ 3000	(220)	422	0.17	5.06

yttria-stabilized zirconia (YSZ) TBCs is mostly independent of geography, operating conditions or degree of exposure and consists of a calcium–magnesium aluminosilicate (CMAS) eutectic ($33CaO-9MgO-13AlO_{1.5}-45SiO_2$) with a low-melting point ($T_m = 1250^{\circ}C$).[31] The effects of CMAS attack on YSZ TBCs have been extensively studied and shown to be highly detrimental on TBC performance and lifetime,[32] as it can induce failure in multiple ways.[26] At high temperatures, CMAS glass can infiltrate pores and cracks and the CTE mismatch between the solid CMAS glass ($\approx 8.1 \times 10^{-6}\,^{\circ}C^{-1}$)[33] and the YSZ ($\approx 11.1 \times 10^{-6}\,^{\circ}C^{-1}$)[34] introduces large compressive residual stresses that can lead to delamination and cold-shock fracture.[33] Additionally, chemical interaction between CMAS and the TBC has been shown to reduce the coating's lifespan, as it destabilizes tetragonal YSZ.[35,36] The mechanism of this transformation involves the dissolution of the metastable tetragonal phase and the reprecipitation of ZrO_2 in the stable monoclinic phase. This mechanism is accompanied by a change in microstructure and a CTE mismatch with the substrate, resulting in a loss of strain tolerance and limiting service lifetimes.

BSAS EBCs showed similar CMAS corrosion problems to TBCs.[37] In freestanding plasma-sprayed specimens with low glass loading (≈ 10 mg/cm^2) and short hold times (1 h) at 1300°C, no substantial reaction of the glass with the coating was observed, but was evident at longer holding times (4 h). Higher loading (60 mg/cm^2) resulted in increased penetration and greater dissolution of BSAS into the molten glass. CMAS penetrated the grain boundaries of the EBC coating and dislodged the BSAS crystallites, which were dissolved into the glass and reprecipitated as a calcium-containing modified celsian phase. At low CMAS loading, precipitates of anorthite ($Ca_2Al_2Si_2O_8$) with Ba and Sr in solid solution and diopside ($CaMgSi_2O_6$) were also detected. The penetration depths observed after 4 h were on the order of typical plasma-sprayed EBC coating thicknesses, making CMAS corrosion a concern regarding lifetimes. However, it is important to note that although the CMAS was shown to penetrate over 200 μm, the loadings used to reach that depth (60 mg/cm^2) were extremely large, and may not be representative of CMAS exposure in typical engine environments.

In addition to altering the EBC composition, CMAS interaction is expected to have an effect on mechanical performance of these multilayer systems due to CTE mismatch effects. Although extensively studied for TBCs, no data exist in the literature for BSAS EBC systems. The aims of the present work were twofold. Firstly, we studied the effect of CMAS on the stress state of BSAS EBC multilayers by means of high-energy X-ray diffraction using a synchrotron source, a technique that has been shown to be well suited for this purpose as it allows stress measurement in multilayers as a function of depth with a spatial resolution of < 50 μm.[38] Not only can the average stress state in the different layers that compose the system be measured with this method but also local stress variations due to partial infiltration and reaction of CMAS can be elucidated. Also, because X-rays provide compositional information, this technique allows the quantification of hexacelsian/celsian fractions within

the BSAS topcoat with spatial resolution, to determine whether the presence of aliovalent cations from the glass affects the hexacelsian-to-celsian transformation kinetics. Secondly, we explored the effect of CMAS exposure on the microstructure and composition of BSAS for glass loadings closer to those commonly found in field tests[36] and for longer exposure times than those reported previously.

II. Experimental Methods

(1) Plasma-Spray Processing

Multilayer systems were deposited using a procedure described previously.[9] In brief, doped BSAS coatings were air plasma sprayed onto coupons (25.4 mm $\times$ 6.4 mm $\times$ 3.5 mm) of melt-infiltrated SiC/SiC ceramic matrix composite that were grit blasted with alumina to improve bonding to the substrate. A bond coat of silicon (Atlantic Equipment Engineers, Bergenfield, NJ) was applied first, followed by a layer of mullite+SAS from a mechanical mixture of 80 wt% mullite and 20 wt% SAS. In a final step, a topcoat of BSAS (HC Stark, Goslar, Germany) was applied. Details of the plasma spraying are proprietary, but are similar to those used by Lee in previous works.[12] The coatings were sprayed into a furnace where the substrate was heated to 1200°C in order to avoid depositing amorphous material that would cause catastrophic failure upon crystallization.[12,13]

Free-standing BSAS bars were plasma sprayed at room temperature for CMAS reaction studies using the procedure described in Harder and Faber.[24] BSAS powders (HC Stark) were sprayed onto grit-blasted steel substrates (55 mm $\times$ 25 mm $\times$ 3 mm), which were ultrasonically cleaned in isopropanol for approximately 5 min before deposition. The deposited material had a final thickness of ≈ 2.5 mm. After the deposition, the coated substrates were cooled at room temperature. Within 2–3 min, the BSAS delaminated cleanly from the substrate due to CTE mismatch between the steel and coating. The freestanding material was cut into 5 mm $\times$ 12 mm bars.

Following a procedure described previously,[24] the freestanding BSAS bars were heat treated at 990°C for 10 h to crystallize the amorphous, as-deposited BSAS into the metastable hexacelsian form. Subsequently, both the multilayer system and the freestanding samples were heat treated at 1300°C for 20 h to promote partial transformation of hexacelsian BSAS into the stable, low-CTE celsian phase, resembling the baseline or Gen 1 EBC as defined by Eaton et al.[39] and Spitsberg and Steibel.[40] The resulting coatings and freestanding bars were approximately 50 wt% celsian.

(2) CMAS Interaction Studies

The interaction of CMAS glass was investigated on both multilayer EBC coatings as well as freestanding BSAS bars. The glass was generated by mixing CaO (Aldrich, St. Louis, MO), MgO (Alfa Aesar, Ward Hill, MA), Al_2O_3 (Praxair, Danbury, CT), and SiO_2 (Alfa Aesar) in ratios consistent with the $33CaO-9MgO-13AlO_{1.5}-45SiO_2$ formula. The mixed powders were ball milled using Al_2O_3 media and water for 24 h. The mixture was dried and heated to 1550°C for 4 h in a Pt crucible to form a glass. The glass was ground with a mortar and pestle and fired a second time at 1550°C for 4 h to ensure a homogeneous mixture. Finally, the glass was removed from the crucible and ground to a fine powder (≤ 50 μm particle size).

Multilayer EBC samples (BSAS/mullite/Si/SiC) with a topcoat composition of ≈ 50 wt% celsian were exposed to the CMAS by mixing 0.5 g of glass with 0.75 g of ethanol to create a mixture that could be subsequently applied to the surface of the coatings until there was a dried loading of 35 mg/cm^2 on the sample, which is close to values usually found in field tests and follows previous studies by Aygun et al.[36] Upon application of the glass, samples were heated to 1300°C for 4, 12, 24, or 48 h with a temperature ramp rate of 10°C/min.

In addition to multilayer EBCs, freestanding BSAS coating material was exposed to CMAS glass. Wells were bored into the material using an ultrasonic drill press (Branson Ultrasonics, Danbury, CT) with a radius of 1.56 mm to a depth of ≈ 1.2 mm. This diameter and depth of the trough were chosen to expose the surface area of the BSAS to 35 mg/cm^2 CMAS when completely filled. After drilling, the freestanding material was also heated to 1300°C for 20 h to promote the hexacelsian-to-celsian transformation. The wells were filled with ≈ 10 mg of CMAS glass, and heat treated at 1300°C for 4, 12, 24, and 48 h with a 10°C/min ramp rate. This approach ensured a constant CMAS loading, in contrast with the multilayer samples where variations in CMAS layer thickness were unavoidable due to coating roughness.

(3) High-Energy X-Ray Diffraction

High-energy X-ray diffraction was used to characterize the crystal structures, phases, and stresses within the multilayer coatings. Experiments were conducted at the 1-ID beam line of the Advanced Photon Source at Argonne National Laboratory. The general setup for transmission diffraction is shown in Fig. 1. High-brilliance, high-energy (80–86 keV) X-rays were produced by a Laue monochromator and were vertically focused using Si refractive lenses. A Mar345 image plate (345 mm diameter, 150 μm pixel size) was set at a distance of 1.2 m from the sample, and a CeO$_2$ standard was used to determine the sample-to-detector distance, beam tilt, and beam center.

Strains were measured in the multilayer EBCs as a function of depth at room temperature. The X-ray beam size was 50 μm × 100 μm, and during each exposure the sample was translated horizontally 2 mm to increase the number of sampled grains, resulting in a total diffracted volume of 0.35 mm.^{3} Samples were moved vertically with respect to the beam in 25 μm steps to probe the layers and the substrate.

To assess internal and residual stresses, strains were determined from the distortion of the Debye rings using a series of MATLAB programs developed at Sector 1 of the Advanced Photon Source at Argonne National Laboratory. The radii of diffraction planes for the materials were determined using the unit cell dimensions specified by powder diffraction files 00-026-0182 (hexacelsian BSAS), 38-1451 (celsian BSAS), 00-015-0776 (mullite), 38-1454 (celsian SAS), 00-027-1402 (silicon), and 00-029-1129 (β-silicon carbide).[41] Planes were chosen so as not to overlap with any other phases or materials. The reflections used to calculate strains present in the coating materials are shown in Table I. Diffraction peaks were fit to pseudo-Voigt functions to determine the peak centers, which were mapped as a function of azimuth to determine the strain calculated by the equation:

$$\varepsilon_\eta = \frac{r_0 - r_\eta}{r_0} \tag{1}$$

where r_η is the radius at azimuthal angle η, and r_0 is the strain-free radius, which is the measured radius at the strain-free angle (η^*). The latter is calculated from the X-ray elastic constants included in Table I using the following expression, valid under the assumption of a purely biaxial stress state[38]:

$$\sin^2 \eta^* = \frac{2\nu_{hkl}}{1 + \nu_{hkl}} \tag{2}$$

Error in the strain measurements were computed based on statistical error from peak fitting and standard errors from the biaxial fit of lattice parameter versus azimuth. The strains in the in-plane (ε_{11}) and the out-of-plane (ε_{33}) directions, corresponding to $\eta = 0°$ and $\eta = 90°$, were used to determine the in-plane equibiaxial stress ($\sigma_\parallel$) by[42]

$$\sigma_\parallel = \left(\frac{E}{1 + \nu}\right)\varepsilon_{11} + \frac{\nu E}{(1 - 2\nu)(1 + \nu)}(2\varepsilon_{11} + \varepsilon_{33}) \tag{3}$$

The X-ray elastic constants, E_{hkl}, ν_{hkl}, and the CTE, α, included in Table I were measured *in situ* at the synchrotron beam line; details of the measurement technique were published in a previous work.[9] In brief, α was determined by measuring changes in *d*-spacing as a function of temperature in crushed, plasma-sprayed BSAS, SAS, silicon, and mullite, as well as in the SiC/SiC coupon. Elastic constants E_{hkl} and ν_{hkl} were determined by *in situ* loading experiments on freestanding plasma-sprayed bars in compression, where Eq. (1) was used to determine strains parallel and perpendicular to the loading direction, as a function of applied stress. Elastic constants for all materials comprising the multilayers were determined in this manner, except for the SiC/SiC substrate, where an average of the Voigt and Reuss models from single crystal elastic constants[43] was used.

The ratio of celsian-to-hexacelsian phases in wt% in the EBCs was determined from the acquired X-ray diffraction patterns as a function of depth and exposure time to CMAS. The 2D patterns were radially integrated to obtain profiles of diffracted intensity versus 2θ that were then fit using the Rietveld method as implemented in the EXPGUI/GSAS package.[44,45] Starting structures were obtained from the ICSD database (references 97411–hexacelsian BSAS, 40530–celsian BSAS, 78793–SAS[46]). Initial values for diffractometer geometry and peak profiles were obtained by refining a pattern from the CeO$_2$ standard. These parameters were later refined along with lattice parameters and phase fractions in patterns from the BSAS topcoat.

(4) Microstructure Characterization

Representative cross sections of the multilayers before and after CMAS interaction at 1300°C for different times were prepared using conventional metallographic procedures for observation in the scanning electron microscope (SEM S3400, Hitachi High Technologies, Tokyo, Japan) using backscattered electrons (BSE) for compositional contrast. Sections of the freestanding bars were also prepared in this manner. From these samples, thin lamellae in areas representative of the observed features were prepared using a dual-beam focused ion beam/SEM (Helios NanoLab, FEI, Hillsboro, OR) and mounted onto copper grids using a micromanipulator (Autoprobe 200, Omniprobe, Dallas, TX). These lamellae were then observed under the transmission electron microscope (TEM, Jeol 2100F) in both conventional and scanning modes. X-ray energy-dispersive spectra and maps were acquired, and selected area diffraction (SAD) patterns were obtained for identification of precipitated phases in the BSAS/CMAS interaction zone. ICSD structure references used for indexing of SAD patterns were those mentioned in Section III *(3)* and 78793 (anorthite).

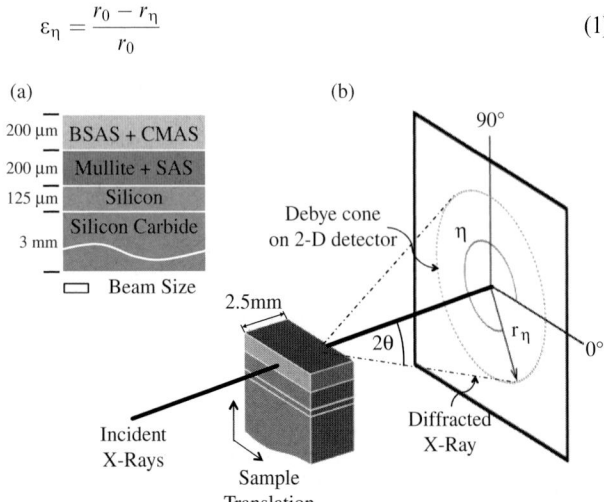

Fig. 1. Schematic for the transmission high-energy X-ray diffraction experiments. (a) To-scale sketch of the multilayer environmental barrier coating studied, showing typical coating thicknesses and diffraction beam size. (b) Experimental setup at beamline 1-ID showing the diffraction geometry used.

3. Results and Discussion

(1) Microstructure of As-Deposited Multilayer EBCs

A representative cross section of the microstructure of plasma-sprayed EBCs multilayers studied in this work is shown in Fig. 2, for a sample heat treated at 1300°C for 20 h. The roughness of the interfaces, due to the plasma-spray process, is easily seen. The average thickness of the layers were close to those reported in Table I although, due to the mentioned roughness, thickness varied as much as ± 50 μm in some parts of the coating. The topcoat layer corresponds to BSAS, which contains a significant amount of porosity. The contrast that can be seen is due to an inhomogeneous distribution of Ba and Sr, changing the local Ba/Sr ratio in adjacent grains. The intermediate layer consists of a mixture of mullite (dark contrast) and 20 wt% SAS. The silicon bondcoat serves to enhance adhesion between the intermediate layer and the SiC/SiC substrate, which can be seen in the bottom part of Fig. 2. In the composite, the SiC fibers are woven and run parallel to the layer interfaces both along and normal to the micrograph.

(2) CMAS Exposure Effects on Multilayer BSAS EBCs

Stress as a function of depth, determined by X-ray diffraction, is shown in Fig. 3 for a multilayer sample that had been exposed to 1300°C for 20 h. The stresses observed in all layers are consistent with those reported previously in this system.[9] The variation of stress was likely due to lack of uniformity in the coating thicknesses as well as the presence of pores.

For samples exposed to CMAS, stresses were measured in each layer of the coating as well as the SiC substrate, but the differences in the stress magnitude of these layers were nearly negligible over the course of the CMAS exposure. The BSAS topcoat in the current study was most affected by the exposure

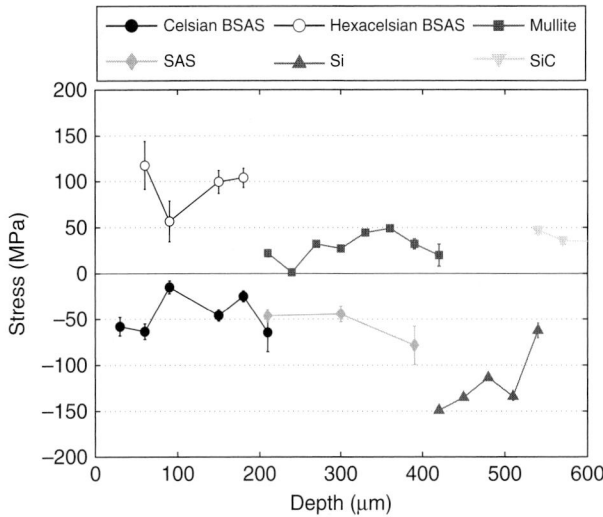

Fig. 3. Biaxial residual stress profile as a function of depth in the BSAS multilayer EBC system after heat treatment at 1300°C for 20 h, measured by transmission X-ray diffraction.

to CMAS. In Fig. 4, the stress in the BSAS topcoat is plotted as a function of depth in each of the samples. The stress at the surface of the BSAS became increasingly compressive with CMAS exposure time. The elevated stress was a result of the interaction of the glass with the BSAS and not the heat treatment, as it was reported that samples of this EBC system that were thermally cycled to 1300°C for 100 h exhibited a constant compressive stress of ≈30 MPa across the coating layer.[9] The elevated compressive stress at the surface of samples exposed to CMAS was likely caused by the CTE of the residual glass (or byproducts that were formed from the interaction) to be larger than the topcoat. The CTE of amorphous CMAS was estimated by Chen to be 8.1×10^{-6}°C^{-1}.[33] This is similar to the CTE of hexacelsian BSAS (8.37×10^{-6} °C^{-1}), but much larger than the celsian BSAS (4.28×10^{-6} °C^{-1}).

The microstructure of the BSAS topcoat also changed with exposure to the CMAS. The surface was roughened considerably during the hexacelsian-to-celsian transformation due to the more faceted nature of the precipitated celsian grains. Additionally, the glass incorporated some of the BSAS material in the form of precipitates, as shown in Fig. 5. At 4 h, a significant number of elongated precipitates appear to grow from the BSAS/glass interface with varied levels of BSE contrast, suggesting a nonhomogeneous composition. A different, smaller type of precipitate can be seen with dark BSE contrast at the

Fig. 2. Cross section of the BSAS multilayer EBC system on an SiC/SiC melt infiltrated composite, after heat treatment at 1300°C for 20 h.

Fig. 4. Biaxial residual stress profiles as a function of depth in the BSAS topcoat after exposure to CMAS at 1300°C for 0, 4, 24, and 48 h, measured by transmission X-ray diffraction.

Fig. 5. Microstructure of the BSAS topcoat in a multilayer EBC system after exposure to CMAS glass at 1300°C for different times. From left to right and top to bottom: 0, 4, 24, and 48 h.

BSAS/glass interface. EDS was used to determine that these precipitates contain Ba, Sr, Al, Ca, and trace amounts of Mg. One explanation for their presence is that the glass attacks the BSAS coating and incorporates it into the molten glass, which crystallizes from the melt as it cools. Similar precipitates were observed in TEM studies by Grant *et al.*[37] who identified them as celsian, anorthite, or diopside. The BSAS coating also showed a "reaction zone" at the interface with the glass. This region exhibited a more faceted grain and pore structure, in addition to a more homogeneous contrast in the BSE images. Reaction time did not affect the morphology of these precipitates, but longer exposure times resulted in fewer and smaller precipitates, due to the small amount of unincorporated glass remaining and longer holding times. Indeed, the 48 h sample exhibited much less residual glass on the surface, suggesting that the CMAS was incorporated into the coating during the exposure.

To investigate the effect of the CMAS glass on the crystal structure of BSAS, the weight percents of celsian and hexacelsian phases in the topcoat the multilayer samples were determined by Rietveld fitting of 1D patterns calculated by integration of Debye ring radii from $\eta = 0$–$360°$. The weight percent celsian was determined as a function of depth in approximately 25 μm increments from the glass on the surface to the interface with mullite layer and shown in Fig. 6. The results of these phase fraction measurements were compared with published values for samples that underwent identical heat treatments without CMAS exposure,[24] represented as horizontal lines. The celsian content is increased at the top of the coating and at the BSAS/mullite interface, and lower in the center of the layer, suggesting that this transformation is favored at the surfaces and interfaces. This was also observed in literature results, as BAS and BSAS were shown to transform at free surfaces and at interfaces before the center of the bulk.[47,48] BSAS material that was exposed to CMAS demonstrated an accelerated transformation rate. The effect was pronounced near the surface of the sample exposed for 4 h. The weight percent celsian at the surface was $\approx 25\%$ higher than the expected average value, and was much higher than the remainder of the sample, which was close to the condition without the glass. The microstructure results in Fig. 5 also support this, as faceted grain and pore structure in the reaction zone more closely resemble the celsian BSAS. In the sample exposed for 24 h, the celsian content is elevated over a depth of ≈ 125 μm into the sample, which suggested deeper CMAS infiltration than the 4 h sample. After 48 h, the topcoat shows a constant celsian content close to 100%, which was $\approx 10\%$ higher than the heat treating alone would have provided. In these CMAS exposures, the data suggest that the exposure of the glass accelerated the hexacelsian-to-celsian

transformation in BSAS. Several mechanisms or combinations thereof could be responsible for the enhancement of this transformation rate. In YSZ-based TBC coatings, CMAS glass attacks by a dissolution–reprecipitation mechanism. Metastable tetragonal YSZ is dissolved into solution, and upon cooling, is reprecipitated as the stable monoclinic phase with an equiaxed microstructure.[35,36] If BSAS is attacked by CMAS in a similar way, the metastable (hexacelsian) and stable (celsian) material would be dissolved into the glass, and would reprecipitate as the stable celsian crystal structure. This mechanism was suggested by Grant *et al.*[37] from results from TEM studies on CMAS-reacted BSAS, and corroborated here. BSAS was dissolved into the CMAS glass and reprecipitated as a modified phase with the incorporation of Ca and a different Ba: Sr ratio.

Another mechanism that could accelerate the transformation rate of hexacelsian to celsian is the incorporation of Mg and Ca cations into the hexacelsian lattice. There is evidence in the literature that the integration of some additives to BAS material accelerates the conversion from hexacelsian to celsian.[49,50] The incorporation of these smaller cations, for which CMAS may act as a source, may reduce the transformation energy by sub-

Fig. 6. Volume fraction of celsian phase in the BSAS topcoat of the studied multilayer EBC system after high-temperature exposure to CMAS at 1300°C for 0, 4, 24, and 48 h. Symbols represent experimental values determined from Rietveld refinements of diffraction patterns obtained by high-energy X-ray diffraction. Horizontal lines correspond to the average measured fraction of celsian phase in BSAS samples that underwent identical heat treatments without CMAS exposure.[24]

stitutionally doping for the Ba and Sr ions in the BSAS. The doped lattice would lower the activation energy and enhance the transformation rate. In any case, such a mineralizing effect will be convoluted with the main solution–reprecipitation mechanism.

(3) CMAS Exposure Effects on Freestanding BSAS

Freestanding BSAS samples were exposed to CMAS at 1300°C for 4, 12, 24, and 48 h, and the microstructures of the bottom of the wells are shown in Fig. 7. The sample exposed for 4 h contained a large amount of residual glass on the surface, but heating for 12 h reduced the remaining glass considerably, with almost none remaining after 24 or 48 h. The BSE images show a reaction zone that extends into the BSAS similar to what was seen in the coatings in Fig. 5, but the penetration depth was greater due to the increased crack density of the bulk BSAS samples, which had a greater number of pathways than the multilayer coatings for the glass to infiltrate. The grain and pore structures in the reaction zone were also more faceted than the underlying material. In the 4 h sample, the reaction zone was small, on the order of ≈ 10 μm from the CMAS/BSAS interface. After 12 h, the reaction zone was ≈ 30 μm, and after 24 h the reaction zone grew to ≈ 50 μm. Heating for 48 h did not extend the zone any further than the 24 h sample due to the complete incorporation of the glass after 24 h.

TEM observations confirmed that the precipitates observed at the early stages of CMAS/BSAS interaction had a modified celsian structure that differed slightly in lattice parameters but showed essentially the same SAD patterns as the celsian present in the as-sprayed bars. Figure 8 shows an STEM image of a representative precipitate growing into the glass after 4 h of CMAS exposure, accompanied by EDS compositional maps and an SAD pattern. The diffraction pattern could only be indexed as a celsian structure. Although interplanar angles and spacings calculated from SAD patterns differed slightly from the BSAS celsian phase, they were not compatible with other possible structures such as anorthite or diopside. The EDS maps additionally show the presence of Ba- and Ca-rich regions in the precipitate, incorporating little or no Mg. Strontium is found not only in the precipitate but also in the glass, suggesting that Ca could be substituting Sr in the precipitates. Barium was not observed in the glass near the precipitate, suggesting that BSAS

is dissolved into the glass and reprecipitated as a Ca-doped BAS phase while the Sr remains in the glass.

Precipitates with SAD patterns consistent with anorthite were found to be present in very small amounts at the BSAS/CMAS interface (seen in Fig. 7 as dark, faceted phases), and EDS measurements showed that some Ba was present in the structure. This observation is again consistent with precipitates observed in previous works at higher CMAS loads,[37] although, contrary to that case, no diopside was found. The absence of diopside could be related to the amount of CMAS available for reaction. Combining our results with the observations of Grant and colleagues, we can conclude that anorthite, diopside and celsian precipitate from molten CMAS glass at loadings between 10 to 35 mg/cm^2. At loadings between 35 and 60 mg/cm^2, celsian and anorthite are present as precipitates, while for CMAS loadings of 60 mg/cm^2 and higher, only the celsian phase is found.

IV. Summary

We have studied exposure effects of CMAS deposits at 1300°C on both the composition and residual stresses of plasma-sprayed BSAS multilayer EBCs, for different exposure times. CMAS reacts with the topcoat dissolving the BSAS topcoat preferentially along the grain boundaries and dislodging the crystallites, roughening the BSAS–CMAS interface. Owing to the higher CTE of the CMAS glass, the result is a composite layer where the residual stress changes from uniform compression to an increasingly compressive stress gradient reaching values as high as ≈ -160 MPa after 48 h of exposure to 35 mg/cm^2 of CMAS. The residual stresses in the underlying layers and SiC/SiC substrates were unaffected by CMAS exposure. The average composition of the topcoat was evaluated by Rietveld fitting, and it was found that the presence of aliovalent atoms such as Ca and Mg accelerated the hexacelsian-to-celsian phase transformation in BSAS. CMAS reacts with BSAS following a solution–reprecipitation mechanism by which BSAS grains are dissolved into the molten glass. Ba and Sr atoms are diffused away from the CMAS/BSAS interface where they precipitate as a modified celsian phase that contains Ca in solid solution. No diopside precipitates were found for loadings of 35 mg/cm^2, but a small number of anorthite grains were identified using electron diffraction in the TEM.

Fig. 7. Microstructure in a freestanding plasma-sprayed BSAS bar after exposure to CMAS glass at 1300°C for different times. From left to right and top to bottom: 4, 12, 24, and 48 h.

Fig. 8. Composition EDS maps from a modified celsian precipitate observed after 4 h of CMAS exposure, STEM image of the studied area, and selected area electron diffraction pattern from the precipitate, with indexing consistent with a celsian structure.

Acknowledgments

Microscopy images were collected at the NUANCE Center at Northwestern University, which is supported by the MRSEC program of the National Science Foundation (DMR-0520513). J. Ramirez-Rico was partially funded by the Spanish Ministry of Science and Technology under grant MAT2007-30141-E.

References

[1]N. S. Jacobson, "Corrosion of Silicon-Based Ceramics in Combustion Environments," *J. Am. Ceram. Soc.*, **76** [1] 3–28 (1993).

[2]K. N. Lee, "Current Status of Environmental Barrier Coatings for Si-Based Ceramics," *Surf. Coat. Technol.*, **133**, 1–7 (2000).

[3]E. J. Opila and R. E. Hann, "Paralinear Oxidation of CVD SiC in Water Vapor," *J. Am. Ceram. Soc.*, **80** [1] 197–205 (1997).

[4]E. J. Opila, "Variation of the Oxidation Rate of Silicon Carbide with Water-Vapor Pressure," *J. Am. Ceram. Soc.*, **82** [3] 625–36 (1999).

[5]E. J. Opila, J. L. Smialek, R. C. Robinson, D. S. Fox, and N. S. Jacobson, "SiC Recession Caused by SiO₂ Scale Volatility Under Combustion Conditions: Ii, Thermodynamics and Gaseous-Diffusion Model," *J. Am. Ceram. Soc.*, **82** [7] 1826–34 (1999).

[6]R. C. Robinson and J. L. Smialek, "SiC Recession Caused by SiO Scale Volatility Under Combustion Conditions: I, Experimental Results and Empirical Model," *J. Am. Ceram. Soc.*, **82** [7] 1817–25 (1999).

[7]J. L. Smialek, R. C. Robinson, E. J. Opila, D. S. Fox, and N. S. Jacobson, "SiC and Si₃N₄ Recession Due to SiO₂ Scale Volatility Under Combustor Conditions," *Adv. Compos. Mater.*, **8** [1] 33–45 (1999).

[8]K. T. Faber, C. M. Weyant, B. Harder, J. Almer, and K. Lee, "Internal Stresses and Phase Stability in Multiphase Environmental Barrier Coatings," *Int. J. Mater. Res.*, **98** [12] 1188–95 (2007).

[9]B. J. Harder, J. Almer, C. M. Weyant, K. N. Lee, and K. T. Faber, "Residual Stress Analysis of Multilayer Environmental Barrier Coatings," *J. Am. Ceram. Soc.*, **92** [2] 452–9 (2009).

[10]B. J. Harder, J. Almer, K. N. Lee, and K. T. Faber, "*In Situ* Stress Analysis of Multilayer Environmental Barrier Coatings," *Powder Diffr.*, **24** [2] 94–8 (2009).

[11]K. N. Lee, D. S. Fox, and N. P. Bansal, "Rare Earth Silicate Environmental Barrier Coatings for SiC/SiC Composites and Si₃N₄ Ceramics," *J. Eur. Ceram. Soc.*, **25** [10] 1705–15 (2005).

[12]K. N. Lee, R. A. Miller, and N. S. Jacobson, "New-Generation of Plasma-Sprayed Mullite Coatings on Silicon-Carbide," *J. Am. Ceram. Soc.*, **78** [3] 705–10 (1995).

[13]K. N. Lee and R. A. Miller, "Development and Environmental Durability of Mullite and Mullite/YSZ Dual Layer Coatings for SiC and Si₃N₄ Ceramics," *Surf. Coat. Technol.*, **86** [1–3] 142–8 (1996).

[14]M. Moldovan, C. M. Weyant, D. L. Johnson, and K. T. Faber, "Tantalum Oxide Coatings as Candidate Environmental Barriers," *J. Therm. Spray Technol.*, **13** [1] 51–6 (2004).

[15]C. M. Weyant, K. T. Faber, J. D. Almer, and J. V. Guiheen, "Residual Stress and Microstructural Evolution in Tantalum Oxide Coatings on Silicon Nitride," *J. Am. Ceram. Soc.*, **88** [8] 2169–76 (2005).

[16]C. M. Weyant, K. T. Faber, J. D. Almer, and J. V. Guiheen, "Residual Stress and Microstructural Evolution in Environmental Barrier Coatings of Tantalum Oxide Alloyed with Aluminum Oxide and Lanthanum Oxide," *J. Am. Ceram. Soc.*, **89** [3] 971–8 (2006).

[17]K. N. Lee, J. I. Eldridge, and R. C. Robinson, "Residual Stresses and Their Effects on the Durability of Environmental Barrier Coatings for SiC Ceramics," *J. Am. Ceram. Soc.*, **88** [12] 3483–8 (2005).

[18]D. Bahat, "Kinetic Study of Hexacelsian–Celsian Phase Transformation," *J. Mater. Sci.*, **5** [9] 805–10 (1970).

[19]N. P. Bansal and M. J. Hyatt, "Crystallization Kinetics of BaO–Al₂O₃–SiO₂ Glasses," *J. Mater. Res.*, **4** [5] 1257–65 (1989).

[20]N. P. Bansal, "Solid State Synthesis and Properties of Monoclinic Celsian," *J. Mater. Sci.*, **33** [19] 4711–5 (1998).

[21]N. P. Bansal and C. H. Drummond, "Kinetic-Study on the Hexacelsian–Celsian Phase-Transformation—Comment," *J. Mater. Sci. Lett.*, **13** [6] 423–4 (1994).

[22]N. P. Bansal and C. H. Drummond, "Kinetics of Hexacelsian-To-Celsian Phase Transformation in SrAl₂Si₂O₈," *J. Am. Ceram. Soc.*, **76** [5] 1321–4 (1993).

[23]K. T. Lee and P. B. Aswath, "Synthesis of Hexacelsian Barium Aluminosilicate by a Solid-State Process," *J. Am. Ceram. Soc.*, **83** [12] 2907–12 (2000).

[24]B. J. Harder and K. T. Faber, "Transformation Kinetics in Plasma-Sprayed Barium- and Strontium-Doped Aluminosilicate (BSAS)," *Scr. Mater.*, **62** [5] 282–5 (2010).

[25]A. G. Evans, D. R. Clarke, and C. G. Levi, "The Influence of Oxides on the Performance of Advanced Gas Turbines," *J. Eur. Ceram. Soc.*, **28** [7] 1405–19 (2008).

[26]S. Krämer, S. Faulhaber, M. Chambers, D. R. Clarke, C. G. Levi, J. W. Hutchinson, and A. G. Evans, "Mechanisms of Cracking and Delamination within Thick Thermal Barrier Systems in Aero-Engines Subject to Calcium–Magnesium–Alumino-Silicate (CMAS) Penetration," *Mater. Sci. Eng. A—Struct.*, **490** [1–2] 26–35 (2008).

[27]J. Kim, M. G. Dunn, A. J. Baran, D. P. Wade, and E. L. Tremba, "Deposition of Volcanic Materials in the Hot Sections of Two Gas-Turbine Engines," *J. Eng. Gas. Turb. Power*, **115** [3] 641–5 (1993).

[28]J. L. Smialek, F. A. Archer, and R. G. Garlick, "Turbine Airfoil Degradation in the Persian-Gulf-War," *J. Min. Met. Mater. S*, **46** [12] 39–41 (1994).

[29]D. J. Dewet, R. Taylor, and F. H. Stott, "Corrosion Mechanisms of ZrO₂–Y₂O₃ Thermal Barrier Coatings in the Presence of Molten Middle-East Sand," *J. Phys. IV*, **3** [C9] 655–63 (1993).

[30]F. Stott, D. J. Dewet, and R. Taylor, "Degradation of Thermal-Barrier Coatings at Very High-Temperatures," *MRS Bull.*, **19** [10] 46–9 (1994).

[31]M. P. Borom, C. A. Johnson, and L. A. Peluso, "Role of Environmental Deposits and Operating Surface Temperature in Spallation of Air Plasma Sprayed Thermal Barrier Coatings," *Surf. Coat. Technol.*, **86** [1–3] 116–26 (1996).

[32]C. Mercer, S. Faulhaber, A. G. Evans, and R. Darolia, "A Delamination Mechanism for Thermal Barrier Coatings Subject to Calcium–Magnesium–Alumino-Silicate (CMAS) Infiltration," *Acta Mater.*, **53** [4] 1029–39 (2005).

[33]X. Chen, "Calcium–Magnesium–Alumina–Silicate (CMAS) Delamination Mechanisms in EB-PVD Thermal Barrier Coatings," *Surf. Coat. Technol.*, **200** [11] 3418–27 (2006).

[34]A. G. Evans, D. R. Mumm, J. W. Hutchinson, G. H. Meier, and F. S. Pettit, "Mechanisms Controlling the Durability of Thermal Barrier Coatings," *Prog. Mater. Sci.*, **46** [5] 505–53 (2001).

[35]S. Krämer, J. Yang, C. G. Levi, and C. A. Johnson, "Thermochemical Interaction of Thermal Barrier Coatings with Molten CaO–MgO–Al₂O₃–SiO₂ (CMAS) Deposits," *J. Am. Ceram. Soc.*, **89** [10] 3167–75 (2006).

[36]A. Aygun, A. L. Vasiliev, N. P. Padture, and X. Ma, "Novel Thermal Barrier Coatings that are Resistant to High-Temperature Attack by Glassy Deposits," *Acta Mater.*, **55** [20] 6734–45 (2007).

[37]K. M. Grant, S. Krämer, J. P. A. Lofvander, and C. G. Levi, "CMAS Degradation of Environmental Barrier Coatings," *Surf. Coat. Technol.*, **202** [4–7] 653–7 (2007).

[38]J. Almer, U. Lienert, R. L. Peng, C. Schlauer, and M. Oden, "Strain and Texture Analysis of Coatings Using High-Energy X-Rays," *J. Appl. Phys.*, **94** [1] 697–702 (2003).

[39]H. E. Eaton, W. P. Allen, N. S. Jacobson, N. P. Bansal, E. J. Opila, J. L. Smialek, K. N. Lee, I. T. Spitsberg, H. Wang, P. J. Meschter, and K. L. Luthra, "Article Having Silicon-Containing Substrate and Barrier Layer and Production Thereof"; US Patent 6,387,456, 2001.

[40]I. Spitsberg and J. Steibel, "Thermal and Environmental Barrier Coatings for SiC/SiC CMCs in Aircraft Engine Applications," *Int. J. Appl. Ceram. Technol.*, **1** [4] 291–301 (2004).

[41]International Centre for Diffraction Data, Newton Square, PA, JCPDS, 2008.

[42]I. C. Noyan and J. B. Cohen, *Residual Stress: Measurement by Diffraction and Interpretation*. Springer-Verlag, New York, 1986.

[43]K. H. Hellwege (ed.), *Elastic, Piezoelectric and Related Constants of Crystals*, Vol. 11. Springer-Verlag, Berlin, 1979.

[44]A. C. Larson and R. B. Von-Dreele, "General Structure Analysis System (GSAS)"; Los Alamos National Laboratory Report, LAUR 86-748, Los Alamos, NM, 2000.

[45]B. H. Toby, "EXPGUI, a Graphical User Interface for GSAS," *J. Appl. Crystallogr.*, **34**, 210–3 (2001).

[46] International Centre for Structure Data, Karlsruhe, ICSD, 2010.

[47]J. I. Eldridge and K. N. Lee, "Phase Evolution of BSAS in Environmental Barrier Coatings," *Cer. Eng. Sci. Proc.*, **22** [4] 383–90 (2001).

[48]M. J. Hyatt and N. P. Bansal, "Crystal Growth Kinetics in $BaO \cdot Al_2O_3 \cdot 2SiO_2$ and $SrO \cdot Al_2O_3 \cdot 2SiO_2$ Glasses," *J. Mater. Sci.*, **31** [1] 172–84 (1996).

[49]K. T. Lee and P. B. Aswath, "Kinetics of the Hexacelsian to Celsian Transformation in Barium Aluminosilicates Doped with CaO," *Int. J. Inorg. Mater.*, **3** [7] 687–92 (2001).

[50]K. T. Lee and P. B. Aswath, "Role of Mineralizers on the Hexacelsian to Celsian Transformation in the Barium Aluminosilicate (BAS) System," *Mat. Sci. Eng. A—Struct.*, **352** [1–2] 1–7 (2003).   □

1.87 mL/min for 12 h; and 1 mL/min for 20 h. After the experiments, the depths of erosion were measured by profilometry and optical interferometry.

III. Results

(1) Temperature Calibration

Results of the temperature calibration are summarized in Fig. 2. In the set-up used in these experiments, the length of capillary tube in the hot zone of the furnace is not sufficient to allow equilibration of the temperatures of the water vapor and the furnace atmosphere, so that the temperature of the vapor jet is lower than the furnace temperature by an amount that is sensitive to the flow rate. Moreover, because this temperature difference increases with increasing flow rate, the velocity of the vapor jet at given furnace temperature does not increase monotonically with the water flow rate (Fig. 2(b)).

An estimate of the position at which the liquid water vaporized could usually be obtained from a slight discoloration visible on the inside of the capillary tube after the test run (from trace amounts of contaminants in the water). The distances between this position and the exit of the capillary were approximately 12 and 17 cm for flow rates of 1.0 and 1.87 mL/min, corresponding to a residence time of the accelerating gas in the capillary of the order of 10^{-3} s.

(2) Recession Measurements and Observations

A sharp wedge-shaped SiC test sample was observed as shown in Fig. 3(a), after exposure for 20 h in a jet of mean velocity 260 m/s and temperature 1160°C (furnace temperature 1500°C). Severe recession occurred both in the position of the leading edge (originally a straight line) and in grooves along the initially flat polished faces of the wedge. The positions of the deepest grooves coincide with flow at the periphery of the water vapor jet.

From the profilometer data shown in Fig. 3(b), the recession depth in the central region of the jet flow is $\sim 9\ \mu m$, and is roughly constant within a distance of ~ 4 mm from the leading edge. The grooves at the edge of the flow are deeper by a factor of 3 or more near the leading edge (~ 27 and 30 μm) and become shallower with increasing distance downstream. The formation of grooves of this depth on both sides of the wedge-shaped test samples would cause the leading edge of the wedge to recede by ~ 500 μm. This is close to the observed recession of the leading edge in Fig. 3(c), thus indicating that material removal in this region is predominantly from the flat faces of the wedge rather than from the stagnation point at the leading edge itself. This conclusion is further supported by the observation that the leading edge remains sharp in the grooves.

Examination by SEM and EDS analysis indicated that there was a continuous SiO_2 layer outside the region of jet flow. However, there was no detectable evidence for the oxide layer within the deep grooves, while in the central region of the jet flow, discontinuous patches of thin oxide were detected. Elemental maps of silicon, oxygen, carbon, and aluminum from the region of Fig. 3(c) are shown in Fig. 4 (note the absence of oxygen along the grooves, reduced oxygen signal at the central region of jet flow, and complementary regions of high and low oxygen and carbon signals). The presence of a small concentration of aluminum outside the region of the jet flow is indicative of formation of volatile aluminum oxy-hydroxides[19–21] by the reaction of water vapor in the surrounding gas with the alumina furnace hardware. There was no evidence of Al within the region exposed to the jet.

Independent measurements of recession depths and oxide thickness were obtained using optical interference microscopy. Three sets of fringes are visible in the white light interference micrograph shown in Fig. 3(d): one set is from interference between reflections from the microscope reference mirror and the top surface (oxide) of the test specimen; a second set from interference between the reference mirror and the SiO_2–SiC interface; and the third set from interference between the top of the oxide layer and the oxide–SiC interface. For a flat surface (i.e., before exposure to the water vapor jet), the first two sets of fringes would be straight and horizontal in Fig. 3(d). The downward displacement of these fringes, visible in Fig. 3(d), is proportional to the recession depth of the surface, while the separation of the two sets of fringes is proportional to the thickness of the oxide layer. The near-vertical fringes from interference between the top and bottom of the oxide layer also give a measure of the layer thickness. Measurements from these fringes as well as measurements from interference micrographs obtained using monochromatic light indicated that the thickness of the oxide layer is ~ 4 μm outside the region impacted by the jet (right side of the micrograph) and the thickness decreases continuously toward the center of the groove, where the oxide thickness is below the detection limit of the interference measurements (~ 30 nm). In the central region of the jet flow, a thin oxide layer (thickness < 200 nm) was detected in patches separated by regions where no oxide was detected.

Higher magnification views of the specimen in Fig. 3 are shown in Fig. 5. Large differences in the surface texture are evident in various regions: the region outside the vapor jet, which is covered with a layer of glassy SiO_2, is relatively smooth, whereas the region within the grooves is much rougher and pitted, with no evidence of glassy SiO_2 detected either from the appearance of the SEM image or from EDS analysis. In the central region of the vapor jet, a mixture of surface textures are evident, circular depressions with rough pitted surfaces surrounded by a smoother thin film of glassy SiO_2. Many of the circular depressions showed radial texture that might suggest an association with cristobalite spherulites.[35,36] However, no evidence of cristobalite crystals was seen in the pitted surfaces.

Fig. 2.   (a) Temperature calibration curves for various furnace temperatures and water flow rates. (b) Vapor jet temperatures and velocities for various furnace temperatures and water flow rates. Speed of sound in water vapor, v_s, indicated for comparison.

Fig. 3. SiC test specimen after exposure for 20 h in water vapor jet with velocity 260 m/s, temperature 1160°C. (a) Before removing from holder used during test: grooves coincide with flow of the edge of the jet. (b) Erosion depths measured from profilometer data. (c) SEM image with superimposed profilometer trace along broken line. (d) Optical interference micrograph (white light) from area indicated in (c) showing three sets of interference fringes.

Erosion patterns similar to Fig. 3 were observed under several conditions of jet velocity and temperature, although the relative depths of the central region and the edge grooves varied with conditions. At lower velocity (160 m/s) and higher temperature (1350°C), the side grooves were not as distinct, while at high velocity their depth increased with velocity and exposure time. However, in all cases, regions similar to Fig. 5(a) were observed in the side grooves, with rough, pitted surfaces and no evidence of glassy SiO_2 detectable by SEM imaging or EDS analysis. Similar rough pitted surfaces were also observed in the tests in which the jet was incident on a flat polished surface inclined at an angle of 45° to the flow. An example is shown in Fig. 6(a).

IV. Discussion

(1) Flow Conditions in Water Vapor Jet

The main characteristics of the flow conditions in the water vapor jet can be deduced from standard fluid dynamics relations. Within the capillary, the flow is characterized by the Reynolds number[37]:

$$Re_d = \frac{\rho \, d \, v}{\mu} \qquad (6)$$

where v, ρ, and μ are the velocity, density, and viscosity of the fluid, and d is the capillary diameter. The flow is laminar for low values of Re_d, with the onset of turbulence occurring at $Re_d = 2300$ and fully developed turbulent flow at $Re_d > 10\,000$. For the range of conditions of the water vapor jet at temperatures up to $\sim 1400°C$ ($d = 1$ mm, $\rho \sim 1.3 \times 10^{-4}$ g/cc, $\mu \sim 5 \times 10^{-4}$ g/cm·s, and $v < 300$ m/s), Eq. (6) gives $Re_d < 1000$. Therefore, the flow at the exit of the capillary is laminar in all of the experiments.

The fluid velocity for laminar flow in a tube varies across the diameter of the tube, approaching a steady-state parabolic profile after flowing a distance greater than the hydrodynamic entry length, x_o. For a fluid injected into the tube with uniform velocity profile and flowing with uniform mean velocity, x_o is given by [32]

$$\frac{x_o}{d} \approx 0.05 Re_d \qquad (7)$$

With $d = 1$ mm and $Re_d < 1000$, as in the present experiments, Eq. (7) gives $x_o < 50$ mm. The observations in Section III(*1*) indicate that evaporation occurred at a distance of ~ 120–170 mm from the exit of the capillary. At the evaporation position,

(a)

(b)

Fig. 6. Comparison of erosion surface textures produced in SiC by (a) turbulent region of water vapor jet incident at an angle of 45° to specimen surface; and (b) H_2/O_2 combustion plume in rocket nozzle conditions (surface temperature ~1600°C); inset showing higher magnification image from area indicated.

furnaces, the time required to approach steady state is >10 000 h. In any practical experiment, the erosion rates are negligible and the kinetics is dominated by the parabolic growth of the oxide layer. Moreover, the limiting oxide thickness becomes sufficiently large that spalling can occur.[8]

At the other extreme of high velocity and pressure, as experienced in industrial gas turbines, the time to reach steady state is only a few minutes, the erosion rates are high (~1 μm/h) and the oxide thickness is very small (<0.1 μm). Similar erosion rates and oxide thicknesses were observed in the central laminar region of the water vapor jet in Section III. Conditions are even more severe in H_2/O_2 rocket combustion environments, which experience higher pressures and velocities of water vapor (and higher temperatures), as indicated in Fig. 7. In these cases, the small oxide thickness (<10 nm in Fig. 7(b)) may lead to breakup of the oxide layer (as possibly seen in the turbulent region of the

water vapor jet in Figs. 5(b) and 6(a)) and a change in the rate-controlling mechanism from volatility of SiO_2 to the direct reaction of SiC or Si_3N_4 with water vapor. Several possibilities were proposed by Opila[8]: (i) rate controlled by reaction of SiC (or Si_3N_4) with water to give SiO_2, with the oxide being swept away as soon as it forms, (ii) "active" volatilization similar to the active oxidation that occurs for SiO_2 formers at low partial pressures of O_2, where $SiO(g)$ is formed rather than SiO_2. In this case, an analogous reaction would be SiC reacting directly with water vapor to give $Si(OH)_4(g)$, without the intermediate SiO_2 formation step. Evidence for a transition in behavior to a direct reaction was reported[8,33] for Si_3N_4 vanes after exposure in industrial turbines. The observations in Figs. 5(b) and 6(a) of deeply pitted surfaces in regions of turbulent flow of the water vapor jet are suggestive of a transition in mechanism. Also shown for comparison in Fig. 6(b) is erosion damage on the surface of an SiC test specimen exposed to high-speed H_2/O_2 combustion products from a rocket nozzle.[‡] A similar rough surface is evident with pitting at multiple scales. Even if a very thin oxide layer exists in these cases, with thickness below the detection level of SEM analysis, it is clear that the erosion process is more complex than a simple model of diffusion through a laminar planar boundary layer.

Because these conditions cannot be readily obtained in a laboratory environment, there is very little understanding of the details of the mechanism. However, there is evidence for highly accelerated erosion rates in this regime.[8,33] The upper limit for erosion rate in this case is given by the Langmuir equation for free evaporation into a vacuum.[3] This limit is indicated in Fig. 7(a). It is evident from Fig. 7(a) that extrapolation of Eq. (10) to conditions of H_2/O_2 rocket engines, where the water vapor pressure is substantially >1 atm, gives erosion rates approaching the Langmuir limit. Extrapolation for conditions of the water vapor jet ($pH_2O = 1$ atm) with laminar flow indicates that a significantly higher jet velocity (~10^5 m/s) would be required to reach the Langmuir limit.

(3) Comparison with Measured Recession Rates

Along the line x_1 in Fig. 1(c), where the flow is laminar, the measured recession rates can be compared with the analysis summarized in the previous sections. Two conditions of velocity and temperature were used in the experiments: (A) velocity of 260 m/s at temperature of 1160°C; and (B) velocity of 160 m/s at temperature of 1350°C.

The temperature dependence (Eq. (10)) enters through the parameter C_1, which can be written in the form

$$C_1 = \beta\, T^n \exp\left(\frac{-\Delta H}{R\,T}\right) \tag{13}$$

where the exponent n is close to $-1/4$, ΔH is the reaction enthalpy of the volatile species (in this case $Si(OH)_4$), and β is a temperature-independent function of molecular weights, collision cross sections, viscosity, and the length dimension that appears in the Reynolds number for flow over the specimen surface (see Appendix B). The temperature variation due to the term T^n is negligible compared with the influence of uncertainties in the exponential term. Thermodynamic measurements by Jacobson et al.[38] over the temperature range 1000–1400 K, within which $Si(OH)_4$ is the dominant reaction product, give $\Delta H = 54.6$ kJ/mol, whereas the HPBR erosion data used in the calibration of the constant C_1 in Eq. (10)[4,8] gave an activation energy of 108 kJ/mol, a result that was interpreted as implying the presence of appreciable amounts of $SiO(OH)_2$ as well the predominant $Si(OH)_4$. Steady-state recession rates predicted from Eqs (10) and (13) with the latter value of activation energy and with the calibration at $T = 1316$°C mentioned above are shown in Fig. 8 at the temperatures of the water jet experiments.

[‡]O. Sudre, unpublished work on SiC composite exposed to testing at Cell 22 facility at NASA Glenn Research Center.

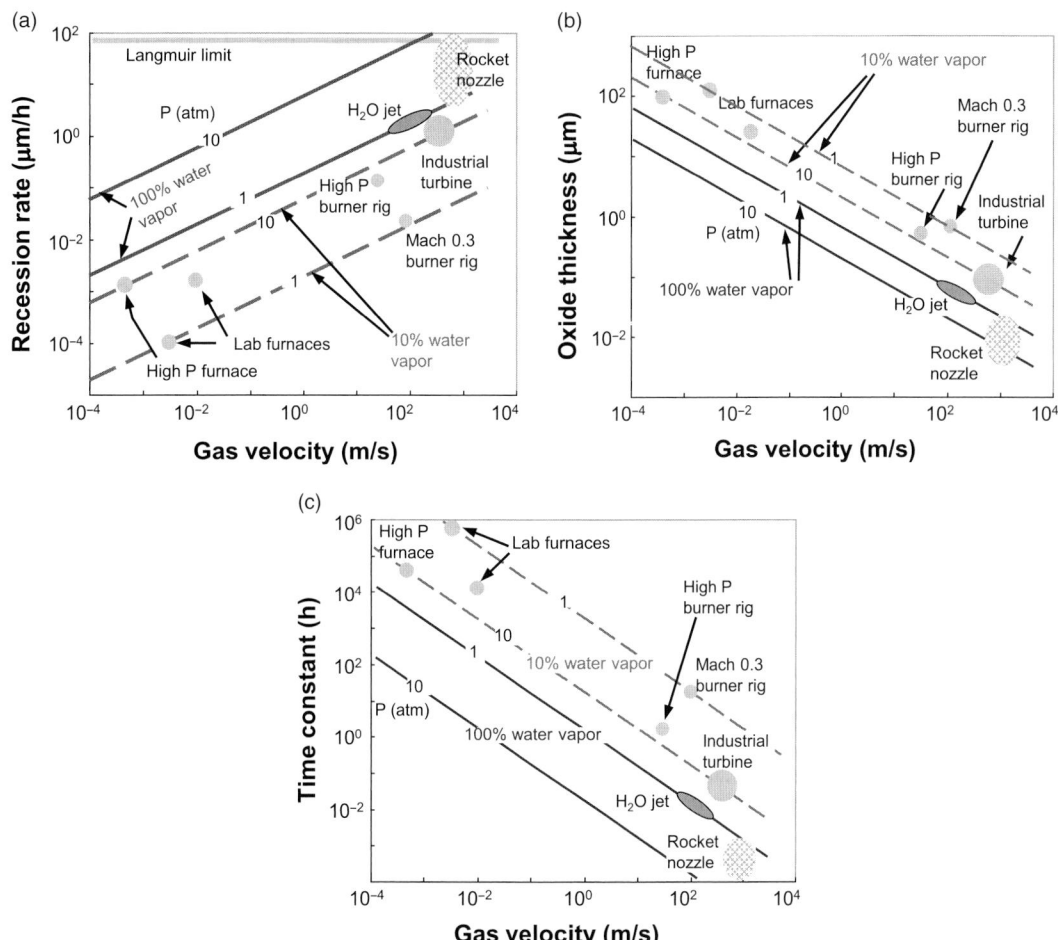

Fig. 7. Plots from Eqs (10–12) comparing erosion of SiC in flowing water vapor (blue solid lines) and in hydrocarbon combustion (red broken lines, 10% water vapor) at temperature of 1316°C, assuming laminar flow over a flat plate and volatilization of $Si(OH)_4$: (a) steady-state recession rate, (b) limiting oxide thickness, and (c) time taken to reach steady state. Conditions corresponding to several test facilities, including the high-speed water vapor jet, are indicated and compared with typical conditions for industrial turbine engines and rocket nozzles (note temperatures are generally higher in rocket nozzles).

Also shown in Fig. 8 are the measured recession rates. The recession rates on the line x_1 in Fig. 1(c) (~ 0.5–0.8 µm/h for condition A and 0.6–0.7 µm/h for condition B) were smaller by a factor of 2 to 3 than the predicted values, while the recession

Fig. 8. Comparison of measured recession rates at two test conditions A ($v_{av} = 260$ m/s, $T = 1160$°C) and B ($v_{av} = 160$ m/s, $T = 1350$°C) with predictions from Eq. (10) using thermodynamic data ($\Delta H = -108$ kJ/mol) from burner rig testing.[4,8] Dotted lines indicate approximate shift of predicted curves that would result from referencing the calibration of the constant C_1 in Robinson and Smialek[4] to the measured gas temperature (1416°C) rather than the measured test specimen temperature (1316°C).

rates in the grooves at the turbulent periphery of the jet were slightly higher than the predicted value for laminar flow (~ 1.5 µm/h). These differences likely relate to different flow conditions in the HPBR tests used to calibrate the constant C_1 in Eq. (10)[4,8]: (i) the test surfaces in the HPBR tests were oriented at an angle $\sim 45°$ to the flow, whereas those in the present experiments were parallel to the flow; and (ii) the gas temperatures in the HPBR tests were higher by $\sim 100°$C than the measured specimen temperatures used in the calibration, whereas the temperatures referenced in the present tests were measured values for the water vapor jet. If the calibration from the HPBR tests were to be referenced to the gas temperature rather than the test specimen temperature (i.e., 1416°C rather than 1316°C), the predicted erosion rates at the temperatures of the water jet tests would be lowered by approximately 40% (dotted lines in Fig. 8), giving close agreement with the measured erosion rates for condition A and reducing the discrepancy for condition B. Additional uncertainty is introduced through differences in the length parameter in the Reynolds number and Sherwood number in Eqs (B-1) and (B-2) and the likelihood that the water vapor jet contains some $Si(OH)_4$ from reaction with the capillary wall, which would act to retard the volatilization reaction on the test specimen surface. Given these uncertainties, the measured recession rates appear to be reasonably consistent with the analysis of Opila *et al.*[5,8]

V. Conclusions

The method demonstrated here provides a means for efficient laboratory scale studies of corrosion of materials and coatings by water vapor, under conditions that are representative of the

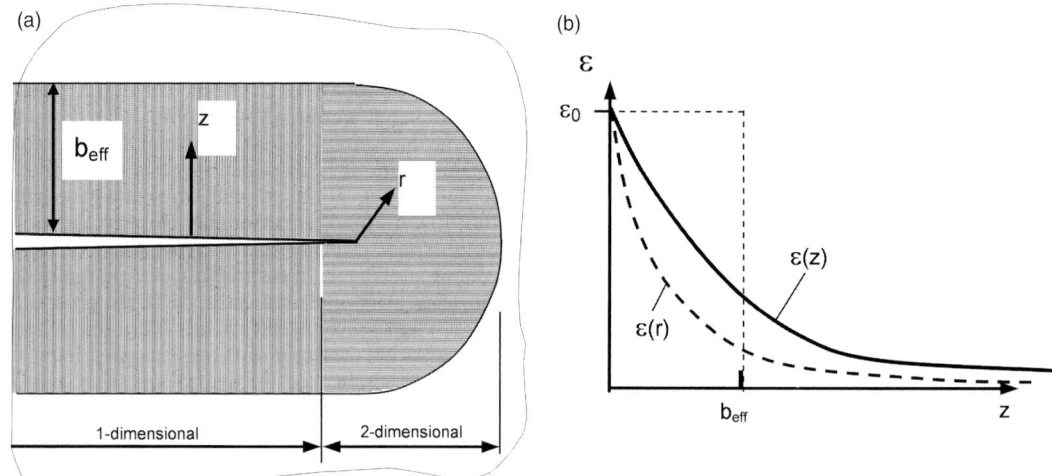

Fig. 1. (a) Diffusion zone in the absence of an external load, (b) swelling profile under assumption of a one-dimensional (solid curve) and axial-symmetric two-dimensional water diffusion (dashed curve); effective zone thickness b_{eff} in the one-dimensional case.

the dilution of water concentration due to the increase of the area elements $dA \propto r\,dr$ in the two-dimensional problem and use $dA \propto dz$ from the one-dimensional model. Also, the continuously varying swelling strain $\varepsilon(r)$ of Eq. (4) is replaced by a step-shaped dependency with a change at the characteristic contour height b_{eff} (Fig. 1(b)) resulting in the same shielding stress-intensity factor.

(1)   Swelling Zone for Unloaded Cracks

The volume expansion at crack surfaces must result in an intrinsic shielding stress-intensity factor K_{sh}. In the case of an unloaded crack, the hydrostatic stress term in Eq. (3) disappears. Because of the Arrhenius dependence of the diffusivity on temperature, Eq. (1), the diffusion process becomes more important as the temperature is increased, even though no stress-enhancement takes place. The shielding stress-intensity factor, K_{sh}, after a time, t, of water contact is given by[36]

$$K_{\text{sh}} = -0.25 \frac{\varepsilon_0 E}{1-\nu} \sqrt{b_{\text{eff}}} \qquad (7)$$

with an effective water layer thickness, b_{eff}, of

$$b_{\text{eff}} \cong \sqrt{D(0)t} \qquad (8)$$

which is roughly the distance from the surface at which the strain has dropped to one-half of the surface value (Fig. 1(b)). Finally, it holds that

$$K_{\text{sh}} \cong -0.25 \frac{\varepsilon_0 E}{1-\nu} [D(0)t]^{1/4} \qquad (9)$$

(2)   Swelling Zone Near the Tip of a Nonpropagating Crack Under Load

The hydrostatic stress is given by the trace of the stress tensor,

$$\sigma_{\text{h}} = \frac{1}{3}(\sigma_{rr} + \sigma_{\varphi\varphi} + \sigma_{zz}) \qquad (10)$$

Because very high tensile stresses occur in the vicinity of a crack tip, a water containing zone must rapidly extend from the tip to the surrounding material during crack growth. As the crack moves, the zone extends along the sides of the crack at the same velocity relative to the crack tip as the crack is moving in the test material.

If K_{I} denotes the stress-intensity factor, the near-tip stress field for the stationary crack in the absence of a T-stress term[36] reads[‡‡]

$$\sigma_{ii} = \frac{K_{\text{tip}}}{\sqrt{2\pi r}} g_{ii}(\varphi) \qquad (11)$$

where r is the crack-tip distance. In Eq. (11), g_{ii} are well-known geometric functions depending on the polar angle φ.[37] The hydrostatic stress under plane strain conditions is given by

$$\sigma_{\text{h}} = \frac{2}{3}(1+\nu) \frac{K_{\text{tip}}}{\sqrt{2\pi r}} \cos(\varphi/2) \qquad (12)$$

Introducing Eq. (12) into Eq. (3) yields location-dependent diffusion coefficients. The introduction of such diffusion coefficients into the problem makes the resulting equations highly intractable, because Eqs. (2) and (4) are correct only for constant diffusion coefficients. The authors are unaware of an exact analytical solution for this diffusion problem. Therefore, to simplify the problem, we assumed that the two-dimensional diffusion problem at the crack tip can be represented by the one-dimensional problem. Therefore, the erfc-relation, Eq. (4), was used for the near-tip zones in the presence of stresses.

Using Eqs. (12), (4), (3) and (1), an equation for the swelling strain can be derived:

$$\varepsilon(r,\varphi,t) = \varepsilon_0 \cdot \text{erfc}\left[\frac{r}{2\sqrt{D(0)t}} \times \exp\left(-\frac{1}{3}(1+\nu)K_{\text{tip}} \frac{\cos(\varphi/2)}{\sqrt{2\pi r}} \frac{\Delta V_{\text{w}}}{R\Theta} \right) \right] \qquad (13)$$

The contour for an effective swelling strain of $\varepsilon(r,\varphi,t) = 0.5\varepsilon_0$ results in

$$\frac{r}{2\sqrt{D(0)t}} \exp\left(-\frac{1}{3}(1+\nu)K_{\text{tip}} \frac{\cos(\varphi/2)}{\sqrt{2\pi r}} \frac{\Delta V_{\text{w}}}{R\Theta} \right)$$
$$= 0.477 \cong 0.5 \qquad (14)$$

with a solution of

[‡‡]The stress distribution around a crack tip is usually expressed in the form of a power series in $r^{n/2}$, where r is the distance from the crack tip. The series starts with $n = -1$. The coefficient of the first term is the stress-intensity factor; the coefficient of the second term of the series is the T-stress. See Fett[36] for further discussion of this subject.

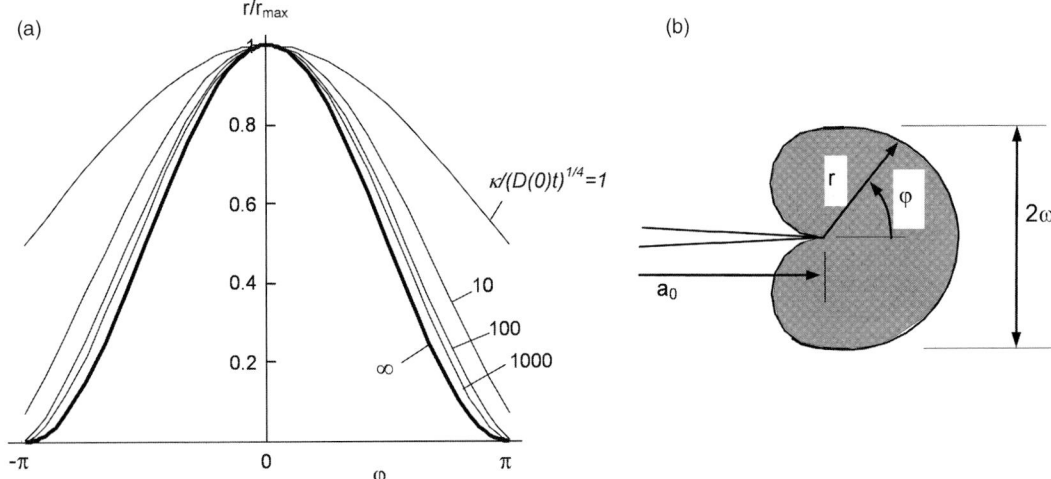

Fig. 2. (a) Diffusion zone radius normalized on the maximum value as a function of the polar angle φ, (b) shape of the swelling zone.

$$r = \frac{\kappa^2 \cos^2(\varphi/2)}{4\left(\text{PLog}\left[\dfrac{\kappa \cos(\varphi/2)}{2[D(0)t]^{1/4}}\right]\right)^2} \qquad (15)$$

with the abbreviation for κ:

$$\kappa = \frac{(1+v)}{3}\frac{K_{tip}}{\sqrt{2\pi}}\frac{\Delta V_w}{R\Theta} \qquad (16)$$

"PLog" stands for the Lambert W function or *product log function*, i.e., the solution $x = f(y)$ of the equation $y = x \exp(x)$ (in *Mathematica*[38] called ProductLog).[‡] Contour radii for constant swelling are plotted in Fig. 2(a) with $\kappa/(D(0)t)^{1/4}$ as the parameter.

In the case of strongly stress-enhanced diffusion, i.e., $\kappa/(D(0)t)^{1/4} \gg 1$, Eq. (15) tends to

$$\frac{r}{r_{max}} = \cos^2(\varphi/2), \quad r_{max}$$
$$= \kappa^2\left(2\text{Plog}\left[\frac{\kappa}{2[D(0)t]^{1/4}}\right]\right)^{-2} \qquad (17)$$

This can be seen from Fig. 2(a) where the thick curve represents the limit case for $\kappa/(D(0)t)^{1/4} \to \infty$ in K_{sh}. The shape of the related swelling zone is shown in Fig. 2(b). The result is similar to that for the phase-transformation zones in zirconia ceramics. But in contrast to those zones, the size of the swelling zones is time-dependent.

The height of the swelling contours for an arbitrary $\kappa/(D(0)t)^{1/4}$ results in very complicated expressions. For reasons of simplicity, the contour radius for $\varphi = 0$, r_{max}, may therefore be used as a characteristic size parameter. According to the analysis by McMeeking and Evans[39] this type of zone must result for $\kappa/(D(0)t)^{1/4} \to \infty$ in $K_{sh} = 0$.

IV. Shielding Term for Growing Cracks

A crack growing at a constant subcritical velocity, v, develops an extended zone of biaxial compression on the newly formed crack faces. The zone forms from the crack tip and as the crack moves forward, it passes through the compressive zone leaving half the compressive zone along each fracture surface. Figure 3(a) shows the swelling zone after a crack increment of Δa. For a very sim-

plified analysis, the continuously varying swelling strain $\varepsilon(r)$ may be replaced by a step-shaped dependency with a change at the characteristic contour height $\omega = b$ (Fig. 3(b)), defined by the same shielding stress-intensity factor.

In this normalized representation, $K_{sh,max}$ denotes the shielding for an infinitely long zone, $\Delta a/\omega \to \infty$. Without placing great demands on accuracy we can represent the K_{sh}–Δa dependency, numerically given by McMeeking and Evans,[39] by the approximation:

$$\frac{K_{sh}}{K_{sh,max}} \cong \tanh\left[0.757\left(\frac{\Delta a}{\omega_{eff}}\right)^{3/4}\right]^{2/3} \qquad (18)$$

with an effective zone height $\omega_{eff} \approx r_{max}$, Eq. (17) (Fig. 3(b)).

The time available for the formation of the swelling zone is roughly $t = r_{max}/v$. Instead of Eq. (17), this now results in the implicit equation:

$$r_{max} = \kappa^2\left(2\text{Plog}\left[\frac{\kappa}{2[D(0)r_{max}/v]^{1/4}}\right]\right)^{-2} \qquad (19)$$

The solution in terms of the power law of subcritical crack growth, which holds far from the thermodynamic threshold for crack growth[12]

$$v = AK_{tip}^n \qquad (20)$$

is then

$$r_{max} = \frac{\kappa^2}{\text{PLog}^2\left[\kappa\sqrt{\dfrac{v}{D(0)}}\right]}, \qquad (21)$$
$$\kappa = \frac{(1+v)}{3}\frac{(v/A)^{1/n}}{\sqrt{2\pi}}\frac{\Delta V_w}{R\Theta}$$

It has to be reemphasized that this maximum zone size is only reached if a crack has grown a sufficiently large increment of $\Delta a > r_{eff}$ at a constant rate.

The maximum shielding stress-intensity factor increases with decreasing crack growth rate

$$K_{sh,max} \cong -0.22\frac{\varepsilon_0 E}{1-v}\frac{\kappa}{\text{PLog}\left[\kappa\sqrt{\dfrac{v}{D(0)}}\right]} \qquad (22)$$

and held under load and then repropagated under a much higher load. The stress-intensity factor applied to the crack tip at the point of arrest was 0.245 MPa·$\sqrt{m}$; the load was maintained for 80 days in water at room temperature, then increased to 0.48 MPa·$\sqrt{m}$ to restart the crack. Crack motion was immediate upon increasing the load on the specimen.

Examination of a crack profile by comparing the two fracture surfaces after completing the fracture (Fig. 7(a)) indicates that with the exception of a small area near the crack tip, *a*, the two surfaces matched over the length of the crack that had been opened and exposed to the water. An interesting observation in Fig. 7(a) is the overlap of the two fracture surfaces near the crack tip, *a*. One possible explanation of the overlap might be the occurrence of silica precipitate from the crack tip solution and its deposition at the crack tip before final fracture.[16] This was the interpretation given for the observation in Wiederhorn *et al.*.[41] We now discount this possibility for two reasons. First, the experiment was carried out in water, and the specimen had to be dried before any studies could be carried out on the fracture surface. In drying the fracture surface, it was wiped with high-quality tissue paper, which would have removed any precipitation deposited on the surface during the experiment. In our previous experience, gels that precipitate on the fracture surface are easily removed by such a procedure.

Second, an AFM scan of the apparent excess material at the crack arrest point using the tapping mode™ indicated no shift in the phase image, which suggests that the viscoelastic/elastic properties of the excess material were close to that of the original glass. If the excess material represented a dried precipitate, the elastic constants would have been lower because of internal porosity and a phase shift would have been observed in the phase image.

Based on these considerations, we interpret the excess material ahead of the actual crack tip as a water diffusion zone where swelling by the volumetric strain, ε_0, took place. After complete cracking of the specimen, each half of this region can expand in the direction normal to the new crack face. This results in a volume expansion and in an overlapping of the surface profiles. Figure 7(b) gives measurements of the overlapping displacements with a maximum value of about 6 nm.

In order to study this effect, a finite element (FE) modelling of the heart-shaped zone according to Fig. 2(b) was performed. The case of a half-space was realized by a plate 60ω wide and 30ω high including 1400 elements and 4300 nodes. Solid continuum elements (8-node bi-quadratic) in plane strain were chosen and the computations carried out with ABAQUS version 6.8. The volume strain was replaced by the equivalent thermal problem by heating the inner of the zone with a temperature of ΔT keeping the temperature outside the zone constant (e.g., at

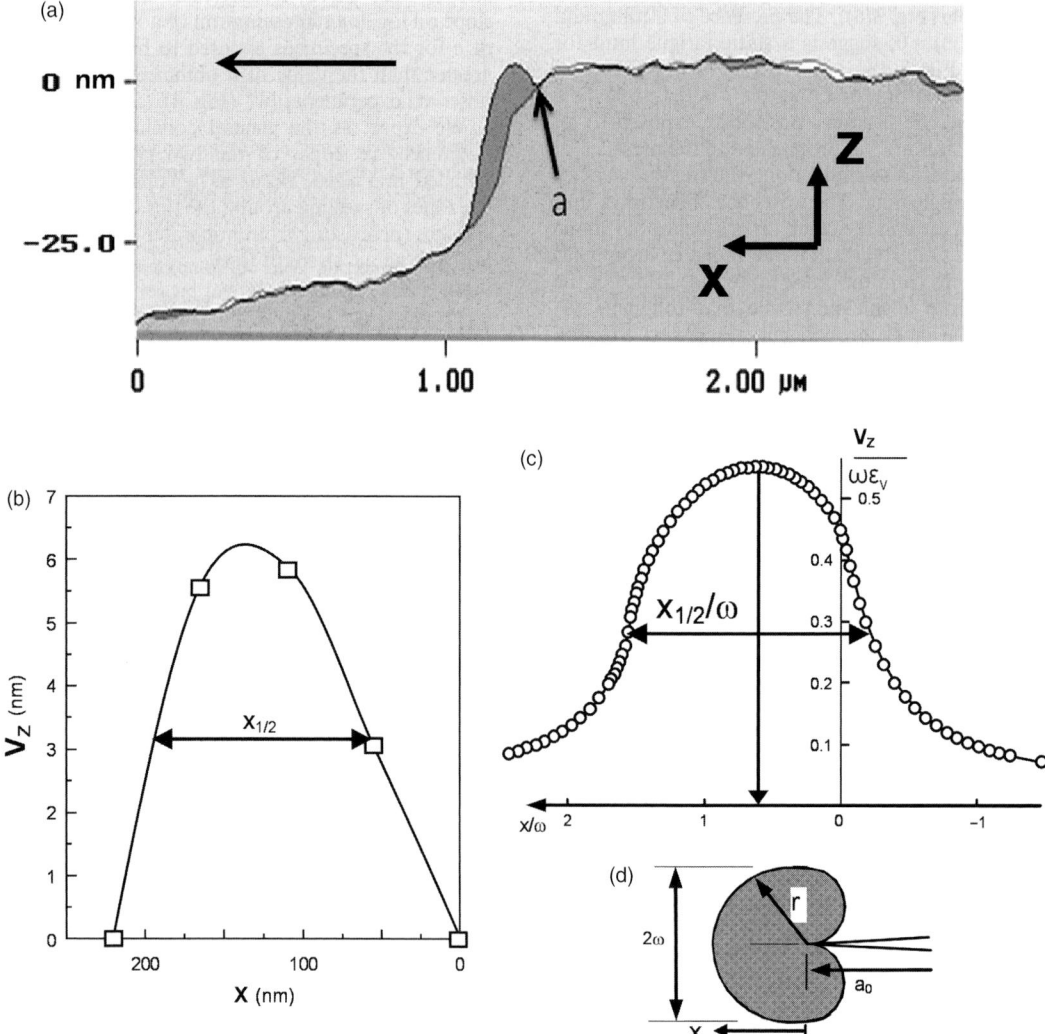

Fig. 7. (a) Overlapping of crack faces for an arrested crack after final fracture.[41] The arrow gives the direction of crack growth, (b) crack–face displacement, V_z, from (a), (c) finite element result for the surface displacements due to a volumetric swelling strain ε_v, (d) the cardioid shape of the region of the glass penetrated by water. The width of the cardioid is 2ω and the distance from the crack tip in front of the crack is *x* (geometric data are defined in the bottom schematic).

$T = 0$). This results in the volume strain

$$\varepsilon_v = 3\alpha\Delta T \qquad (28)$$

The displacements are shown in normalized representation in Fig. 7(c). The maximum displacement from FE-modelling was found at 0.6ω ahead of the crack tip to be $v_y/(\omega\varepsilon_v) \cong 0.55$. From Figs. 7(b) and (c) we can conclude that $\omega\varepsilon_0 \cong 11$ nm.

The width at half of maximum displacement, $x_{1/2}/\omega$, is (see Fig. 7(c)) $x_{1/2}/\omega = 1.8$. The corresponding measured value of $x_{1/2}$ is $x_{1/2} \approx 140$ nm (see Fig. 7(b)). From these two results we obtain: $\omega \cong 80$ nm $\Rightarrow \varepsilon_0 = 13.7\%$, or, a linear strain of about 4.6%, which is equivalent to a biaxial compression stress of $\sigma_x = \sigma_y = -4000$ MPa, Eq. (5).

The height of the zone (identical with $r \times \sin\varphi$ for $\varphi \cong \pi/3$) and an estimate of ΔV_w can be determined from Eq. (17). Inserting the values of $D_0 = 10^{-21}$ m^2/s ^{25} into Eq. (17) yields $\omega \cong 140$ nm for $\Delta V_w = 30 \times 10^{-6}$ m^3/mol, $\omega \cong 110$ nm for $\Delta V_w = 20 \times 10^{-6}$ m^3/mol and $\omega \cong 80$ nm for $\Delta V_w = 10 \times 10^{-6}$ m^3/mol. The latter value of ω is in good agreement with the value resulting from the evaluation of Fig. 7(b). This calculation suggests that $\Delta V_w = 10 \times 10^{-6}$ m^3/mol, a value not that far from the molar volume of water, $\Delta V_m = 18 \times 10^{-6}$ m^3/mol. Thus, a systematic investigation of the zone size at different holding loads, K, holding times t, and temperatures, Θ, provides an independent tool for the determination of the activation volume, ΔV_w, for the diffusion of water in silica glass.

VI. Summary/Conclusions

In this paper, we investigated the effect of water diffusion into silica glass fracture surfaces on the strength and static fatigue of this glass. The approach used is identical to that used by McMeeking and Evans[39] in their calculation of the effect of a phase transformation on the strength and toughening of zirconium oxide. The toughness of zirconium oxide was enhanced substantially by a volume expansion of the material around the crack tip, caused by a stress-induced phase transformation in the zirconium oxide. In the case of silica glass, the volume expansion is the result of swelling when water diffuses into the fracture surface of the glass and the glass around the crack tip. The swelling is time-dependent and so the toughening of the glass increases with time. The theory is able to predict an increase of strength when testing is carried out in liquid nitrogen; a small change in the slope of the static fatigue curve as a result of the zone of swelling around the crack tip, and an overlapping of the fracture surfaces by about 6 nm in the near vicinity of the crack tip due to water diffusion into the highly strained region surrounding the crack tip. A further benefit of this study was the development of a new method of measuring the activation volume for the diffusion of water through silica glass.

References

[1]C. R. Kurkjian, P. K. Gupta, R. K. Brow, and N. Lower, "The Intrinsic Strength and Fatigue of Oxide Glasses," *J. Noncrystal. Solids*, **316**, 114–24 (2003).

[2]F. W. Preston, "The Mechanical Properties of Glass," *J. Appl. Phys.*, **13**, 623–34 (1942).

[3]R. E. Mould, "The Strength of Inorganic Glasses"; pp. 119–49 in *Fundamental Phenomena in the Materials Sciences, Vol. 4, Fracture of Metals, Polymers and Glasses*, Edited by L. J. Bonis, J. J. Duga, and J. J. Gilman. Plenum Press, New York, 1967.

[4]B. A. Proctor, I. Whitney, and J. W. Johnson, "The Strength of Fused Silica," *Proc. R. Soc. A*, **297** [1451] 534–57 (1967).

[5]V. M. Sglavo and D. J. Green, "Fatigue Limit in Fused Silica," *J. Eur. Ceram. Soc.*, **21**, 561–7 (2001).

[6]S. M. Wiederhorn and L. H. Bolz, "Stress Corrosion and Static Fatigue of Glass," *J. Am. Ceram. Soc.*, **53** [10] 543–8 (1970).

[7]S. M. Wiederhorn, H. Johnson, A. M. Diness, and A. H. Heuer, "Fracture of Glass in Vacuum," *J. Am. Ceram. Soc.*, **57** [8] 336–41 (1974).

[8]S. Sakaguchi and Y. Hibino, "Fatigue in Low-Strength Silica Optical Fibres," *J. Mater. Sci.*, **19**, 3416–20 (1984).

[9]M. Muraoka and H. Abé, "Subcritical Crack Growth in silica Optical Fibers in a Wide Range of Crack Velocities," *J. Am. Ceram. Soc.*, **79** [1] 51–7 (1996).

[10]T. I. Suratwala and R. A. Steele, "Anomalous Temperature Dependence of Sub-Critical Crack Growth in Silica Glass," *J. Non-Crystalline Sol.*, **316**, 174–82 (2003).

[11]S. M. Wiederhorn, S. W. Freiman, E. R. Jr. Fuller, and C. J. Simmons, "Effect of Water and Other Dielectrics on Crack Growth," *J. Mater. Sci.*, **17**, 3460–78 (1982).

[12]B. R. Lawn, *Fracture of Brittle Solids*, 2nd edition, Cambridge University Press, Cambridge, 1993.

[13]T. A. Michalske and S. W. Freiman, "A Molecular Mechanism for Stress Corrosion in Vitreous Silica," *J. Am. Ceram. Soc.*, **66** [4] 284–8 (1983).

[14]K. Han, M. Ciccotti, and S. Roux, "Measuring Nanoscale Stress Intensity Factors with an Atomic Force Microscope," *EPL*, **89**, 66003 (2010).

[15]G. Pezzotti and A. Leto, "Contribution of Spatially and Spectrally Resolved Cathodo-luminescence to Study Crack-Tip Phenomena in Silica Glass," *Phys. Rev. Lett.*, **103**, 175501 (2009).

[16]Y. Bando, S. Ito, and M. Tomozawa, "Direct Observation of Crack Tip Geometry of SiO$_2$ Glass by High-Resolution Electron-Microscopy," *J. Am. Ceram. Soc.*, **67**, C36–7 (1984).

[17]R. J. Charles and W. B. Hillig, *Symposium on Mechanical Strength of Glass and Ways of Improving It*. Florence, Italy, September 25–29, 1961, pp. 511–27. Union Scientifique Continentale du Verre, Charleroi, Belgium, 1962.

[18]S. Ito and M. Tomozawa, "Crack Blunting of High-Silica Glass," *J. Am. Ceram. Soc.*, **65** [8] 368–71 (1982).

[19]K. Hirao and M. Tomozawa, "Kinetics of Crack Tip Blunting of Glasses," *J. Am. Ceram. Soc.*, **70** [1] 43–8 (1987).

[20]K. Hirao and M. Tomozawa, "Dynamic Fatigue of Treated High-Silica Glass: Explanation by Crack Tip Blunting," *J. Am. Ceram. Soc.*, **70** [6] 377–82 (1987).

[21]B. R. Lawn, K. Jakus, and A. C. Gonzalez, "Sharp vs Blunt Crack Hypotheses in the Strength of Glass: A Critical Study Using Indentation Flaws," *J. Am. Ceram. Soc.*, **68** [1] 25–34 (1985).

[22]D. H. Roach, S. Lathabai, and B. R. Lawn, "Interfacial Layers in Brittle Cracks," *J. Am. Ceram. Soc.*, **71** [2] 97–105 (1988).

[23]T. A. Michalske and B. C. Bunker, "Effect of Surface Stress on Stress Corrosion of Silicate Glass"; pp. 3689–701 in *Advances in Fracture Research*, Vol. 6, Edited by K. Salama, K. Ravi-Chandler, D. M. R. Tablin, and P. Ramao Rao. Pergamon Press, New York, 1989.

[24]J.-P. Guin, S. M. Wiederhorn, and T. Fett, "Crack-Tip Structure in Soda-Lime–Silicate Glass," *J. Am. Ceram. Soc.*, **88** [3] 652–9 (2005).

[25]T. Fett, J. P. Guin, and S. M. Wiederhorn, "Stresses in Ion-Exchange Layers of Soda-Lime–Silicate Glass," *Fatigue Fract. Eng. Mater. Struct.*, **28**, 507–14 (2005).

[26]T. A. Michalske, "The Stress Corrosion Limit: Its Measurement and Implications"; pp. 277–89 in *Fracture Mechanics of Ceramics, Vol. 5, Surface Flaws, Statistics, and Microcracking*, Edited by R. C. Bradt, A. G. Evans, D. P. H. Hasselman, and F. F. Lange. Plenum Press, New York, 1977.

[27]E. Gehrke, Ch. Ullner, and M. Hähnert, "Fatigue Limit and Crack Arrest in Alkali-Containing Silicate Glasses," *J. Mater. Sci.*, **26**, 5445–55 (1991).

[28]R. H. Doremus, "Diffusion of Water in Silica Glass," *J. Mater. Res.*, **10** [9] 2379–89 (1995).

[29]K. M. Davis and M. Tomozawa, "Water Diffusion into Silica Glass: Structural Changes in Silica Glass and their Effect on Water Solubility and Diffusivity," *J. Non-Cryst. Solids*, **185**, 203–20 (1995).

[30]A. Zouine, O. Dersch, G Walter, and F. Rauch, "Diffusivity and Solubility of Water in Silica Glass in the Temperature Range 23–200°C," *Phys Chem. Glasses: Eur. J. Glass Sci Technol. B*, **48** [2] 85–91 (2007).

[31]A. Oehler and M. Tomozawa, "Water Diffusion into Silica Glass at a Low Temperature Under High Water Vapor Pressure," *J. Non-Cryst. Glasses*, **347**, 211–9 (2004).

[32]M. Nogami and M. Tomozawa, "Effect of Stress on Water Diffusion in Silica Glass," *J. Am. Ceram. Soc.*, **67**, 151–4 (1984).

[33]M. I. Gorbacheva and R. M. Zaoints, "Moisture Expansion of Various Crystalline and Amorphous Phases," *Steklo Keram.*, **11**, 18–9 (1974).

[34]J. Thurn, "Water Diffusion Coefficient Measurements in Deposited Silica Coatings by the Substrate Curvature Method," *J. Non-Cryst. Solids*, **354**, 5459–65 (2008).

[35]W. M. Kuschke, H. D. Carstanjen, N. Pazarkas, D. W. Plachke, and E. Arzt, "Influence of Water Absorption by Silicate Glass on the Strains in Passivated Al Conductor Lines," *J. Electron. Mater.*, **27**, 853–7 (1998).

[36]T. Fett, *Stress intensity factors, T-stresses, Weight functions, IKM 50*. Universitätsverlag Karlsruhe, Karlsruhe, Germany, 2008.

[37]D. Munz and T. Fett, *Ceramics, Mechanical Properties, Failure Behaviour, Materials Selection*. Springer, Berlin, 1999.

[38]Wolfram Research. *Mathematica*. Wolfram Research, Champaign, IL, USA.

[39]R. M. McMeeking and A. G. Evans, "Mechanics of Transformation-Toughening in Brittle Materials," *J. Am. Ceram. Soc.*, **65**, 242–6 (1982).

[40]D. B. Fraser, "Acoustic Properties of Vitreous Silica," *J. Appl. Phys.*, **39** [13] 5868–78 (1968).

[41]S. M. Wiederhorn, J.-P. Guin, and T. Fett, "The Use of Atomic Force Microscopy to Study Crack Tips in Glass," *Met. Mater. Trans.*, **42A**, 267–78 (2011). □

J. Am. Ceram. Soc., **94** [S1] S204–S214 (2011)
DOI: 10.1111/j.1551-2916.2011.04516.x
© 2011 The American Ceramic Society

A Constitutive Description of the Inelastic Response of Ceramics

Vikram S. Deshpande[†]

Department of Engineering, University of Cambridge, Cambridge CB2 1PZ, U.K.

E. A. Nell Gamble, Brett G. Compton, Robert M. McMeeking, Anthony G. Evans, and Frank W. Zok

Materials Department, University of California, Santa Barbara, 93106 California

The objective of the article is to present a unified model for the dynamic mechanical response of ceramics under compressive stress states. The model incorporates three principal deformation mechanisms: (i) lattice plasticity due to dislocation glide or twinning; (ii) microcrack extension; and (iii) granular flow of densely packed comminuted particles. In addition to analytical descriptions of each mechanism, prescriptions are provided for their implementation into a finite element code as well as schemes for mechanism transitions. The utility of the code in addressing issues pertaining to deep penetration is demonstrated through a series of calculations of dynamic cavity expansion in an infinite medium. The results reveal two limiting behavioral regimes, dictated largely by the ratio of the cavity pressure p to the material yield strength σ_Y. At low values of p/σ_Y, cavity expansion occurs by lattice plasticity and hence its rate diminishes with increasing σ_Y. In contrast, at high values, expansion occurs by microcracking followed by granular plasticity and is therefore independent of σ_Y. In the intermediate regime, the cavity expansion rate is governed by the interplay between microcracking and lattice plasticity. That is, when lattice plasticity is activated ahead of the expanding cavity, the stress triaxiality decreases (toward more negative values) which, in turn, reduces the propensity for microcracking and the rate of granular flow. The implications for penetration resistance to high-velocity projectiles are discussed. Finally, the constitutive model is used to simulate the quasi-static and dynamic indentation response of a typical engineering ceramic (alumina) and the results compared to experimental measurements. Some of the pertinent observations are shown to be captured by the present model whereas others require alternative approaches (such as those based on fracture mechanics) for complete characterization.

I. Introduction

CERAMICS have been used extensively for armor protection. When confined within a metallic medium, their exceptional dynamic strength in compression causes impacting projectiles to deform and erode. Absent the confinement, the tensile stresses induced in the ceramic cause extensive cracking that eliminates the benefit. The practical challenge is to conceive designs that maintain the confinement for a sufficient duration to realize the full benefit. Current design practice has been developed through extensive testing but has limited scope given the large design space; the most efficient approach would incorporate numerical simulations.

A. Heuer—contributing editor

Manuscript No. 28828. Received October 27, 2010; approved February 19, 2011.
This work is financially supported by the Office of Naval Research through a Multi-disciplinary University Research Initiative Program on "Cellular Materials Concepts for Force Protection," Prime Award No. N00014-07-1-0764.
†Author to whom correspondence should be addressed. e-mail: vsd@eng.cam.ac.uk

Various computational tools have been developed and dynamic constitutive models devised that characterize the response of ceramics to extreme loads.[1–6] In general, two approaches have been used:

(i) Phenomenological damage mechanics, wherein the ceramic is regarded as an elastic–plastic material subject to a notional damage process that reduces the strength as the deformation proceeds.

(ii) Micromechanically motivated models that incorporate aspects of the physics governing compressive damage and plastic deformation.

Each of these approaches has its benefits and draw-backs. The micromechanical models give insight into the mechanisms at play but are computationally too expensive to be used in large-scale structural simulations. On the other hand, while the phenomenological approaches are suited to large-scale computations, they require extensive calibration and typically have a regime of applicability that is limited to scenarios resembling the calibration schemes. Among the phenomenological models, the one that is most complete is that devised by Johnson and Holmquist.[5] The model incorporates a phenomenological law inspired by the response of ceramics impacted at high velocity under highly confined conditions. It embodies a Drucker–Prager yield surface[7] that evolves with damage through the effective plastic strain (analogous to that in the Johnson–Cook model[8] for metals). The representation has multiple coefficients and exponents that require calibration through dynamic measurements. It does not incorporate any microstructural characteristics (such as grain size) or normative material properties (such as toughness and hardness). Nevertheless, when a material has been calibrated in accordance with the proposed protocol, projectile penetration can be simulated with adequate fidelity.[9] The evident limitation is that, for each candidate ceramic, the coefficients and exponents must be recalibrated, because they are not connected to basic microstructure/property relationships.

Recently, Deshpande and Evans[10] (subsequently referred to as DE) devised a micromechanically motivated model that is amenable to large-scale dynamic structural computations. Two specific inelastic phenomena have been included in this model. The first is lattice plasticity due to dislocations and twins, characterized by the von Mises stress relative to the (rate-dependent) flow strength of the material. This property can be probed using hardness measurements conducted with a spherical indenter. The second is damage in the form of microcracks that emanate from preexisting flaws. They evolve subject to a combination of local deviatoric and hydrostatic stresses. The microcracks have dimensions and separations that scale with the grain size. They extend in a manner dictated by the short crack toughness of the ceramic. The constitutive law embodying these mechanisms has been used to successfully simulate the influence of normative material properties, including grain size, hardness, and toughness, on the extents of plastic deformation and damage that occur in polycrystalline alumina upon impact of a hard spherical projectile[11] and upon quasi-static penetration by a hard sphere.[12] The limitation of the foregoing representation is that the mechanical response beyond the point at which the damage becomes saturated (namely, the microcracks have coalesced) has yet to be addressed. The present article establishes an approach for representing the behavior of a damage-saturated ceramic and combines with the foregoing lattice plasticity and damage models to produce a unified model that can be used to simulate deep penetration.

The fully comminuted ceramic created within the highly confined environment beneath an impacting projectile is in a unique state. It consists of grain-sized particles bounded by narrow separations between them (Fig. 1) and has a relative density very close to unity because the volume occupied by the separations is small. This state differs from that found in particulate materials such as soils and in powder compacts, which have relative densities around 60%. When subjected to a combination of compressive and shear stress, the particles can slide and rotate past one another: a phenomenon hereafter referred to as *granular plasticity*.

The article is organized as follows: it begins with a description of the unified model, its physical interpretation, and the associated constitutive law. Thereafter, preliminary features of the constitutive response are elucidated via the predicted stress/strain response of a prototypical hard ceramic (namely alumina) under varying levels of stress triaxiality. Subsequently, two specific loading scenarios are examined and, where possible, used to interpret phenomena found experimentally. The first is the expansion of a spherical cavity in an infinite medium. The objective is to characterize the development and propagation of elastic, plastic, and damage waves as well as understand the interplay between plasticity and damage. The second considers quasi-static and dynamic penetration. Here comparisons are made of the predictions of the original DE model and the current constitutive law, with emphasis on the damage evolution pattern and the load versus displacement response.

II. Development of the Constitutive Model

The inelastic deformation of ceramic polycrystals occurs in accordance with the foregoing three primary mechanisms (Fig. 1) proceeding partially in series and partially in parallel. We envisage the ceramic with a population of preexisting flaws or heterogeneities that, under stress, can extend into microcracks. Before microcracks form, plastic deformation can only occur by dislocation slip or twinning: termed *lattice plasticity*. Its onset is characterized by the von Mises yield criterion, depicted in effective stress (σ_e)−mean stress (σ_m) space as a horizontal line (Fig. 2). When microcracks grow and coalesce, the ceramic transitions into a granular medium comprising narrowly

separated granules with dimensions dictated typically by the grain size. The fully comminuted ceramic has a small shear yield strength governed by friction between the particles. Its yield criterion has a form similar to Drucker–Prager for granular media. The transition between these two yield surfaces with increasing microcrack density is depicted in Fig. 2. At each level of damage, D, yield is characterized by a sloping line in σ_e−σ_m space; as described below, we shall assume that all of these lines intercept the lattice plasticity line at the same location. Upon full comminution, inelastic deformation can occur by one of two mechanisms: granular plasticity of the comminuted particles or lattice plasticity within the particles themselves.

To devise a mechanism-inspired model from the foregoing physical picture, the following four elements must be incorporated:

(i)   A lattice plasticity model.
(ii)   A criterion for the evolution of damage due to microcracking.
(iii)   The transition between the yield surfaces for lattice and granular plasticity.
(iv)   The incorporation of elasticity (with lattice and granular plasticity) to give the overall constitutive description.

(1) The Constituent Models

(A) Lattice Plasticity: In structural ceramics, lattice plasticity occurs in accordance with two limits: at low strain rates, resistance from the lattice governed by the Peierls stress and, at high rates, phonon drag.[13] With this in mind, we approximate the effective plastic strain rate $\dot{\varepsilon}_e^{pl}$ versus the von Mises effective stress σ_e relationship by

$$\frac{\dot{\varepsilon}_e^{pl}}{\dot{\varepsilon}_0} = \begin{cases} \left(\dfrac{\dot{\varepsilon}_0}{\dot{\varepsilon}_t}\right)^{(1-n)/n}\left(\dfrac{2\sigma_e}{\sigma_0}-1\right) & 2\sigma_e > \sigma_0\left[\left(\dfrac{\dot{\varepsilon}_t}{\dot{\varepsilon}_0}\right)^{1/n}+1\right] \\[3mm] \left(\dfrac{2\sigma_e}{\sigma_0}-1\right)^n & \sigma_0 < 2\sigma_e \leq \sigma_0\left[\left(\dfrac{\dot{\varepsilon}_t}{\dot{\varepsilon}_0}\right)^{1/n}+1\right] \\[3mm] 0 & 2\sigma_e \leq \sigma_0 \end{cases}$$

$$(1)$$

where $\sigma_0(\varepsilon_e^{pl})$ is the flow stress at an equivalent plastic strain ε_e^{pl}, while $\dot{\varepsilon}_0$ and n are the reference strain rate and strain rate sensitivity exponent, respectively, and $\dot{\varepsilon}_t$ is the transition strain-rate above which plasticity becomes phonon drag limited. Note that the exponent n sets the strain rate sensitivity of the ceramic in the low strain rate regime. Strain hardening is characterized by a conventional power law, notably:

$$\sigma_0 = \frac{\sigma_Y}{2}[1 + (\varepsilon_e^{pl}/\varepsilon_Y)^M] \qquad (2)$$

where σ_Y is the uniaxial yield strength, ε_Y the plastic strain at which $\sigma_0 = \sigma_Y$, and M the strain hardening exponent. Associated flow is assumed with a flow potential, $G_p \equiv \sigma_e$. The plastic strain rate is then

$$\dot{\varepsilon}_{ij}^{pl} = \dot{\varepsilon}_e^{pl}\frac{\partial G_p}{\partial \sigma_{ij}} \qquad (3)$$

(B) Microcrack Evolution: The model envisages an array of f microcracks per unit volume, subject to principal stresses σ_1 and σ_3 and growing in an otherwise elastic medium (Fig. 3). Each microcrack develops from an initial flaw, radius a, by means of two wings, length ℓ, extending parallel to X_1 (Fig. 3). The inclined flaws are subject to Coulomb friction, with friction coefficient μ. The radius and separation of the flaws scale with the grain size as $a = g_1 d$ and $1/f^{1/3} = g_2 d$, respectively, where g_1 and g_2 are parameters. The initial and current levels of damage are expressed respectively as $D_0 = (4/3)\pi(\alpha a)^3 f$ and $D = (4/3)\pi(\ell+\alpha a)^3 f$, where $\alpha \equiv \tan\psi$ is a shape factor to account for the angle of the initial flaw as shown in Fig. 3. The

Fig. 1. The inelastic mechanisms included in the constitutive model for ceramics.

Polycrystal ensemble

σ_1

σ_2

Slip

Dislocations, twins

Lattice plasticity

Microcracking (0 < D < 1)

Granular plasticity (D = 1)

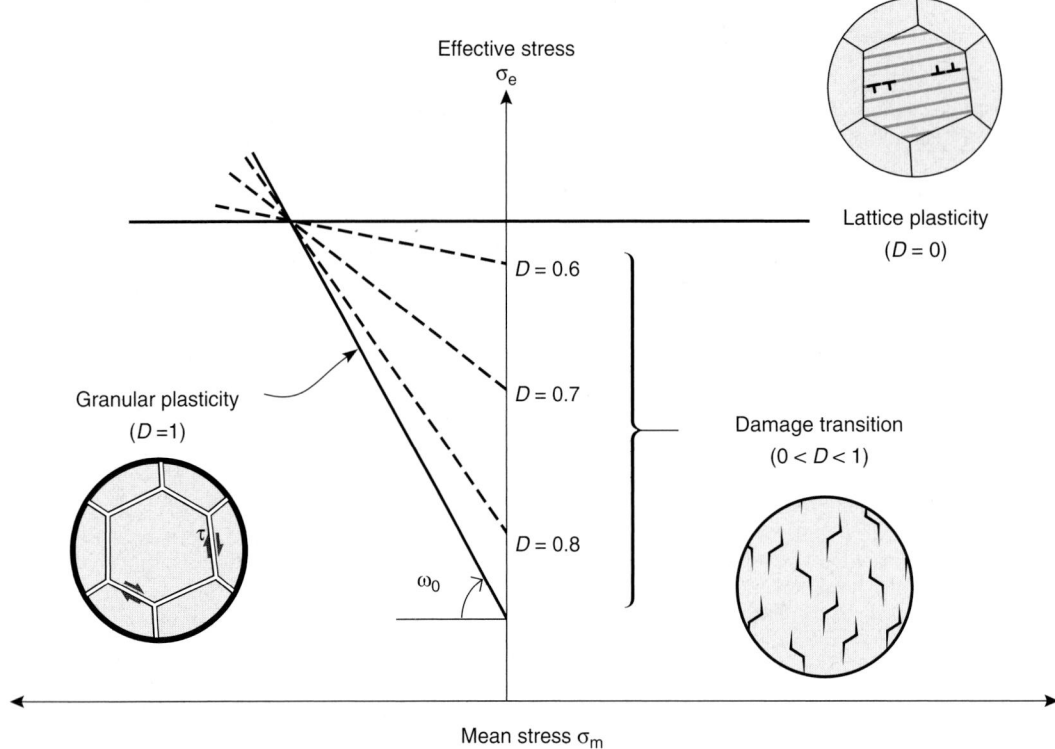

Fig. 2. Schematic of the yield surface transition from lattice to granular plasticity.

development of the microcracks is governed by the mode I stress intensity, K_I, at the tip of the wing cracks. In turn, K_I depends on the stress triaxiality, $\lambda \equiv \sigma_m/\sigma_e$, in accordance with three behavioral regimes (*I, II, III*) with four associated nonlinear functions of damage and friction (*A, B, C,* and *E,* defined in the Appendix A). In *Regime I* ($\lambda \le -B/A$), the cracks are shut, with $K_I = 0$. In *Regime II*, frictional slip occurs along the initial flaw surfaces, with

$$K_I/\sqrt{\pi a} = A\sigma_m + B\sigma_e \qquad (4)$$

In *Regime III*, $\lambda \ge AB/[C^2 - A^2]$, contact along the faces of the initial flaw is lost, where upon

$$K_I/\sqrt{\pi a} = (C^2\sigma_m^2 + E^2\sigma_e^2)^{1/2} \qquad (5)$$

The crack growth rate, $\dot{\ell}$ is related to the stress intensity factor by

$$\dot{\ell} = \begin{cases} \min\left[\dot{\ell}_0(K_I/K_{IC})^m, \sqrt{G/\rho_0}\right] & D < 1 \\ 0 & \text{otherwise} \end{cases} \qquad (6)$$

where K_{IC} is the mode I (short-crack) fracture toughness, m is a rate exponent, and $\dot{\ell}_0$ the reference crack growth rate at $K_I = K_{IC}$. The crack speed is limited by the shear wave speed $\sqrt{G/\rho_0}$, where G and ρ_0 are the shear modulus and density of the intact ceramic, respectively. When $D \to 1$, the microcracks coalesce, causing $K_I \to \infty$ (in *Regimes II and III*), thus requiring that the microcracking model be supplanted by the ensuing model for granular plasticity. This is where the new model deviates from that originally proposed by Deshpande and Evans.[10]

(C) Granular Plasticity: The comminuted ceramic is modeled as a granular medium using a nonassociated, visco-plastic, Drucker–Prager type constitutive law, with an effective

stress, $\hat{\sigma}$, defined as

$$\hat{\sigma} \equiv \frac{\sigma_e + (\tan\omega)\sigma_m}{1 - \tan\omega/3} \qquad (7)$$

where ω is the friction angle. The factor in the denominator ensures that, under a uniaxial compressive stress σ, the effective stress $\hat{\sigma} \equiv |\sigma|$.

The effective stress is used to define an effective strain rate $\dot{\varepsilon}_e^g$, motivated by the following considerations. Under compressive hydrostatic stress states ($\sigma_m < 0$), the flow of the comminuted ceramic is expected to follow a Bagnold-type granular flow law,[14] wherein the effective stress for the medium scales with the square of the effective strain rate. This relation arises because the stresses generated in the medium at high-strain rates are not associated with interparticle friction but, rather, are due to repeated collisions between particles. During these collisions, both the momentum change per collision and the number of collisions per unit time are proportional to the strain rate, resulting in the quadratic scaling. In contrast, for tensile hydrostatic stress states ($\sigma_m > 0$), particle interactions are negligible, violating the assumptions that underpin Bagnold flow. These features suggest an overstress model of the form:

$$\frac{\dot{\varepsilon}_e^g}{\dot{\varepsilon}_s} = \begin{cases} \left(\dfrac{\hat{\sigma}}{\Sigma_c} - 1\right)^s & \hat{\sigma} > \Sigma_c \\ 0 & \text{otherwise} \end{cases} \qquad (8a)$$

where Σ_c is the uniaxial compressive strength of the ceramic and $\dot{\varepsilon}_s$ a reference strain rate (specified subsequently in Section II(2)). Consistent with Bagnold scaling, the exponent is selected to be $s = 0.5$ when $\sigma_m \le 0$: whereupon (8a) gives the Bagnold scaling $\dot{\varepsilon}_e^g \sim \sqrt{\hat{\sigma}}$ in the limit $\hat{\sigma} \gg \Sigma_c$. Conversely, when $\sigma_m > 0$, s needs to be large (typically $s = s_0 \ge 10$) to ensure that $\hat{\sigma}$ does not significantly exceed the uniaxial strength, Σ_c, thereby preventing the comminuted ceramic from attaining unreasonably high-tensile stresses. To ensure that the strain rate $\dot{\varepsilon}_e^g$ is continuous around $\sigma_m = 0$, a piecewise function for s is specified, with

Fig. 3. Schematic of a microcracked solid containing an array of wing cracks.

the transition from $s = 0.5$ to s_0 occurring over the range $-\Sigma_c \leq \sigma_m \leq \Sigma_c$ and given by

$$s = \begin{cases} s_0 & \sigma_m \geq \Sigma_c \\ s_0 + \frac{1}{2}(s_0 - 0.5)\left(\dfrac{\sigma_m - \Sigma_c}{\Sigma_c}\right)^q & 0 \leq \sigma_m < \Sigma_c \\ 0.5 + \frac{1}{2}(s_0 - 0.5)\left(\dfrac{\sigma_m + \Sigma_c}{\Sigma_c}\right)^q & -\Sigma_c \leq \sigma_m < 0 \\ 0.5 & \text{otherwise} \end{cases} \tag{8b}$$

where the exponent q governs the transition rate.

To complete the characterization, a flow rule is specified. Experiments by Bagnold[14] and others suggest that, when deformation is governed by particle collisions, flow is nonassociated such that the flow potential is not determined by $\hat{\sigma}$. To ensure this generality, a flow surface G_g is defined as

$$G_g \equiv \begin{cases} \sigma_e + (\tan\delta)\sigma_m & \sigma_m < 0 \\ \sqrt{(\varepsilon\Sigma_t)^2 + \sigma_e^2} + \dfrac{\Sigma_c}{\Sigma_t}\sigma_m & \text{otherwise} \end{cases} \tag{9a}$$

where Σ_t is the hydrostatic tensile strength, given by setting $\hat{\sigma} = \Sigma_c$ and $\sigma_m/\sigma_e = 1/3$ in Eq. (7)

$$\Sigma_t \equiv \frac{\Sigma_c(1 - \tan\omega/3)}{\tan\omega} \tag{9b}$$

The quantity δ is the flow dilatation angle of the comminuted ceramic defined such that, when $\delta = 0$, the flow is volume conserving under compressive hydrostatic stress states. The parameter ε is a nondimensional regularization parameter (taken as 0.01) used to ensure that the flow is purely dilatational under hydrostatic tension. The granular strain rate follows from the flow surface G_g as

$$\dot{\varepsilon}_{ij}^g = \frac{\dot{\varepsilon}_e^g}{\varsigma}\frac{\partial G_g}{\partial\sigma_{ij}} \tag{10a}$$

where the normalization parameter is

$$\varsigma \equiv \begin{cases} \sqrt{\dfrac{3}{2} + \dfrac{1}{3}(\tan\delta)^2} & \sigma_m < 0 \\ \sqrt{\dfrac{(3/2)\sigma_e^2}{(\varepsilon\Sigma_t)^2 + \sigma_e^2} + \dfrac{1}{3}\left(\dfrac{\Sigma_c}{\Sigma_t}\right)^2} & \text{otherwise} \end{cases}. \tag{10b}$$

The factor ς is included in the flow rule so that

$$\left(\frac{1}{\varsigma}\frac{\partial G_g}{\partial\sigma_{ij}}\right)\left(\frac{1}{\varsigma}\frac{\partial G_g}{\partial\sigma_{ij}}\right) \equiv 1 \tag{11}$$

i.e., the magnitude of the effective granular plastic strain rate is set solely by $\dot{\varepsilon}_e^g$ with $\dot{\varepsilon}_e^g \equiv \sqrt{\dot{\varepsilon}_{ij}^g\dot{\varepsilon}_{ij}^g}$, while the direction of plastic straining is given by $\frac{\partial G_g}{\partial\sigma_{ij}}$.

(2) The Transition from Lattice to Granular Plasticity

Now that the basic models have been established, linkages must be provided. Namely, because the granular plasticity model is valid for all states of the comminuted ceramic (over the entire range $D_0 \leq D \leq 1$), evolution equations of the granular plasticity parameters from $D = D_0$ to $D = 1$ are required. These equations are specified through four parameters: (i) the uniaxial compressive strength Σ_c; (ii) the friction angle ω; (iii) the flow dilatation angle δ and (iv) a reference granular strain rate $\dot{\varepsilon}_s$. In its initial intact state (with $D = D_0$), the ceramic behaves as a von-Mises plastic material (with no contribution from granular plasticity). Accordingly, the granular plasticity parameters are chosen to ensure that, when $D = D_0$, the Drucker–Prager model reduces to the von-Mises model and that the granular strain rate is zero. The following evolution equations then ensue.

(i) *Uniaxial compressive strength Σ_c.* As D increases, Σ_c decreases from the plastic yield strength σ_0 to the fully comminuted strength σ_c, specified using

$$\Sigma_c = \sigma_0 - (\sigma_0 - \sigma_c)\left(\frac{D - D_0}{1 - D_0}\right)^q \tag{12}$$

The exponent q governs the transition rate with respect to D.

(ii) *The friction angle ω.* When the granular strength Σ_c is equal to the plastic yield strength σ_0, the friction angle $\omega = 0$. The angle increases with damage, reaching a value ω_0 upon complete comminution. Its evolution with the current material strength is specified by:

$$\tan\omega = \frac{(\sigma_0 - \Sigma_c)\tan\omega_0}{\sigma_0 - \sigma_c(1 - \tan\omega_0/3) - \Sigma_c\tan\omega_0/3} \tag{13a}$$

Thus, the granular yield surface evolves by pivoting about a coordinate in stress space defined by $\sigma_e = \sigma_0$ and

$$\sigma_m = \frac{\sigma_0 - \sigma_c}{\tan\omega_0} + \frac{\sigma_c}{3} \tag{13b}$$

This evolution is depicted in Fig. 2.

(iii) *The flow dilatation angle δ.* Because the intact ceramic behaves as an incompressible plastic medium, the flow dilatation angle should vary between $\delta = 0$ at $D = D_0$ to that for the fully equation of motion is given by

$$\frac{\partial}{\partial r}(r^2 \sigma_r) - 2r\sigma_\theta = \rho \ddot{u} r^2 \tag{26}$$

where r denotes the radial position of a material point in the current configuration, u is the radial displacement of that point, the overdot denotes differentiation with respect to time, and ρ is the material density in the current configuration. Equation (26) is solved using an updated Lagrangian scheme, detailed in Deshpande and Evans.[10] In the FE calculations, the outer radius of the medium being analyzed is taken to be sufficiently large so as to ensure that the elastic wave does not reach the outer boundary over the time-span of the calculations reported here, thereby simulating an infinite medium. Unless otherwise specified, the computations use the property values for alumina (summarized in Section II(4)). The focus of subsequent numerical results and discussion is on the early time response, wherein dynamic effects dominate. For the expansion of a spherical cavity in an infinite *elastic* medium, Hopkins[17] has shown that the stress distributions near the vicinity of the cavity are well approximated by the corresponding quasi-static result at times $\bar{t} \equiv c_e t / a_0 > 4$, where c_e is the p-wave speed. Thus, all results presented here are for times $\bar{t} < 4$.

(2) Effect of Pressure

The predicted temporal variation of the radial displacement U/a_0 of the cavity surface is plotted against time $\bar{t}$ in Fig. 6 for three levels of the cavity pressure. The cavity displacement increases approximately linearly with time for the three pressure levels considered here. We proceed to investigate in further detail the cases in which $p_0 = 12$ and 24 GPa.

First consider the case where $p_0 = 12$ GPa. The spatial distributions of damage D, effective lattice plastic strain ε_e^{pl} and effective granular plastic strain ε_e^{g} are plotted in Fig. 7(a) at three select times. Note that ε_e^{pl} and ε_e^{g} are the time integrals of the strain rates given by Eqs. (1) and (8a), respectively. The distributions are plotted as a function of R/a_0, where R is the radial coordinate in the undeformed configuration. Full comminution of the ceramic (i.e., $D = 1$) first occurs at a finite distance from the cavity surface, viz. at $R/a_0 \approx 1.5$. The damage zone then expands radially inwards while simultaneously expanding outwards. However, while the damage zone continues to expand outwards for the duration of the computations, the progression of the damage zone toward the surface to the cavity halts at time $\bar{t} \approx 1.4$ when the inner boundary of the damage zone reaches $R/a_0 \approx 1.1$, i.e., the damage zone does not extend up to the cavity surface. We shall show subsequently that this feature is due to the fact that the large lattice plastic strains near the cavity

surface reduce the stress triaxiality and thereby reduce the propensity for microcracking. The spatial distributions of ε_e^{pl} included in Fig. 7(a) show that, for $\bar{t} > 1.4$, the lattice plastic strain rises sharply for $R/a_0 < 1.1$, i.e., the region over which damage does not occur. Correspondingly, the distribution of granular plastic strains clearly show that significant granular plasticity only occurs over the region where $D = 1$. Note that the total inelastic strains (i.e., lattice plus granular plastic strains) are continuous across the boundary of the damage zone near the surface of the cavity; while ε_e^{g} drops to zero across the boundary where the damage variable transitions from $D = 1$ to $D = D_o$, ε_e^{pl} rises sharply in order to keep the total inelastic strain approximately continuous.

At the higher applied cavity pressure of $p_0 = 24$ GPa, the qualitative picture discussed above remains unchanged. However, the higher pressure means that the stress triaxiality near the cavity surface is further reduced and hence microcracking initiates even further away from the cavity surface at $R/a_0 \approx 2.2$ and expands *only* away from the cavity surface, where the stress triaxiality increases. The discontinuity in the value of ε_e^{pl} across the damage zone interface near the surface of the cavity is now more obvious. However, similar to the $p_0 = 12$ GPa case, the total inelastic strains are approximately continuous across this interface, with granular plasticity inside the damage zone replaced by lattice plasticity.

Insights into the effects of lattice plasticity on stress triaxiality (which in turn affect microcracking) are gleaned from computations in which the damage is artificially suppressed, by setting $K_{IC} = \infty$. The results are plotted in Fig. 8 for three times and three values of yield strength, selected such that $p_0/\sigma_Y = 1.71$, 1.5, and 0 when $p_0 = 24$ GPa. For the purely elastic case ($p_0/\sigma_Y = 0$), the triaxiality is at its lowest (most negative) at the elastic wave front (as noted by Hopkins[17]). Thus, absent lattice plasticity, damage initiates at the cavity surface, where both the stresses and the stress triaxiality are greatest, and propagates radially outward, trailing the elastic wave front. When lattice plasticity is activated, as in the cases of $p_0/\sigma_Y = 1.71$ and 1.5, plasticity initiates on the cavity surface, where the deviatoric stress is highest, and subsequently spreads radially outward. This plasticity results in a reduction in the stress triaxiality in the plastic zone, as evident upon comparison of the results for $p_0/\sigma_Y = 1.71$ and 1.5 with those of the elastic case. The magnitude of this effect increases with increasing plastic strain. Moving radially outward from the cavity surface, the plastic strain diminishes and hence its effect on stress triaxiality decreases. Eventually, the triaxiality reaches the value corresponding to the elastic case at the elastic/plastic interface. Because of the opposing effects of elastic wave propagation and plasticity on stress triaxiality—the elastic wave reducing triaxiality with increasing R/a_0 and the near-surface plasticity reducing triaxiality as $R/a_0 \to 1$—the triaxiality attains a *maximum value at a finite distance* away from the cavity surface. For cases where damage is activated (such as that shown in Fig. 7), damage initiates at or near this location and then spreads radially both inward and outward. A corollary of the results presented in Fig. 8 is that the stress triaxiality decreases with increasing values of p_0/σ_Y, due to the increased levels of lattice plasticity. This is evident from comparisons of the results for $p_0/\sigma_Y = 1.71$ and 1.5 in Fig. 8. The reductions in the stress triaxiality for higher values of p_0/σ_Y result in damage initiating further away from the cavity surface as observed in Fig. 7(b).

(3) The Interplay Between Plasticity and Damage

The results in the preceding section demonstrate the interplay between lattice plasticity and stress triaxiality. Its connection to damage development and cavity expansion emerges from a series of computations for a cavity pressure $p_0 = 12$ GPa and three values of yield strength σ_Y (all other material properties being kept fixed at their reference values, including the finite fracture toughness). The variation in the displacement of the cavity surface U/a_0 with time $\bar{t}$ for each of these cases is plotted in

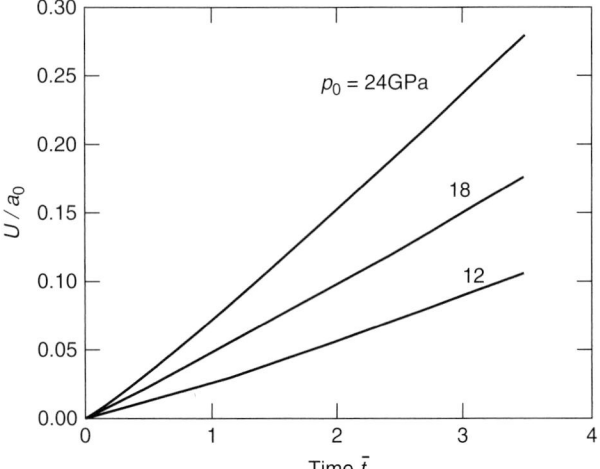

Fig. 6. The evolution of the radial displacement of the cavity surface with time. Results are for alumina with reference properties.

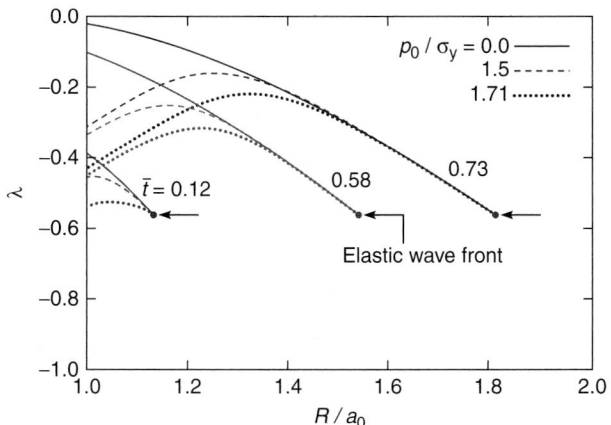

Fig. 7. Spatial distributions of damage D, effective lattice plastic strain ε_e^{pl} and effective granular plastic strain ε_e^g for applied cavity pressures, p_0, of (a) 12 GPa and (b) 24 GPa. Results are shown for alumina with reference properties at three select times $\bar{t}$.

Fig. 9(a). Interestingly, the cavity expansions are approximately equal for the $\sigma_Y = 5$ and 9 GPa cases (indeed, slightly lower for $\sigma_Y = 5$ GPa) but significantly higher for $\sigma_Y = 3$ GPa. In order to

Fig. 8. Evolution of stress triaxiality associated with dynamic spherical cavity expansion in an elastic–plastic (nondamaging) alumina.

parameterise this information we plot the cavity displacement U_c/a_0 at a time $\bar{t} = 3$ against yield strength σ_Y in Fig. 9(b) for cavity pressures $p_0 = 12$ and 24 GPa. Three regimes emerge: one is dominated by damage, at high values of σ_Y, wherein U_c/a_0 is independent of σ_Y; a second dominated by lattice plasticity, wherein U_c/a_0 increases with decreasing σ_Y; and an intermediate regime in which lattice plasticity and damage occur simultaneously. Thus, there exists a critical value of σ_Y above which U_c does not decrease any further. This clearly shows that increasing the yield strength of the ceramic indefinitely will not bring any further performance benefits as the response becomes dominated by the microcracking deformation mode. However, the critical value of σ_Y increases with increasing applied pressure; for the parameters considered here, the critical yield strength increases from $\sigma_Y \approx 5$ to 8 GPa as p_0 increases from 12 to 24 GPa.

V. Indentation

The fidelity of the model has been assessed in two companion studies (B. Compton *et al.*, unpublished data).[15] Here we highlight some of their results in order to illustrate the applicability of the model. All tests were performed on an armor-grade

Fig. 9. (a) The evolution of radial displacement of the cavity surface with time for an applied cavity pressure $p_0 = 12$ GPa. (b) Effects of yield strength and pressure on the radial cavity displacement U_c at time $\bar{t} = 3$.

alumina (Corbit 98, produced by Industri Bitossi, Sovigliana vinci, Florence, Italy) with an average grain size $d = 3$ μm. It consists of 98% alumina and minor amounts of a glassy phase, the latter being situated predominantly at triple grain junctions. Details of the properties of this material and the model parameters (listed in Section II(4)) as well as the procedures used to obtain their values are in.[15]

(1) Quasi-static Indentation

The quasi-static penetration resistance was probed using a spheroconical diamond indenter with a 200 μm tip radius and 120° cone angle, mounted on a conventional servo-hydraulic mechanical test system.[15] Displacements were measured with submicrometer accuracy using a one-armed extensometer positioned on the alumina surface. Additionally, corrections for machine compliance and elastic (Hertzian) deformation of the indenter were used to remove displacements external to the indentation site. Corresponding FE computations were performed using the current model and the original DE model. (The parameter set of the DE model is a subset of the current model, i.e., all parameters of the current model expect those associated with granular plasticity. To make a fair comparison between the current and DE models we use the values parameters listed in Section II(4) in the calculations with the DE model as well.)

Figure 10 shows a comparison between predictions and measurements of both the load–displacement response and the damage immediately under the indenter at an applied load of 200 N. The current model is shown to accurately predict both the extent of the damage immediately under the indenter and the load versus penetration response. By contrast, the DE model predicts significantly greater damage and unrealistically large surface uplift around the indent periphery. This large damage manifests as large inelastic displacements during loading and hence the DE model also does not predict the measured load versus penetration response with an adequate level of accuracy. These discrepancies of the DE model are a consequence of the elastic dilatation associated with the cracks but not associated with granular plasticity.

(2) Dynamic Indentation

Constrained alumina tiles, each 50 mm × 50 mm × 12 mm thick, were impacted by 7.14 mm diameter steel spheres at velocities in the range 300–800 ms^{-1} (B. Compton *et al.*, unpublished data).

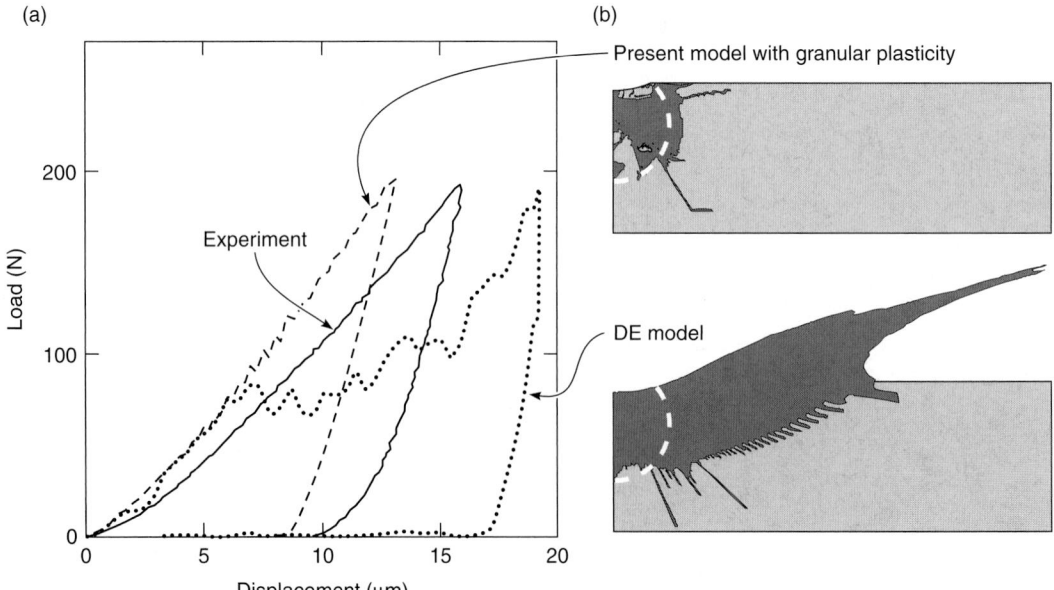

Fig. 10. (a) Comparisons between predicted and measured quasi-static indentation responses. (b) Corresponding subsurface damage at an applied load of 200 N. Predictions are shown for both the current and the DE models with the area shaded dark corresponding to fully microcracked ceramic. The dashed lines indicate the extent of the observed commutated zone (Adapted from[15]).

Fig. 11. Comparisons between predicted and measured subsurface damage in alumina resulting from impact of a 304 stainless-steel (SS) sphere at an incident velocity of 750 ms^{-1}. The dark shaded regions are fully damaged whereas the light shaded regions are undamaged. Predictions are shown for both the current and the DE models. Three cases are considered: no face-sheet, 0.5-mm-thick 304 SS face-sheet and 1-mm-thick 304 SS face-sheet.

Three situations were considered: (i) tiles with no face sheet; (ii) tiles with a 0.5 mm thick 304 stainless-steel face sheet; and (iii) tiles with a 1-mm-thick 304 stainless-steel face sheet. After the tests, the tiles were stained with a blue dye and impregnated with an epoxy resin in order to contain the comminuted ceramic. They were then sectioned in half in order to reveal the damage pattern directly beneath the impact site. Comparisons between the observations and the predictions (from both current and original DE models) for an impact velocity 750 ms^{-1} are presented in Fig. 11. The current model captures the observations for the three scenarios of present interest with reasonable fidelity, including the observation that the extent of damage is significantly reduced by the presence of the steel face sheets. By contrast, in line with the quasi-static indentation simulations, the DE model overestimates the damage significantly in all cases.

VI. Limitations of the Model

The micromechanical model for ceramics presented in this study has some inherent limitations that fall into two catagories: (i) limitations related to microstructural assumptions and (ii) limitations related to the neglecting of the statistical aspects of ceramic failure. Here we briefly discuss these limitations.

Three inelastic mechanisms have been accounted for in the approach presented here, viz. microcracking from preexisting flaws presumed to be present at grain boundaries, lattice plasticity and granular plasticity. Cracking in polycrystalline ceramics is however known to occur from numerous other sources of heterogeneities such as inclusions and voids,[23] glass present at grain boundaries[24] and surface flaws. The modeling framework presented here is sufficiently general that these mechanisms can be included in the constitutive model, albeit at the cost of greater complexity and additional material parameters that will inevitably require calibration. We thus expect to include these additional mechanisms as additional experimental data that suggests the need to incorporate these mechanisms becomes available.

It is well known that the failure strength of ceramics is both statistical in nature and size dependent. This is often well represented by the Weibull distribution.[25] The statistical effects give rise to the well-known observation that nominally identical ceramic specimens have different mechanical properties including sometime markedly different ballistic responses. To account for these statistical variations, Leavy *et al.*[26] included uncertainty into a deterministic model such as the Johnson and Holmquist[5]

model. This enabled them to not only capture the observed failure patterns in a Brazilian test but including a distribution of strengths also significantly alleviated the mesh dependency problems inherent in these ceramic models that are based on a damage mechanics approach. Inclusion of uncertainty into the present model is a necessary next step in developing a predictive ceramic model for armor applications.

VII. Conclusions

A unified model for the dynamic inelastic response of ceramics under compressive stress states has been developed. It incorporates three inelastic deformation mechanisms: (i) lattice plasticity due to dislocation glide or twinning; (ii) microcrack extension; and (iii) granular flow of densely packed comminuted particles. In addition to analytical descriptions of each of these mechanisms, prescriptions have been provided for their implementation into a FE code as well as schemes for transitioning between mechanisms.

The utility of the code in addressing issues related to deep penetration has been demonstrated through a series of calculations of dynamic cavity expansion in an infinite medium subject to a fixed internal pressure p. The results reveal two limiting behavioral regimes: one dictated by lattice plasticity and another by granular plasticity. The competition between these two deformation mechanisms along with the interplay between plasticity and stress triaxiality lead to an optimal value of yield strength at which the cavity expansion rate is minimized. This result suggests that, once the damage and fracture parameters are specified, an optimal yield strength exists for superior penetration resistance at a prescribed pressure. It remains to be established whether this effect can be exploited in designing ceramics to resist penetration over a targeted range of loadings.

Granular plasticity proves to be an essential ingredient in the constitutive law, as demonstrated by the quasi-static and dynamic indentation studies. When neglected (as in the original DE model), the spatial extent of damage and the magnitude of the strains are strongly overpredicted. Nevertheless, the model fails to capture all features in the indentation response. Most notably, the formation of lateral subsurface cracks during the unloading phase of indentation cannot be predicted. To capture this feature along with the ensuing spallation that commonly accompanies indentation, the model would need to be expanded to include a fracture mechanics-based component for modeling "macro" rather than "micro" cracking.

Appendix A: Summary of Microcrack Growth Coefficients

The coefficients used in Section II(1) are defined as follows. The parameters A and B are given as

$$A \equiv c_1(c_2A_3 - c_2A_1 + c_3) \tag{A-1}$$

and

$$B \equiv \frac{c_1}{\sqrt{3}}(c_2A_3 + c_2A_1 + c_3) \tag{A-2}$$

where

$$c_1 = \frac{1}{\frac{\pi^2}{2^{3/4}}\left[\left(\frac{D}{D_0}\right)^{1/3} - 1 + \beta\sqrt{2}\right]^{3/2}} \tag{A-3}$$

$$c_2 = 1 + 2\left[\left(\frac{D}{D_0}\right)^{1/3} - 1\right]^2\left(\frac{D_0^{2/3}}{1 - D^{2/3}}\right) \tag{A-4}$$

and

$$c_3 = \pi^2\left[\left(\frac{D}{D_0}\right)^{1/3} - 1\right]^2 \tag{A-5}$$

while

$$A_1 = \pi\sqrt{\frac{\beta}{3}}[(1 + \mu^2)^{1/2} - \mu] \tag{A-6}$$

and

$$A_3 = A_1\left\{\frac{(1 + \mu^2)^{1/2} + \mu}{(1 + \mu^2)^{1/2} - \mu}\right\} \tag{A-7}$$

Here β is a coefficient introduced by Ashby and Sammis[16] to convert exact two-dimensional solutions to three-dimensional stress states. The coefficients E and C are defined as

$$E^2 = \frac{B^2C^2}{C^2 - A^2} \tag{A-8}$$

and

$$C \equiv A + \gamma\sqrt{\frac{1}{\sqrt{2}}\left(\frac{D}{D_0}\right)^{1/3}} \tag{A-9}$$

with γ a constant used to match tensile data.

Acknowledgments

E. A. G. was supported by a National Defense Science and Engineering Fellowship. V. S. D. acknowledges ONR for support via award no. N00014-09-1-0573.

References

[1] A. M. Rajendran and D. J. Grove, "Modeling the Shock Response of Silicon Carbide, Boron Carbide and Titanium Diboride," *Int. J. Impact Eng.*, **18**, 611–31 (1996).

[2] G. T. Camacho and M. Ortiz, "Computational Modeling of Impact Damage in Brittle Materials," *Int. J. Solids Struct.*, **33**, 2899–938 (1996).

[3] P. J. Hazell and M. J. Iremonger, "Crack Softening Damage Model for Ceramic Impact and its Application with in a Hydrocode," *J. Appl. Phys.*, **82**, 1088–92 (1997).

[4] H. D. Espinosa, P. D. Zavattieri, and S. K. Dwivedi, "A Finite Deformation Continuum/Discrete Model for the Description of Fragmentation and Damage in Brittle Materials," *J. Mech. Phys. Solids*, **46**, 1909–42 (1998).

[5] G. R. Johnson and T. J. Holmquist, "Response of Boron Carbide Subjected to Large Strains, High Strain Rates, and High Pressures," *J. Appl. Phys.*, **85**, 8060–73 (1999).

[6] C. H. M. Simha, S. J. Bless, and A. Bedford, "Computational Modeling of the Penetration Response of a High-Purity Ceramic," *Int. J. Impact Eng.*, **27**, 65–861 (2002).

[7] D. C. Drucker and W. Prager, "Soil Mechanics and Plastic Analysis or Limit Design," *Quart. Appl. Math.*, **10**, 157–65 (1952).

[8] G. R. Johnson and W. H. Cook, "Fracture Characteristics of Three Metals Subjected to Various Strains, Strain Tatesm Temperatures and Pressures," *Eng. Fract. Mech.*, **21**, 31–48 (1985).

[9] D. W. Templeton, T. J. Holmquist, H. W. Meyer, D. J. Grove, and B. Leavy, "A comparison of ceramic material models"; pp. 299–308 in *Proceedings of the Ceramic Armor by Design Symposium*, Pacific Rim IV International Conference on Advanced Glass and Ceramics, Wailea, Maui, Hawaii, 2001.

[10] V. S. Deshpande and A. G. Evans, "Inelastic Deformation and Energy Dissipation in Ceramics: A Mechanism-Based Dynamic Constitutive Model," *J. Mech. Phys. Solids*, **56**, 3077–100 (2008).

[11] Z. Wei, V. S. Deshpande, and A. G. Evans, "The Influence of Material Properties and Confinement on the Dynamic Penetration of Alumina by Hard Spheres," *J. Appl. Mech.*, **76**, 051305, 8pp (2009).

[12] Z. Wei, A. G. Evans, and V. S. Deshpande, "Mechanisms and Mechanics Governing the Indentation of Polycrystalline Alumina," *J. Am. Ceram. Soc.*, **91**, 2987–96 (2008).

[13] H. J. Frost and M. F. Ashby, *Deformation Mechanism Maps*. Pergamon Press, Oxford, 1982.

[14] R. A. Bagnold, "Experiments on a Gravity-Free Dispersion of Large Solid Particles in a Newtonian Fluid Under Shear," *Proc. R. Soc. Lond.*, **A225**, 49–63 (1954).

[15] E. A. Gamble, B. Compton, V. S. Deshpande, A. G. Evans, and F. W. Zok, "Penetration resistance of an armor ceramic: I- Quasi-static loading," *J. Am. Ceram. Soc.*, (2010) (this issue).

[16] M. F. Ashby and C. G. Sammis, "The Damage Mechanics of Brittle Solids in Compression," *Pure Appl. Geophys.*, **133**, 489–521 (1990).

[17] H. G. Hopkins, "Dynamic Expansion of Spherical Cavities in Metal"; pp. 85–164 in *Progress in Solid Mechanics*, Vol. 1, Edited by I. N. Sneddon, and R. Hill. North-Holland, Amsterdam, 1960.

[18] M. J. Forrestal, D. Y. Tzou, E. Askari, and D. B. Longcope, "Penetration into Ductile Metal Targets with Rigid Spherical-Nose Rods," *Int. J. Impact Eng.*, **16**, 699–710 (1995).

[19] R. W. Macek and T. A. Duffey, "Finite Cavity Expansion Method for Near-Surface Effects and Layering During Earth Penetration," *Int. J. Impact Eng.*, **24**, 239–58 (2000).

[20] D. Durban and R. Masri, "Dynamic Spherical Cavity Expansion in a Pressure Sensitive Elastoplastic Medium," *Int. J. Solids Struct.*, **41**, 5697–716 (2004).

[21] S. Satapathy, "Dynamic Spherical Cavity Expansion in Brittle Ceramics," *Int. J. Solids Struct.*, **38**, 5833–45 (2001).

[22] J. D. Walker, "Analytic model for penetration of thick ceramic targets"; pp. 337–48 in *Proceedings of the ceramic armor by design symposium*, Pacific Rim IV International Conference on Advanced Glass and Ceramics, Wailea, Maui, Hawaii, 2001.

[23] C. G. Sammis and M. F. Ashby, "The Failure of Brittle Porous Solids Under Compressive Stress States," *Acta Metall.*, **34**, 511–26 (1986).

[24] D. R. Clarke, "Grain Boundaries in Polycrystalline Ceramics," *Annu. Rev. Mater. Sci.*, **17**, 57–74 (1987).

[25] W. Weibull, "A Statistical Distribution Function of Wide Applicability," *J. Appl. Mech.*, **18**, 293–7 (1951).

[26] R. B. Leavy, R. M. Brannon, and O. E. Strack, "The Use of Sphere Indentation Experiments to Characterize Ceramic Damage Models," *Appl. Ceram. Technol.*, **7**, 606–15 (2010). □

J. Am. Ceram. Soc., **94** [S1] S215–S225 (2011)
DOI: 10.1111/j.1551-2916.2011.04472.x
© 2011 The American Ceramic Society

Journal

Damage Development in an Armor Ceramic Under Quasi-Static Indentation

Eleanor A. Gamble,[‡] Brett G. Compton,[‡] Vikram S. Deshpande,[§] Anthony G. Evans,[‡] and Frank W. Zok[†,‡]

[‡]Materials Department, University of California, Santa Barbara, California 93106

[§]Engineering Department, Cambridge University, Cambridge CB2 1PZ, U.K.

The objective of the present study is to assess the capabilities of a recently developed mechanism-based model for inelastic deformation and damage in structural ceramics. In addition to conventional lattice plasticity, the model accounts for microcrack growth and coalescence as well as granular flow following comminution. The assessment is made through a coupled experimental/computational study of the indentation response of a commercial armor ceramic. The experiments include examinations of subsurface damage zones along with measurements of residual surface profiles and residual near-surface stresses. Extensive finite element computations are conducted in parallel. Comparisons between experiment and simulation indicate that the most discriminating metric in the assessment is the spatial extent of subsurface damage following indentation. Residual stresses provide additional validation. In contrast, surface profiles of indents are dictated largely by lattice plasticity and thus provide minimal additional insight into the inelastic deformation resulting from microcracking or granular flow. A satisfactory level of correlation is obtained using property values that are either measured directly or estimated from physically based arguments, without undue reliance on adjustable (nonphysical) parameters.

I. Introduction

CERAMIC materials are recognized for their effectiveness in protecting against ballistic threats. Their high hardness enables energy dissipation by plastic deformation and erosion of projectiles during the early stages of impact. But their brittleness precludes their use as primary load-bearing elements within structural systems. Furthermore, ceramics have limited utility in protecting against other dynamic threats such as blast loads. Consequently, in order to satisfy the full spectrum of structural and protective requirements, ceramics will need to be integrated into complex multimaterial systems with metallic alloys and fiber-reinforced polymer composites. With almost limitless configurational possibilities, the most rapid progress in the design of weight-efficient systems will be achieved through numerical simulations coupled with select validation tests. The success of this approach is predicated on the availability of reliable constitutive laws for the constituent materials.

The goal of the present article is to begin assessment of the capabilities of a recently developed mechanism-based model for ceramic deformation. The first version of the model, by Deshpande and Evans,[1] embodied two principal inelastic mech-

anisms: (i) lattice plasticity by dislocation glide or twinning (Fig. 1(a)) and (ii) extension of grain-sized microcracks (Fig. 1(b)). A recent extension of the model has addressed the material behavior after the damage has saturated, that is, once the microcracks have coalesced.[2] In this state, the material can deform by one of two mechanisms: lattice plasticity within the individual particles, or granular plasticity, wherein the particles flow past one another (Fig. 1(c)). The latter mechanism is closely analogous to that which occurs in soils, albeit with a much higher packing density of the constituent particles. The extended version of the Deshpande–Evans (DE) model is detailed in a companion paper in this issue.

The study focuses specifically on the response of a prototypical armor ceramic to *quasi-static indentation* by a hard spheroconical indenter. Such tests can reproduce the high contact pressures and level of confinement characteristic of the early stages of projectile impact. Examinations of the subsurface regions performed in this study and reported previously by Lawn and colleagues[3,4] reveal the presence of a heavily microcracked "quasi-plastic zone" beneath the indenter. The correlation between these microcracked regions and the Mescall zones formed during ballistic impact has been noted by LaSalvia and McCauley.[5] These correlations suggest that the quasi-static indentation tests are well suited to probing some of the pertinent inelastic deformation mechanisms: albeit lacking in the dynamic aspects of the loading. The principal advantage of the quasi-static tests (relative to impact) is their amenability to accurate instrumentation during loading. Secondary benefits include the lack of dynamic spallation or extraneous cracking caused by stress wave reflections in impact tests. Additionally, by using bonded interface specimens[4] (described below), the indentation tests allow for careful observations of subsurface damage, without the need for posttest sectioning and polishing and the associated artifacts that can ensue.

In addition to indentation tests and subsequent microstructural examinations, the experimental portion of the study includes measurement of the residual stresses in the region of the indents as well as the residual surface profiles. The experimental measurements are accompanied by extensive finite element computations using a range of material parameter values. A paper, comparing the predictions of the calibrated constitutive law with the damage and deformation caused by dynamic impact, is in preparation (B. G. Compton *et al.*, unpublished data).

The article is organized in the following way. Section II contains a description of the material and the experimental procedures employed in this study. Section III provides a synopsis of the extended DE model, with emphasis on the inelastic mechanisms and the associated material parameters required for calibration. Experimental measurements and observations are summarized in Section IV. This is followed by presentation of the computational results of the damage zones beneath the indents, in Section V, as well as an assessment of the residual stresses and surface profiles, in Section VI. Implications of the correlations and limitations of the extended DE model are addressed in the final section.

A. Heuer—contributing editor

Manuscript No. 28771. Received October 13, 2010; approved February 5, 2011.
This work was supported by the Office of Naval Research through a Multidisciplinary University Research Initiative Program on "Cellular Materials Concepts for Force Protection," Prime Award No. N00014-07-1-0764.
[†]Author to whom correspondence should be addressed. e-mail: zok@engineering.ucsb.edu

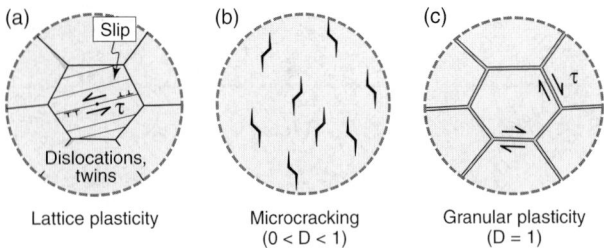

Fig. 1. Mechanisms of inelastic deformation and damage operative in polycrystalline ceramics under compressive stress states.

II. Material and Test Procedures

The material used for the experimental study was a fine-grained armor-grade ceramic (Corbit 98, produced by Industrie Bitossi, Sovigliana Vinci, Italy). It consists of 98% alumina and minor amounts of a glassy phase, the latter being situated predominantly at triple grain junctions. The glassy phase is evident on polished cross-sections in backscatter electron imaging mode in the scanning electron microscope (SEM) (Fig. 2(a)). It is also manifested as triple grain pores following either thermal etching (caused by volatilization of the glass, Fig. 2(b)) or chemical etching (Figs. 2(c) and (d)). The average grain diameter, d, is 3 μm.

A series of Vickers indentation tests was performed over a load range of 0.7–10 kg. The measurements yield a hardness $H = 15.9 \pm 1.8$ GPa and a (long crack) fracture toughness, $K_{IC} = 2.9 \pm 0.3$ MPa$\sqrt{m}$. The latter tests reveal that fracture occurs along an intergranular path, attributable to the presence of a thin grain-boundary glass layer.[6]

The Young's modulus, E, measured by instrumented nanoindentation, is 405 GPa. This value exceeds that reported by the manufacturer (384 GPa) by about 5%. The difference is attributable to the presence of both residual porosity and the compliant glassy phase at the grain boundaries. Neither of these features is probed by the nanoindentation tests. Consequently, the lower value is selected for use in subsequent computations.

Penetration resistance was probed using a spheroconical diamond indenter with a 200 μm tip radius and 120° cone angle. Two specimen geometries were employed. (i) Flat-polished plates were used to measure the plastic indentation stress–strain response. They were also used for measurement of surface profiles through the indents and the near surface residual stresses (described below). (ii) The nature and spatial extent of sub-surface damage were ascertained using bonded interface specimens.[4] To produce such specimens, pairs of polished alumina plates were bonded with a thin (∼5 μm) layer of cyanoacrylate adhesive and clamped together with a pressure of about 10 MPa. One of the orthogonal surfaces was then polished. The clamping pressure was sufficient to maintain sample integrity during polishing and indentation but negligible in comparison to the those required to induce plastic deformation (on the order of 10 GPa). Indentation tests were performed on the latter surface with the indenter placed directly over the bonded interface (Fig. 3). Surface profiles around the indentations made on the bonded interface specimens were indistinguishable from those of corresponding indents on flat-polished samples. The inference is that the presence of the interface does not affect the overall deformation response. Whether some salient features of the deformation mechanisms directly on the bonded surfaces differ from those operating internally (say 10 μm below the bondline) remains to be critically addressed.

Following testing, the samples were separated and cleaned. The internal polished surfaces were coated with gold and examined using Nomarski interference optical microscopy and SEM. In some instances, the region directly beneath the indent was also ion-milled (to a depth of 200 nm) after indentation to remove surface texture resulting from plastic deformation. The latter surface treatment allows more definitive identification of grain-boundary cracks. Additionally, narrow grooves were cut perpendicular to the polished face using focused ion beam (FIB) machining. These grooves were cut both directly beneath the indent and at a remote location; the exposed surfaces were examined in the SEM.

Fig. 2. SEM micrographs of polished sections of Corbit 98: (a) as-polished and imaged in backscatter electron mode (darker regions being the glassy phase); (b) polished and thermally etched at 1400°C for 1 h; and (c, d) polished and etched with phosphoric acid for 4 min at 130°C. Images in (b)–(d) taken in secondary electron mode.

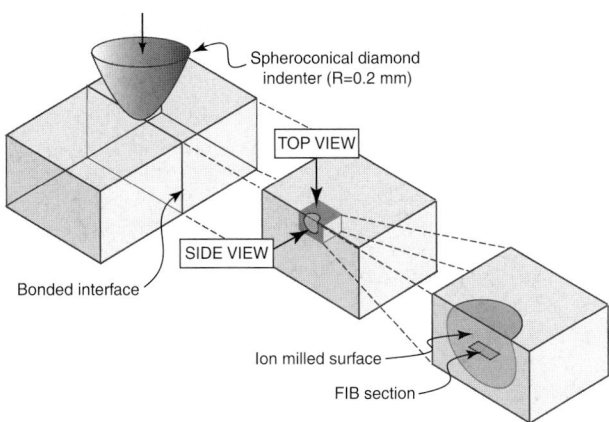

Fig. 3. Schematic of the bonded interface specimen.

Surface profiles on flat-polished specimens in the vicinity of the indents were measured using an optical profilometer (Wyko NT1100, Veeco, Tucson, AZ). Profilometry scans were also used to measure the diameter of the indent crater and, in turn, the nominal hardness (i.e., peak load/residual indent area). A series of such measurements to various peak loads was used to establish the plastic indentation stress–strain response.

Residual hydrostatic stresses in the near-surface vicinity of the indents were measured using a luminescence piezospectroscopy technique.[7,8] The technique involves use of a laser beam in an optical microscope to stimulate luminescence of trace Cr^{3+} ions and a spectrometer to measure the stress-induced frequency shift of the characteristic R_2 luminescence peak. The residual hydrostatic stress is proportional to the frequency shift through the piezospectroscopic coefficient $(7.61 \ (cm \cdot GPa)^{-1})$. The diameter of the beam employed in the present study was about 3 μm and its penetration depth 10 μm. Experiments performed to measure the effective beam diameter and the beam interactions with the subsurface regions of the material are detailed in Appendix A. The corresponding analysis (also in Appendix A) was subsequently used to establish a depth averaging protocol for the residual stresses computed by finite elements for comparison with the experimental measurements.

III. Constitutive Law and Finite-Element Analysis

The extended DE model[2] was used to represent the mechanical response of the alumina. A descriptive synopsis of the model follows and the key formulae are compiled in Appendix B. Graphical representations of the constitutive law are shown in Fig. 4.

(1) Lattice Plasticity

Plasticity is described by the effective stress σ_e and the von Mises yield criterion, with yield stress σ_y, and the subsequent hardening by a conventional power law, characterized by a hardening exponent M. (Strain rate sensitivity is also taken into account, through an overstress model, calibrated by fitting plasticity data for alumina.[9] However, such effects are exceedingly small in ceramics in the quasi-static loading regime at ambient temperature and, for all practical purposes, can be neglected.)

(2) Distributed Microcracking

Two non-dimensional parameters characterize the initial damage state: the normalized initial flaw size, $g_1 \equiv a/d$, and the normalized flaw spacing, $g_2 \equiv 1/(f^{1/3}d)$ (f being the number of microcracks per unit volume). Three behavioral regimes are obtained, delineated by the stress triaxiality, $\lambda \equiv \sigma_m/\sigma_e$ (σ_m being the mean stress) (Fig. 4(a)). In Regime I, preexisting cracks are closed and the shear stress is insufficient to overcome the frictional resistance of the contacting crack surfaces. Conse-

quently, the stress intensity is zero and crack growth cannot occur. This behavior is obtained when the stress triaxiality is less than a critical value, λ_C: dictated largely by the friction coefficient, μ, of the crack surfaces and the current level of microcrack damage. For friction coefficient values typical of engineering ceramics, $\lambda_C \approx -0.5$ to -1. At higher triaxialities, in Regime II, the crack surfaces remain in contact but are able to slide past one another. Sliding leads to a finite stress intensity which drives the formation of wing cracks at the tips of existing flaws. The cracks extend, initially stably, and eventually link to produce a fully comminuted ceramic. The crack growth rate is expected to increase with increasing stress intensity. Following Aeberli and Rawlings,[10] the crack growth rate is taken to scale with $(K_I/K_{IC})^m$ where $m \gg 1$. Furthermore, the growth rate is limited by the shear wave speed. In Regime III, the crack faces are not in contact with one another and thus the cracks can propagate unstably.

(3) Granular Plasticity

Once fully comminuted, the grains can rotate and slide past one another (analogous to deformation of soils). Granular flow is taken to obey the linear Drucker–Prager yield criterion, characterized by a critical strength σ_e^{cr} at $\sigma_m = 0$ and a friction angle ω (defined in Fig. 4(b)). The plastic potential is taken to be linear as well, with dilatational angle, ψ (Fig. 4(b)). Associated flow is obtained by setting $\psi = \omega$, whereupon the plastic strain exhibits significant dilatation. At the other extreme, where $\psi = 0$, the granular plastic strain is purely deviatoric.

Numerical simulations of penetration were performed using ABAQUS/Explicit V6.9-EF1. The use of the explicit analysis allows the same implementation of the constitutive law to be used for simulating both dynamic and quasi-static loading. The indenter was modeled as a rigid surface with a cone angle of 120° and a tip radius of 200 μm (matching that of the indenter used in the experiments). The alumina specimen being indented was represented by a biased mesh (2 μm edge length near the indentation surface and coarser toward its base and sides). Mass scaling was used to obtain a time step of 4 ns. The kinetic energy in the model was monitored to ensure that the stress due to inertial effects was minimal. Surface sliding was allowed to occur in accordance with Coulomb's law with a friction coefficient of 0.4.

To complete a full indentation simulation in a reasonable amount of time, the simulated indenter displacement rate was selected to be 125 mm/s: 750 000 times that used in the experiments (0.17 μm/s). The reference strain and crack growth rates were scaled by the same factor to reproduce the experimental stress-displacement relationship.

Calibration of the constitutive model was performed in the following way. First, the plastic indentation stress–strain response was computed using a range of parameter values, selected to be consistent with available data for alumina, and neglecting damage. The computed curve that best matched the experimental data was used to define σ_y and M. These parameter values were fixed for all subsequent calculations. Select calculations were performed to assess the sensitivity of the indentation stress–strain response to the choice of damage parameters. Additional calculations, both with and without damage, were performed to ascertain the residual indent profiles, the residual hydrostatic stresses and the spatial extent of sub-surface damage. Each of these metrics was compared with the corresponding experimental measurements. The pertinent material parameters and the range of values employed in the present computations are summarized in Table I.

IV. Indentation Deformation and Damage

The indentation stress–strain response is plotted in Fig. 5. A satisfactory fit of the data were obtained with the lattice plasticity model using a yield strength $\sigma_y = 5.75$ GPa (approximately one-third of the Vickers hardness) and a hardening exponent $M = 0.1$. The corresponding indent profiles are shown

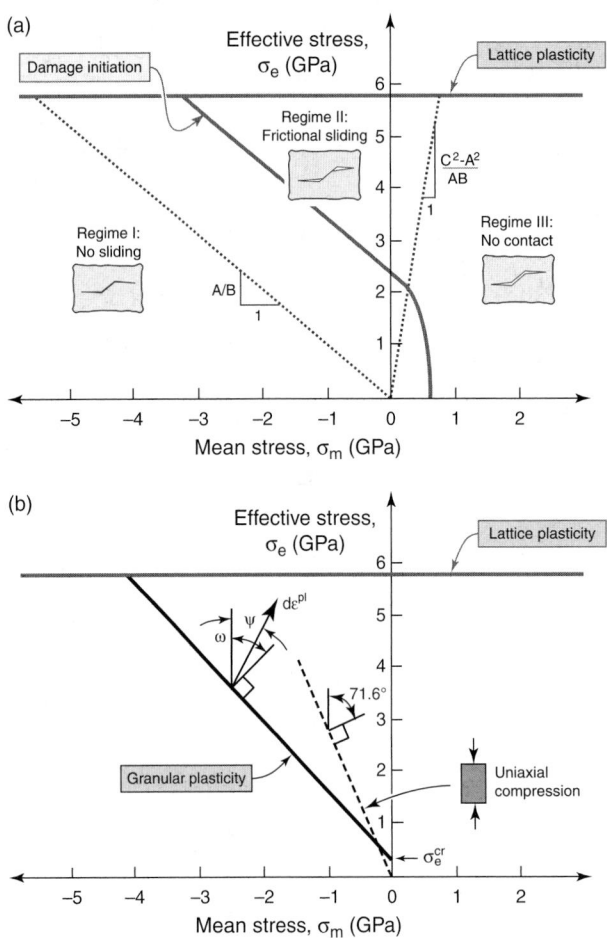

Fig. 4. (a) Conditions for microcracking and lattice plasticity in the alumina of present interest. The dotted lines represent the boundaries between the three microcracking regimes, each depicted by a corresponding schematic. The parameters A, B and C are defined in Appendix B. (b) Schematic of the linear Drucker-Prager yield criterion for granular plasticity. Also shown is the stress trajectory for uniaxial compression. Here $d\varepsilon^{pl}$ is the plastic strain increment.

Table I. Summary of Property Data Used in the Extended DE Model

Parameter	Selected value	Range considered	Units	Description
ρ	3810		kg/m^3	Density
ν	0.22			Poisson's ratio
E	384		GPa	Young's modulus
d	3.0		μm	Grain diameter
σ_Y	5.75	3.0–8.5	GPa	Yield stress
K_{IC}	3.0	2.0–4.5	MPa$\sqrt{m}$	Fracture toughness
M	0.1	0.05–0.2		Strain hardening exponent
m	30			Crack growth rate sensitivity exponent
n	34			Strain rate dependence exponent
i_0	0.01[†]		m/s	Reference crack growth rate
ε_0	0.002			Reference strain
$\dot{\varepsilon}_0$	0.001[†]		1/s	Reference plastic strain rate
g_1	0.25	0.25–0.5		Normalized flaw size
g_2	6	2–12		Normalized flaw spacing
α	0.707			Crack orientation factor
β	0.45			Parameter to ensure 2-D compatibility
γ	2			Crack geometry factor
μ	0.75			Crack coefficient of friction
$\dot{\varepsilon}_t$	10^6		1/s	Transition shear strain rate
ω	70°	50°–70°		Soil friction angle
σ_c	10^6		Pa	Soil uniaxial compressive strength
x_d	5			Soil transition exponent
$\dot{\varepsilon}_{cut-off}$	2·10^6		1/s	Soil transition strain rate
ψ	0°	0°–70°		Dilatational angle

[†]Reference strain rate was scaled by the ratio of the indentation displacement rate to the displacement rate used in the FE computations (750 000), to ensure consistent quasi-static flow stress.

Fig. 5. Measured and computed indentation stress–strain response of alumina. Inset shows typical profilometry scan used to obtain residual diameter as well as indent profile (the latter plotted in Fig. 6). $\sigma_y = 5.75$ GPa and $M = 0.1$ for finite element computations. The key parameter values in the damage model are: $a/d = 1/2$, $K_{IC} = 3$ MPa$\sqrt{m}$, $\beta = 70°$, and $\psi = 0°$.

in Fig. 6. The plasticity model overestimates the maximum indent depths by only about 20% and slightly underestimates the surface uplift around the indent periphery. In light of the extensive subsurface damage (shown below), the agreement between the experimental and computed results is remarkably good. The implication is that the indent profile is dictated largely by the resistance to lattice plasticity and is not a discriminating metric for assessing the damage model.

Optical images of the bonded interface specimens reveal well-delineated subsurface damage zones with a truncated spherical shape (Fig. 7). The diameter of the damage zone scales with the residual indent diameter and coincides with this diameter at the indent surface. SEM examinations of the side surfaces (polished previously to a mirror finish) reveal numerous slip lines within each grain: a consequence of dislocation glide and/or twinning (Fig. 8). Because of the plasticity, the grains undergo a distortion that is manifested as surface rumpling at lower magnifications (such as those of the optical micrographs).

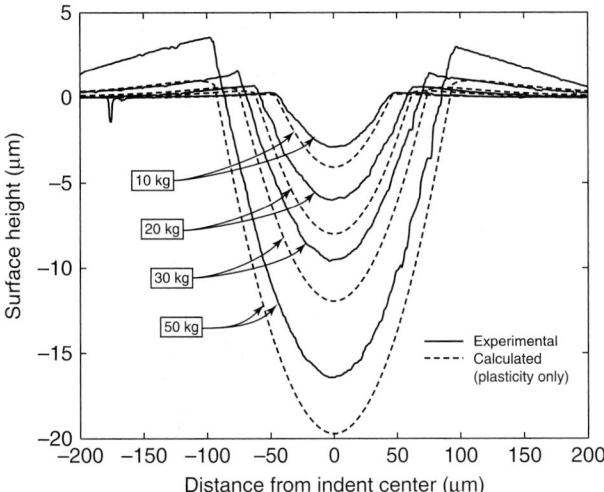

Fig. 6. Measured surface profiles after indentation to various peak loads. Also shown are the results of plasticity calculations.

The SEM images of the internal polished surfaces indicate the presence of grain-boundary microcracks, although the surface distortions preclude a conclusive assessment. They also reveal the presence of macroscopic lateral and radial cracks emanating from the periphery of the damage zone (Fig. 7(b)). Corroborating evidence of such cracks is found in the optical images as well as those obtained by optical profilometry (see, e.g., inset in Fig. 5).

Subsequent ion mill polishing of the polished surfaces yielded the most definitive evidence of microcracking. In the subsurface regions directly beneath an indentation site, the material was fully comminuted (Figs. 9(b) and (c)). That is, virtually *all* grain-boundary facets were cracked. Despite the comminution, the material strength was sufficient to prevent material loss from the indent site during normal specimen handling (including inversion and shaking). SEM images taken in a region remote from the indent revealed preferential removal of the glassy phase at the triple junctions but no relief at grain boundaries (Figs. 9(d) and (e)). Comparisons of images from the two regions re-affirmed that the boundary features near the indent surface are indeed due to microcracks and are not artifacts of the surface preparation procedure. Images of surfaces produced by FIB machining (shown in the lower parts of Figs. 9(c) and (e)) provided additional evidence that all grain boundaries in the near surface regions were cracked.

V. Damage Zone Simulations

An assessment of the extended DE model was made in the following way. First, for properties amenable to direct measurement, a narrow range of values was identified for the computations. For instance, the mode I long-crack fracture toughness of the alumina of present interest is 2.9 MPa$\sqrt{m}$. But, since the microcracks that form beneath the indent are inherently short and generally experience combined mode I/II loadings, the pertinent toughness may differ somewhat from this value. Furthermore, oxide ceramics are susceptible to subcritical cracking in the presence of water vapor, at stress intensities as low as about 2/3 of their fracture toughness. To cover the range of realistic possibilities, the fracture toughness values employed in the present computations span the range of 2.0–4.0 MPa$\sqrt{m}$. All property values are listed in Table I.

A few parameter values are less amenable to measurement, although realistic ranges can be identified. For example, the initial flaw size for intergranular fracture is expected to scale with the grain size. Assuming the existence of flaws on the grain-boundary facets, the flaw size would be $g_1 = a/d \approx 1/4$. Alternatively, if flaws are initiated by slip within the grain, the effective flaw size would be $g_1 \approx 1/2$. Thus the range of interest for the computations is expected to be bound approximately by $1/4 \leq g_1 \leq 1/2$. The flaw spacing parameter is more difficult to estimate *a priori*. But, as demonstrated below, the computed results are insensitive to its value in the domain $3 \leq g_2 \leq 12$.

Because of the high packing density of the comminuted material after microcrack coalescence, the friction angle ω is expected to be greater than that obtained for more loosely packed powders (typically 30°–50°).[11] However, to be consistent with experimental observations[12,13] an upper limit of $\omega = 71.6°$ is set by the requirement that microcracking and granular plasticity occur under uniaxial compression. Otherwise, for $\omega \geq 71.6°$, the only operative mechanism under uniaxial compression would be lattice plasticity: inconsistent with reported observations of compressive failure by microcrack coalescence at stresses below those needed for global lattice plasticity.[14] Thus, in subsequent calculations, ω was varied from 50° to 70°. A mechanistic basis for selecting the dilatational angle ψ is lacking and thus values over the entire parameter range ($0 \leq \psi \leq \omega$) were considered.

The spatial extent of the computed subsurface damage for a 20 kg indent is shown in Fig. 10 for the pertinent range of K_C and g_1, with the other parameters fixed ($\psi = 0°$, $\omega = 70°$, $g_2 = 6$). For comparison, an outline of the experimentally observed damage

Fig. 7. Optical micrographs of top and side views of bonded interface specimens after indentation.

Fig. 8. (a, b) Slip lines and grain-boundary microcracks on polished surfaces of a bonded interface specimen directly beneath the indent (30 kg load). (c) Large lateral crack emanating from periphery of the damaged region.

region is superimposed on the computed results. Both the shape and extent of the damaged region were considered in assessing the degree of correlation between the experimental and computed results. The computed results exhibit an abrupt transition from damaged to undamaged material, closely analogous to that observed in experiments. The abrupt boundary is a manifestation of the extremely small amount of crack extension that is predicted to occur under increasing stress. A closer examination of the present results indicate that the wing cracks are predicted to extend only very short distances (~ 200 nm, depending on the triaxiality) before the local stress reaches a maximum and further growth occurs under rapidly diminishing stress. Consequently, most wing

Fig. 9. (a) SEM images of a bonded interface specimen after removal of about 200 nm of material by ion mill polishing. Virtually all grain-boundary facets contain cracks in the regions directly beneath the indented surface (b, c). FIB machined surface perpendicular to the polished plane reveals similar grain-boundary cracks on all facets (bottom of (c)). In regions remote from the indented surface, image contrast is due solely to selective removal of the glassy phase at triple junctions (d, e). Similar features are evident on the corresponding FIB machined surface (bottom of (e)), without evidence of microcracking.

Fig. 10. Effects of flaw size and fracture toughness on damage beneath a 20 kg indent ($\psi = 0°$, $\beta = 70°$, $g_2 = 6$). Blue indicates intact material ($D = D_0$), while red indicates full comminution ($D = 1$). The white dashed lines correspond to the edge of the experimentally observed sub-surface damage region in Fig. 7(b).

cracks that extend even slightly will propagate unstably to full comminution. One implication is that partially extended wing cracks are unlikely to be observable in the experiments.

Satisfactory agreement between the computed results and experiments was obtained for the combination $K_C = 3.5$ MPa$\sqrt{m}$ and $g_1 = 1/2$. But the optimal combination is not unique. An equally good correlation was obtained for $K_C = 3$ MPa$\sqrt{m}$ and $g_1 = 1/4$. Nevertheless, the range of inferred property values is sufficiently narrow and well within the expected limits. Moreover, the computed damage pattern quickly deviates from the

experimental observations when property values are selected outside this range. For instance, for $K_C = 3.5$ MPa$\sqrt{m}$ and $g_1 = 1/4$, anomalous undamaged regions appear both directly beneath the indent and further down along the centerline. Increasing the fracture toughness to 4 MPa$\sqrt{m}$ in combination with the same flaw size yields a prediction that damage is almost nonexistent.

The effects of flaw spacing, friction angle and dilatation angle on the damage pattern for $K_C = 3$ MPa$\sqrt{m}$ and $g_1 = 1/4$ for the same 20 kg indent are shown in Fig. 11. Among these, only the friction angle has an appreciable effect on the shape of the computed damage zone. Notably, for $\omega \leq 60°$, the boundary between damaged and undamaged material is highly irregular with multiple radial damage bands extending well beyond that observed in the experiments. Selecting $\omega = 70°$ yields acceptable results. This high friction angle (approaching the expected upper limit of 71.6°) is consistent with the high packing density (nearly 100%) of the comminuted grains. The flaw spacing and the dilatation angle play a decidedly secondary role over a realistic range of values. Consequently, further calculations presented here are based on the selections $g_2 = 6$ and $\psi = 0°$.

Finally, the effects of the indent load on both measured and computed damage patterns are shown in Fig. 12 for $K_C = 3$ MPa$\sqrt{m}$, $g_1 = 1/4$ or $1/2$, $\psi = 0°$, $\omega = 70°$, and $g_2 = 6$. The finite element results are in reasonable agreement with the

Fig. 11. Numerical results of the effects of flaw spacing, friction angle and dilatation angle on subsurface damage for a 20 kg indent ($a/d = 1/4$ and $K_C = 3.0$ MPa$\sqrt{m}$).

Fig. 12. Effect of indent load on subsurface damage. Dashed lines indicate the edge of the experimentally observed damage zones.

Fig. 13. Distributions of (a, b) residual hydrostatic stress and (c, d) surface profiles after indentation, for peak loads of (a, c) 10 kg and (b, d) 20 kg. In addition to the experimental measurements, computed curves are plotted for cases in which the ceramic deforms by lattice plasticity alone as well as plasticity accompanied by microcracking. The latter are based on a crack spacing $g_2 = 6$, fracture toughness $K_C = 3$ MPa$\sqrt{m}$, and two values of initial flaw size: $a/d = 1/4$ or $1/2$, as indicated.

experimental observations over the entire load range, with slight underestimation of the damage depth at 10 kg and slight over-estimation at 30 kg.

VI. Residual Stresses and Indent Profiles

As a further assessment of the model, comparisons were made of the residual hydrostatic stress (the numerical results being averaged into the depth according to the method described in Appendix A). The comparisons are plotted on Fig. 13 for indent loads of 10 and 20 kg. Two sets of damage parameter values were considered, differentiated by the initial flaw size ($g_1 = 1/4$ or $1/2$). The other property values were fixed. Also shown are the finite element results assuming that the only inelastic mechanism is lattice plasticity.

The experimental measurements indicate that, within the indent, the residual stress is compressive and essentially uniform across the indent. Although significant scatter exists (attributable to the probe size being comparable to the grain diameter), the average values within the indents are roughly constant (-260 ± 100 MPa) for indent loads of 10 and 20 kg. These levels persist beyond the indent boundary, to a distance of about twice the indent radius. Thereafter, at greater distances from the indent, the stresses gradually approach zero.

The computed residual stress distributions for lattice plasticity alone are inconsistent with the measurements. For instance, the peak computed stresses (near the indent periphery) are about five times the measured values. This discrepancy is a direct consequence of the relaxation from granular plasticity that accompanies microcracking in the experiments but is not taken into account in the plasticity model. Nevertheless, in regions well away from the indent periphery ($\geq$ two indent radii), the plasticity calculations predict a gradual approach to zero stress at a rate comparable to

that of the experimental measurements. This correlation is consistent with the absence of damage in the latter regions.

The computations based on the damage model appear to adequately bind the residual stress measurements within the indent. That is, for $a/d = 1/2$ and an indent load of 10 kg, the magnitude of the average residual stress within the indent (≈ 250 MPa) is underestimated by the model; in contrast, for $a/d = 1/4$, it is overestimated. Both data sets exhibit a local tensile peak at a distance of about twice the indent size, before once again turning compressive and approaching the finite element results based on lattice plasticity alone. These tensile peaks coincide with the edge of the computed damage zone and are due to large tensile hoop stresses in the surrounding elastic material. In reality, these tensile stresses are at least partially relieved by the radial cracks emanating from the indent. Such cracks cannot form in the present computations because of the axisymmetric nature of the model and hence the residual stresses are overestimated. These unrelieved tensile hoop stresses may also lead to over-prediction of the spatial extent of the damage.

For completeness, comparisons were made of the residual surface profiles (Figs. 13(c) and (d)). Within the indents, the damage model yields results similar to those for lattice plasticity alone, reaffirming the notion that the profile is not appreciably affected by the damage. Outside the indent, additional pile-up associated with damage is evident, although the computed results overestimate the measured values.

VII. Concluding Remarks

The results of the extended DE model are in broad agreement with the experimental measurements and observations of quasi-static indentation tests. A particularly striking feature is the correlation obtained using property values that are either measured directly or estimated from physically based arguments, without undue reliance on adjustable (nonphysical) parameters. Specifically, for oxide ceramics, the effective fracture toughness pertinent to microcracking correlates well with the value measured by Vicker's indentations. Furthermore, the inferred initial flaw size falls in the range expected on the basis of grain-boundary facets or grain-sized slip-induced flaws. The inferred value of the friction angle that defines the flow stress for granular plasticity is near the upper limit required to yield damage under uniaxial compression. Such a high value is tentatively attributed to the high packing density of the comminuted material.

The dilatation angle and the flaw spacing have little effect on the predicted damage patterns. Loading conditions under which the these parameters play a significant role have yet to be ascertained. It is surmised that the dilatation angle may be more important in deep penetration experiments, wherein large amounts of confined granular flow are obtained, as described by Shockey *et al.*[15,16] In contrast, in the present calculations, the strains obtained via the granular plasticity mechanism were minimal.

The most discriminating metric in the assessment of the extended DE model is the spatial extent of subsurface damage. Residual hydrostatic stresses provide additional validation. In contrast, surface profiles of the indents are dictated largely by the plastic properties of the ceramic and thus provide minimal insight into the inelastic deformation resulting from microcracking or granular flow.

Fig. A1. Variation in peak intensity with position across a polished epoxy/alumina interface. The lines were calculated assuming a circular probe of uniform intensity, with the radii indicated.

Acknowledgments

E. A. G. was supported by a National Defense Science and Engineering Graduate Fellowship. The authors also acknowledge D.R. Clarke for fruitful discussions regarding residual stress measurement.

Appendix A: Measurement of Piezospectroscopic Interaction Volume

Experiments were conducted to ascertain the response function of the laser probe used for residual stress measurements. They were motivated by the recognition that the probe interrogates a finite volume of material. When stress gradients exist over length scales comparable to the probe dimensions, the true stress distributions can only be inferred through a deconvolution of a series of spectroscopic measurements. Alternatively, if the principal objective is to assess the fidelity of a model for the stress distribution, the computed results can be averaged accordingly for comparison with the experimental data. The latter approach was adopted in the present study.

The laser probe used to stimulate luminescence is focused, nominally to a point, within the area of interest in an optical microscope. In practice, the minimum probe diameter attainable depends on the system optics and the objective magnification and is typically in the range 1–2 µm. The probe also penetrates beneath the surface, to a depth controlled by the distribution in probe intensity with distance from the focal plane. The measurements are further influenced by the probe sensitivity: that is, the efficiency with which each volume element conveys information to the spectrometer. For instance, photons emitted deep below the surface are scattered by the material and yield a lower signal intensity than that emanating from photons emitted near the free surface. Combined, the probe size and probe sensitivity define the probe response function.[17,18]

To determine the size of the probe in the focal plane, a series of fluorescence measurements were taken across the interface between polished alumina and mounting epoxy. At each point, measurements were made of the integrated intensities of the R_1 and R_2 peaks. The translation increment between measurement points was 0.5 µm. The intensity distribution, shown in Fig. A1, was then modeled by assuming that the beam intensity is uniform in the focal plane and that the probe is circular

in shape. The fitting yields a probe diameter of 2–3 µm, virtually the same as the simulation element size in the contact region.

The coupled effects of the probe shape normal to the focal plane and the scattering of emitted photons were ascertained using a modification of the technique described by Lipkin and colleagues[14] and Wan and colleagues.[15] Specifically, spectroscopic measurements were made as the focal plane was translated normal to the sample surface. The variation in integrated intensity of the fluorescence spectrum with focal plane depth is plotted in Fig. A2. The probe intensity normal to the focal plane was assumed to be the sum of two components: a Lorentzian distribution, with a peak at the focal plane, and a constant value, associated with on-axis light. Furthermore, scattering effects are described by a conventional exponential decay function. The integrated intensity of a fluorescence spectrum obtained with the probe focused at depth c below the sample surface is thus given by

$$\frac{I}{I_0} = \int_0^\infty \left[B \cdot \exp(-2\alpha z) + (1-B) \cdot \exp(-2\alpha z) \left(\frac{p^2}{p^2 + (z-c)^2} \right) \right] dz \tag{A-1}$$

where B is the fraction of laser light that is on axis, z is depth below the surface, α is the absorption coefficient, p is a characteristic length describing the probe width along the z-axis and I_0 is a reference intensity. A least squares fit of this relation to the experimental data (Fig. A2) yields parameter values: $B = 0.138$, $\alpha = 0.059 \ \mu m^{-1}$ and $p = 5.25 \ \mu m$.

Appendix B: Constitutive Laws for Ceramic Deformation in the Extended Deshpande-Evans Model

(1) Lattice Plasticity

When the confining pressure is sufficiently high, plasticity occurs by slip and/or twinning. At room temperature, dislocation motion in oxide ceramics such as alumina is limited by the lattice resistance. Furthermore, according to Frost and Ashby,[9] the lattice resistance is largely rate independent until the dislocation velocity reaches a value limited by phonon drag. Here,

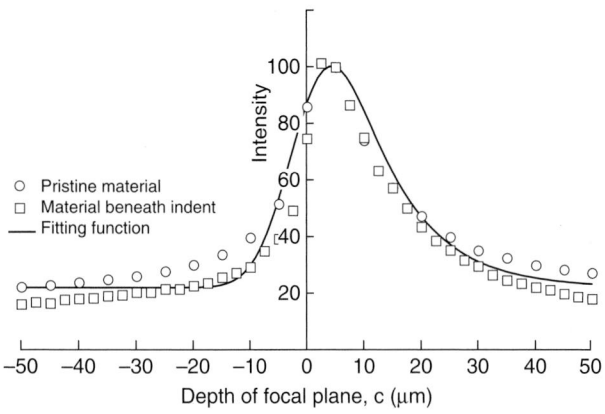

Fig. A2. Variation in peak intensity with focal plane position, for both pristine material and at the center of an indent. The solid line shows the least squares fit of Eq. (A-1).

the plastic strain rate can be described by

$$\frac{\dot{\varepsilon}_e^{pl}}{\dot{\varepsilon}_0} = \begin{cases} \min\left[\left(\frac{2\sigma_e}{\sigma_0} - 1\right)^n, \left(\frac{\dot{\varepsilon}_0}{\dot{\varepsilon}_t}\right)^{(1-n)/n} \left(\frac{2\sigma_e}{\sigma_0} - 1\right) \right] & 2\sigma_e > \sigma_0 \\ 0 & \text{otherwise} \end{cases}$$

(B-1)

where $\dot{\varepsilon}_t$ is the transition strain rate at which dislocation motion becomes limited by phonon drag. To ensure a rate-insensitive response, the strain rate sensitivity exponent, n, is selected to be $\gg 1$. In the present implementation, $n = 30$. The flow stress depends on the current plastic strain via a conventional power law

$$\sigma_0 = \frac{\sigma_y}{2}\left[1 + \left(\frac{\varepsilon_e^{pl}}{\varepsilon_y}\right)^M \right]$$

(B-2)

where ε_y is the yield strain. Associated flow is assumed with a flow potential $G_p \equiv \sigma_e$.

(2) Microcrack Growth

Under compressive stress states, inelastic deformation occurs by stable extension of wing cracks from preexisting flaws (Fig. 1(b)). The stress intensity driving wing crack growth has been analyzed by Ashby and Hallam[19] in two dimensions and extended to three dimensions by Ashby and Sammis.[20] For simplicity, the flaws are assumed to have a uniform spacing, $1/f^{1/3}$, and uniform initial diameter, $2a$. Furthermore, the flaws are assumed to be oriented at 45° to the direction of the maximum principal compressive stress: the orientation that yields the maximum value of K_I. The damage state is characterized by

$$D = \frac{4}{3}\pi(l + \alpha a)^3 f$$

(B-3)

where l is the length of the wing crack and $\alpha a = 1/\sqrt{2}$ is the length of the preexisting flaw in the direction of the wing crack. D increases monotonically from its initial value,

$$D_0 = \frac{4}{3}\pi(\alpha a)^3 f$$

(B-4)

to $D = 1$ when neighboring cracks coalesce.

There are three contributions to the stress intensity at the wing crack tip[17] (i) a sliding stress across the crack faces that wedges open the wing cracks, (ii) a closing stress on the wing crack due to the global compressive hydrostatic stress, and (iii) an internal stress in the uncracked ligament between wing cracks

that balances the wedging stress. The net stress intensity factor at the tip of the wing crack can be expressed in terms of the first two stress invariants as

$$\frac{K_I}{\sqrt{\pi a}} = A\sigma_m + B\sigma_e$$

(B-5)

where

$$A \equiv c_1(c_2 A_3 - c_2 A_1 + c_3)$$

(B-6)

$$B \equiv \frac{c_1}{\sqrt{3}}(c_2 A_3 + c_2 A_1 + c_3)$$

(B-7)

$$c_1 = \frac{1}{\pi^2 \alpha^{3/2}\left[\left(\frac{D}{D_0}\right)^{1/3} - 1 + \frac{\beta}{\alpha} \right]^{3/2}}$$

$$c_2 = 1 + 2\left[\left(\frac{D}{D_0}\right)^{1/3} - 1 \right]^2 \left(\frac{D_0^{2/3}}{1 - D^{2/3}}\right)$$

(B-8)

$$c_3 = 2\alpha^2\pi^2\left[\left(\frac{D}{D_0}\right)^{1/3} - 1 \right]^2$$

$$A_1 = \pi\sqrt{\frac{\beta}{3}}\left[(1 + \mu^2)^{1/2} - \mu \right]$$

(B-9)

$$A_3 = \pi\sqrt{\frac{\beta}{3}}\left[(1 + \mu^2)^{1/2} + \mu \right]$$

These results apply when the crack faces remain in contact but the shear stress is sufficient to overcome friction and cause sliding of the crack faces. This is denoted Regime II.

Two other behavioral regimes are obtained. When the stress triaxiality is sufficiently low (negative), sliding does not occur across the crack faces and hence $K_I = 0$ (regime I). This condition is obtained when

$$\lambda \equiv \frac{\sigma_m}{\sigma_e} \leq -\frac{B}{A}$$

(B-10)

At the other extreme, where the triaxiality is high, the crack faces lose contact with one another (regime III). In this domain, K_I is given approximately by[21,22]

$$\frac{K_I}{\sqrt{\pi a}} = \left(C^2\sigma_m^2 + E^2\sigma_e^2\right)^{1/2}$$

(B-11)

The coefficients C and E are selected to ensure the elastic strains are continuous across the transition between regimes II and III. This requires that

$$E^2 = \frac{B^2 C^2}{C^2 - A^2}$$

(B-12)

and the regime transition occurs when

$$\lambda = \frac{AB}{C^2 - A^2}$$

(B-13)

where C is defined as

$$C \equiv A + \gamma\sqrt{\alpha\left(\frac{D}{D_0}\right)^{1/3}}$$

Crack growth is assumed to occur at a rate $i/i_0 = (K_I/K_{IC})^m$, but limited by the shear wave speed of the uncracked material.

Thus

$$l = \min\left[i_0 (K_I/K_{IC})^m, \sqrt{\frac{G}{\rho}} \right] \tag{B-14}$$

where G and ρ are the shear modulus and the density of the uncracked ceramic. The sensitivity exponent, $m = 34$, and reference crack growth rate, $i_0 = 0.01$ m/s, were obtained by fitting data reported by Aeberli and Rawlings.[11]

For numerical implementation, full comminution was deemed to have occurred when the damage reaches a value $D = 1 - 0.1 D_0$. (Allowing $D = 1$ results in division by 0 in the calculation of c_2.)

(3)   Granular Plasticity

Granular plasticity of the fully comminuted ceramic is taken to follow a linear Drucker–Prager law,[23] with an effective stress defined as

$$\hat{\sigma} \equiv \frac{\sigma_e + (\tan\beta)\sigma_m}{1 - \tan\omega/3} \tag{B-15}$$

where ω is the friction angle. The granular strain rate is given by

$$\frac{\dot{\varepsilon}_e^g}{\dot{\varepsilon}_s} = \begin{cases} \left(\frac{\hat{\sigma}}{\Sigma_c} - 1\right)^s & \hat{\sigma} > \Sigma_c \\ 0 & \text{otherwise} \end{cases} \tag{B-16}$$

where Σ_c is the uniaxial compressive strength of the comminuted ceramic (assumed to be 1 MPa) and $\dot{\varepsilon}_s$ is the reference strain rate. For compressive mean stress states, the effective stress scales with the square of the strain rate in a Bagnold-type flow law giving $s = 1/2$. For $\sigma_m > 0$, s is large (~ 10) to prevent the granular medium from supporting large tensile stresses.

In general, granular flow is expected to occur with some dilatation, dictated by the value of ψ. The granular flow surface is G_g is defined as

$$G_g \equiv \begin{cases} \sigma_e + (\tan\psi)\sigma_m & \sigma_m < 0 \\ \sqrt{(\zeta\,\Sigma_t)^2 + \sigma_e^2} + \frac{\Sigma_c}{\Sigma_t}\sigma_m & \text{otherwise} \end{cases} \tag{B-17}$$

where Σ_t is the hydrostatic tensile strength of the comminuted ceramic and ζ is a nondimensional regularization parameter to ensure purely dilatational flow under hydrostatic tension.

References

[1] V. S. Deshpande and A. G. Evans, "Inelastic Deformation and Energy Dissipation in Ceramics: A Mechanism-Based Constitutive Model," *J. Mech. Phys. Solids*, **56**, 3077–100 (2008).

[2] V. S. Deshpande, E. A. Gamble, B. G. Compton, R. M. McMeeking, A. G. Evans, and F. W. Zok, "A constitutive description of the inelastic response of ceramics," *J. Am. Ceram. Soc.*, 2011 (this issue).

[3] B. R. Lawn, "Indentation of Ceramics with Spheres: A Century after Hertz," *J. Am. Ceram. Soc.*, **81** [8] 1977–94 (1998).

[4] F. Guiberteau, N. P. Padture, and B. R. Lawn, "Effect of Grain Size on Hertzian Contact Damage in Alumina," *J. Am. Ceram. Soc.*, **77** [7] 1825–31 (1994).

[5] J. C. LaSalvia and J. W. McCauley, "Inelastic Deformation Mechanisms and Damage in Structural Ceramics Subjected to High-Velocity Impact," *Int. J. Appl.Ceram. Technol.*, **7** [5] 595–605 (2010).

[6] C. A. Powell-Dogan and A. H. Heuer, "Devitrification of the Grain Boundary Glassy Phase in a High-Alumina Ceramic Substrate," *J. Am. Ceram. Soc.*, **77** [10] 2593–8 (1994).

[7] Q. Ma and D. R. Clarke, "Stress Measurement in Single-Crystal and Polycrystalline Ceramics Using Their Optical Fluorescence," *J. Am. Ceram. Soc.*, **76** [6] 1433–40 (1993).

[8] J. He and D. R. Clarke, "Determination of the Piezospectroscopic Coefficients for Chromium-Doped Sapphire," *J. Am. Ceram. Soc.*, **78** [5] 1347–53 (1995).

[9] H. J. Frost and M. F. Ashby, *Deformation Mechanism Maps*. Pergamon Press, Oxford, U.K., 1982.

[10] K. E. Aeberli and R. D. Rawlings, "Effect of Simulated Body Environments on Crack Propagation in Alumina," *J. Mater. Sci. Lett.*, **2**, 215–20 (1983).

[11] T. W. Lambe and R. V. Whitman, *Soil Mechanics*. John Wiley & Sons Inc, New York, 1969.

[12] A. Nash, "Compressive Failure Modes of Alumina in Air and Physiological Media," *J. Mater. Sci.*, **18**, 3571–7 (1983).

[13] M. Adams and G. Sines, "Determination of Biaxial Compressive Strength of a Sintered Alumina Ceramic," *J. Am. Ceram. Soc.*, **59** [7] 300–4 (1976).

[14] J. Lankford, "Compressive Strength and Microplasticity in Polycrystalline Alumina," *J. Mater. Sci.*, **12**, 791–6 (1977).

[15] D. A. Shockey, A. H. Marchand, S. R. Skaggs, G. E. Cort, M. W. Burkett, and R. Parker, "Failure Phenomenology of Confined Ceramic Targets and Impacting Rods," *J. Impact Eng.*, **9** [3] 263–75 (1990).

[16] D. A. Shockey, J. W. Simons, and D. R. Curran, "The Damage Mechanism Route to Better Armor Materials," *Int. J. Appl. Ceram. Technol.*, **7** [5] 566–73 (2010).

[17] D. M. Lipkin and D. R. Clarke, "Sample-Probe Interactions Inspectroscopy: Sampling Microscopic Property Gradients," *J. Appl. Phys.*, **77** [5] 1855–63 (1995).

[18] K. Wan, W. Zhu, and G. Pezzotti, "Determination of In-Depth Probe Response Function Using Spectral Perturbation Methods," *J. Appl. Phys.*, **98**, 1–7 (2005).

[19] M. F. Ashby and S. D. Hallam, "The Failure of Brittle Solids Small Cracks Under Compressive Containing Stress States," *Acta Metall.*, **34** [3] 497–510 (1986).

[20] M. F. Ashby and C. G. Sammis, "The Damage Mechanics of Brittle Solids in Compression," *Pure Appl. Geophys.*, **133** [3] 489–521 (1990).

[21] J. R. Bristow, "Microcracks, and the Static and Dynamic Elastic Constants of Annealed and Heavily Cold-Worked Metals," *J. Phys. D Appl. Phys.*, **11** [2] 81–5 (1960).

[22] B. Budiansky and R. J. O'Connell, "Elastic Moduli of a Cracked Solid," *Int. J. Solids Struct.*, **12** [2] 81–97 (1976).

[23] D. C. Drucker and W. Prager, "Soil Mechanics and Plastic Analysis or Limit Design," *Q. Appl. Mathematics*, **10** [2] 157–65 (1952). □

J. Am. Ceram. Soc., **94** [S1] S226–S235 (2011)

DOI: 10.1111/j.1551-2916.2011.04432.x

© 2011 The American Ceramic Society

Journal

Large Plastic Deformation in High-Capacity Lithium-Ion Batteries Caused by Charge and Discharge

Kejie Zhao, Matt Pharr, Shengqiang Cai, Joost J. Vlassak, and Zhigang Suo[†]

School of Engineering and Applied Sciences, Kavli Institute for Bionano Science and Technology, Harvard University, Cambridge, Massachusetts 02138

Evidence has accumulated recently that a high-capacity electrode of a lithium-ion battery may not recover its initial shape after a cycle of charge and discharge. Such a plastic behavior is studied here by formulating a theory that couples large amounts of lithiation and deformation. The homogeneous lithiation and deformation in a small element of an electrode under stresses is analyzed within nonequilibrium thermodynamics, permitting a discussion of equilibrium with respect to some processes, but not others. The element is assumed to undergo plastic deformation when the stresses reach a yield condition. The theory is combined with a diffusion equation to analyze a spherical particle of an electrode being charged and discharged at a constant rate. When the charging rate is low, the distribution of lithium in the particle is nearly homogeneous, the stress in the particle is low, and no plastic deformation occurs. When the charging rate is high, the distribution of lithium in the particle is inhomogeneous, and the stress in the particle is high, possibly leading to fracture and cavitation.

I. Introduction

LITHIUM-ION batteries are being developed to achieve safe operation, high capacity, fast charging, and long life. Each electrode in a lithium-ion battery is a host of lithium.[1] Lithium diffuses into and out of the electrode when the battery is charged and discharged. If a particle of the electrode material is charged or discharged slowly and is unconstrained by other materials, the distribution of lithium in the particle is nearly homogeneous, and the particle expands or contracts freely, developing no stress. In practice, however, charge and discharge cause a field of stress in the particle when the distribution of lithium is inhomogeneous,[2-5] when the host contains different phases,[6] or when the host is constrained by other materials.[7] The stress may cause the electrode to fracture,[6-10] which may lead the capacity of the battery to fade.[11-14]

Lithiation-induced fracture not only occurs in current commercial lithium-ion batteries, but is also a bottleneck in developing future lithium-ion batteries.[15,16] For example, of all known materials for anodes, silicon offers the highest theoretical specific capacity—each silicon atom can host up to 4.4 lithium atoms. By comparison, in commercial anodes of graphite, every six carbon atoms can host up to one lithium atom. However, silicon is not used in anodes in commercial lithium-ion batteries, mainly because the capacity of silicon fades after a small number of cycles—a mode of failure often attributed to lithiation-induced fracture.

Recent experiments, however, have shown that the capacity can be maintained over many cycles for silicon anodes of small feature sizes, such as thin films,[17] nanowires,[18] and porous structures.[19] When silicon is fully lithiated, the volume of the material swells by ~300%.[15] For anodes of small feature sizes, evidence has accumulated recently that this lithiation-induced strain can be accommodated by plastic deformation. For instance, cyclic lithiation causes 50-nm-thick silicon films to develop undulation without fracture.[17] Likewise, cyclic lithiation causes surfaces of silicon nanowires to roughen.[18] These studies suggest that lithiated silicon undergoes plastic deformation. Furthermore, during cyclic lithiation of an amorphous silicon thin film attached to a substrate, the measured relation between the stress in the film and the state of charge exhibits pronounced hysteresis.[20] This observation indicates that the lithiated silicon film deforms plastically when the stress exceeds a yield strength. At this writing, the atomistic mechanism of plastic flow of lithiated silicon is unclear, but is possibly due to a "lubricating effect" of lithium. Cyclic lithiation and delithiation can also cause silicon to grow cavities.[21]

This paper develops a theory of finite plastic deformation of electrodes caused by charge and discharge. Plastic deformation has not been considered in most existing theories of lithiation-induced deformation.[2-10] The works that do include plasticity have been limited to small deformation.[22,23] Here, we analyze the homogeneous lithiation and deformation in a small element of an electrode within nonequilibrium thermodynamics, stipulating equilibrium with respect to some processes, but not others. The element is assumed to undergo plastic deformation when the state of stress reaches a yield condition. We then combine large plastic deformation and diffusion to analyze the lithiation of a spherical particle of an electrode. When a small particle is charged and discharged slowly, the stress is small. When the particle is charged and discharged quickly, the stress is high, potentially leading to fracture or cavitation.

II. Nonequilibrium Thermodynamics of Coupled Lithiation and Deformation

Figure 1 illustrates an element of an electrode subject to a cycle of charge and discharge. The element is small in size, so that fields in the element are homogeneous. In the reference state (Fig. 1(a)), the element is a unit cube of a host material, free of lithium and under no stress. When the element is connected to a reservoir of lithium at chemical potential μ and is subject to stresses s_1, s_2, and s_3, as illustrated in Fig. 1(b), the element absorbs a number C of lithium atoms, becomes a block of sides λ_1, λ_2, and λ_3, and gains in free energy W.

By definition, s_1, s_2, and s_3 are nominal stresses—forces acting on the element in the current state divided by the areas of the element in the reference state. The true stresses, σ_1, σ_2, and σ_3, are forces per unit areas of the element in the current state. The true stresses relate to the nominal stresses by $\sigma_1 = s_1/(\lambda_2\lambda_3)$, $\sigma_2 = s_2/(\lambda_3\lambda_1)$, and $\sigma_3 = s_3/(\lambda_1\lambda_2)$.

Associated with small changes in the stretches, $\delta\lambda_1$, $\delta\lambda_2$, and $\delta\lambda_3$, the forces do work $s_1\delta\lambda_1 + s_2\delta\lambda_2 + s_3\delta\lambda_3$. Associated with small change in the number of lithium atoms, δC, the chemical potential does work $\mu\delta C$. Thermodynamics dictates that the

R. McMeeking—contributing editor

Manuscript No. 28712. Received October 2, 2010; approved January 4, 2011.
This work is supported by the National Science Foundation through a grant on Lithium-ion Batteries (CMMI-1031161).
[†]Author to whom correspondence should be addressed. e-mail: suo@seas.harvard.edu

combined work should be no less than the change in the free energy:

$$s_1\delta\lambda_1 + s_2\delta\lambda_2 + s_3\delta\lambda_3 + \mu\delta C \geq \delta W \tag{1}$$

The work done minus the change in the free energy is the dissipation. The inequality (1) means that the dissipation is non-negative with respect to all processes. The object of this section is to examine the large amounts of lithiation and deformation by constructing a theory consistent with the thermodynamic inequality (1).

As illustrated in Fig. 1(b), when the unit cube of the host is lithiated under stresses, the deformation is anisotropic: the unit cube will change both its shape and volume. For example, a thin film of an electrode constrained on a stiff substrate, upon absorbing lithium, deforms in the direction normal to the film, but does not deform in the directions in the plane of the film.

The material deforms by mechanisms of two types: inelastic and elastic. Inelastic deformation involves mixing and rearranging atoms. Elastic deformation involves small changes of the relative positions of atoms, retaining the identity of neighboring atoms as well as the concentration of lithium. When the stresses are removed and the reservoir of lithium is disconnected, the material element will retain part of the anisotropic deformation (Fig. 1(c)). The phenomenon is reminiscent of plasticity of a metal. The remaining deformation is characterized by three stretches λ_1^i, λ_2^i, and λ_3^i, which we call inelastic stretches. The part of deformation that disappears upon the removal of the stresses is characterized by three stretches λ_1^e, λ_2^e, and λ_3^e, which we call elastic stretches. The total stretches are taken to be the products of the two types of the stretches

$$\lambda_1 = \lambda_1^e\lambda_1^i, \quad \lambda_2 = \lambda_2^e\lambda_2^i, \quad \lambda_3 = \lambda_3^e\lambda_3^i \tag{2}$$

Similar multiplicative decomposition is commonly used to describe elastic–plastic deformation of metals[24] and polymers,[25] as well as growth of tissues.[26]

We characterize the state of the material element by a total of seven independent variables: λ_1^e, λ_2^e, λ_3^e, λ_1^i, λ_2^i, λ_3^i and C.

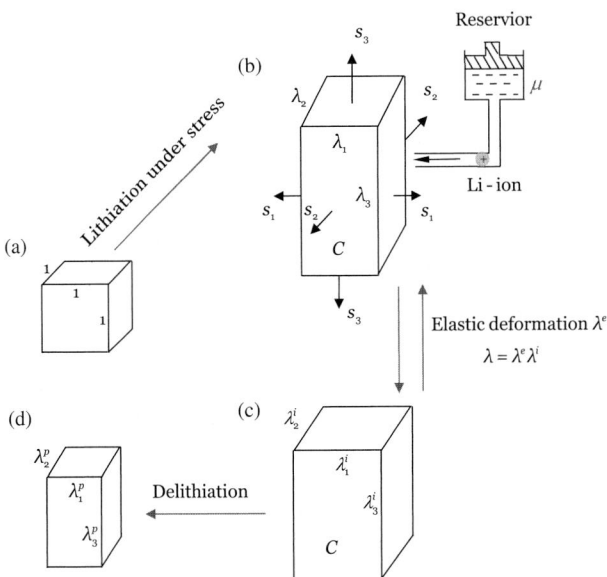

Fig. 1. After a cycle of lithiation and delithiation, an electrode material may not recover its initial shape. (a) In the reference state, an element of an electrode material is a lithium-free and stress-free unit cube. (b) Subject to forces and connected to a reservoir of lithium, the material element absorbs lithium, and undergoes anisotropic deformation. (c) When the stresses are removed and the reservoir of lithium is disconnected, the material element unloads elastically, and the remaining inelastic deformation is anisotropic. (d) After the material element desorbs lithium under no stress, the lithium-free host becomes a rectangular block.

Plasticity of a metal and inelasticity of an electrode differ in a significant aspect. While plastic deformation of a metal changes shape but conserves volume, inelastic deformation of an electrode changes both volume and shape. We decompose the inelastic stretches by writing

$$\lambda_1^i = \Lambda^{1/3}\lambda_1^p, \quad \lambda_2^i = \Lambda^{1/3}\lambda_2^p, \quad \lambda_3^i = \Lambda^{1/3}\lambda_3^p \tag{3}$$

Here, Λ is the volume of the material element after the removal of the stresses, namely,

$$\lambda_1^i\lambda_2^i\lambda_3^i = \Lambda \tag{4}$$

Inelastic shape change of the material element is described by λ_1^p, λ_2^p, and λ_3^p (Fig. 1(d)). By the definition of (3) and (4), the plastic stretches do not change volume, namely,

$$\lambda_1^p\lambda_2^p\lambda_3^p = 1 \tag{5}$$

We will call λ_1^p, λ_2^p, and λ_3^p the plastic stretches. In our terminology, inelastic deformation includes the changes in both volume and shape, while plastic deformation involves only the change in shape.

A combination of (2) and (3) gives $\lambda_1 = \lambda_1^e\lambda_1^p\Lambda^{1/3}$. Taking the logarithm of both sides of this equation, we write $\log\lambda_1 = \log\lambda_1^e + \log\lambda_1^p + \log\Lambda^{1/3}$. The quantity $\log\lambda_1$ is the natural strain, $\log\lambda_1^e$ the elastic part of the natural strain, and $\log\lambda_1^p$ the plastic part of the natural strain.

Most existing theories of lithiation-induced deformation do not consider plasticity. In effect, these theories assume that, after a cycle of lithiation and delithiation, the material element recovers its initial shape. Such an assumption disagrees with experimental observations of lithiation of large-capacity hosts, such as silicon, as discussed in the Introduction and in several recent papers.[22,23,27] This paper will allow plasticity, and will describe rules to calculate large plastic deformation.

The state of the element can be characterized by an alternative list of seven independent variables: λ_1^e, λ_2^e, λ_3^e, λ_1^p, λ_2^p, Λ, and C. To progress further, we make the following simplifying assumptions. The inelastic expansion of the volume is taken to be entirely due to the absorption of lithium, and is a function of the concentration of lithium:

$$\Lambda = \Lambda(C) \tag{6}$$

This function is taken to be characteristic of the material, and is independent of the elastic and plastic stretches. Equation (6) eliminates Λ from the list of independent variables, so that the state of the material element is characterized by six independent variables: λ_1^e, λ_2^e, λ_3^e, λ_1^p, λ_2^p, and C.

Following the theory of plasticity, we assume that the free energy of the material element is unaffected by the plastic stretches. Thus, the free energy is a function of four variables:

$$W = W(\lambda_1^e, \lambda_2^e, \lambda_3^e, C) \tag{7}$$

This assumption is understood as follows. The plastic stretches characterize inelastic shape change, involving rearranging atoms without changing the concentration of lithium. This rearrangement of atoms may dissipate energy, but does not alter the amount of free energy stored in the host materials. The situation is reminiscent of shear flow of a liquid of small molecules—the free energy is stored in molecular bonds, independent of the amount of flow. For a work-hardening metal, however, plastic strain will change the free energy stored in the material, for example, by creating more dislocations. Thus, assuming (7) amounts to a stipulation that plastic strains do not create such microstructural changes in lithiated electrodes.

Rewriting the inequality (1) in terms of the changes in the six independent variables, λ_1^e, λ_2^e, λ_3^e, λ_1^p, λ_2^p, and C, we obtain that

$$
\left(\sigma_1\lambda_1\lambda_2\lambda_3 - \frac{\partial W}{\partial \log \lambda_1^e}\right)\delta \log \lambda_1^e + \left(\sigma_2\lambda_1\lambda_2\lambda_3 - \frac{\partial W}{\partial \log \lambda_2^e}\right)
$$
$$
\times \delta \log \lambda_2^e + \left(\sigma_3\lambda_1\lambda_2\lambda_3 - \frac{\partial W}{\partial \log \lambda_3^e}\right)\delta \log \lambda_3^e
$$
$$
+ \left(\mu - \frac{\partial W}{\partial C} + \Omega\sigma_m\right)\delta C
$$
$$
+ \lambda_1\lambda_2\lambda_3\left[(\sigma_1 - \sigma_3)\delta \log \lambda_1^p + (\sigma_2 - \sigma_3)\delta \log \lambda_2^p\right] \geq 0
\tag{8}
$$

where $\sigma_m = (\sigma_1 + \sigma_2 + \sigma_3)/3$ is the mean stress, and $\Omega = \lambda_1^e\lambda_2^e\lambda_3^e d\Lambda(C)/dC$ is the volume per lithium atom in the host.

Each of the six independent variables represents a process to evolve the material. The processes take place at different rates. We will adopt a commonly used simplifying approach. Say that we are interested in a particular time scale. Processes taking place faster than this time scale are assumed to be instantaneous. Processes taking place slower than this time scale are assumed to never occur. In the present problem, the particular time scale of interest is the time needed for a particle of an electrode material of a finite size to absorb a large amount of lithium. This time is taken to be set by the diffusion time scale.

Elastic relaxation is typically much faster than diffusion. We assume that the material element is in equilibrium with respect to the elastic stretches, so that in (8) the coefficients associated with $\delta\lambda_1^e$, $\delta\lambda_2^e$, and $\delta\lambda_3^e$ vanish:

$$
\sigma_1 = \frac{\partial W}{\lambda_1\lambda_2\lambda_3\partial \log \lambda_1^e}, \quad \sigma_2 = \frac{\partial W}{\lambda_1\lambda_2\lambda_3\partial \log \lambda_2^e},
$$
$$
\sigma_3 = \frac{\partial W}{\lambda_1\lambda_2\lambda_3\partial \log \lambda_3^e}
\tag{9}
$$

We further assume that the material element is in equilibrium with respect to the concentration of lithium, so that in (8) the coefficient of δC vanishes:

$$
\mu = \frac{\partial W(\lambda_1^e, \lambda_2^e, \lambda_3^e, C)}{\partial C} - \Omega\sigma_m
\tag{10}
$$

The free energy is adopted in the following form:

$$
W = W_0(C) + \Lambda G\left[(\log \lambda_1^e)^2 + (\log \lambda_2^e)^2 + (\log \lambda_3^e)^2\right.
$$
$$
\left. + \frac{v}{1-2v}(\log \lambda_1^e\lambda_2^e\lambda_3^e)^2\right]
\tag{11}
$$

where G is the shear modulus, and v Poisson's ratio. Equation (11) can be interpreted as the Taylor expansion in terms of elastic strains. We have assumed that the elastic strains are small, and only retain terms up to those that are quadratic in strains. We have neglected any dependence of elastic moduli on the concentration of lithium. A combination of (9) into (11) shows that the stresses relate to the elastic strains as

$$
\sigma_1 = 2G\left(\log \lambda_1^e + \frac{v}{1-2v}\log \lambda_1^e\lambda_2^e\lambda_3^e\right)
$$
$$
\sigma_2 = 2G\left(\log \lambda_2^e + \frac{v}{1-2v}\log \lambda_1^e\lambda_2^e\lambda_3^e\right)
\tag{12}
$$
$$
\sigma_3 = 2G\left(\log \lambda_3^e + \frac{v}{1-2v}\log \lambda_1^e\lambda_2^e\lambda_3^e\right)
$$

A combination (10) and (11) expresses the chemical potential of lithium as

$$
\mu = \frac{dW_0(C)}{dC} - \Omega\sigma_m
\tag{13}
$$

In writing (12) and (13), we have neglected the terms quadratic in the elastic strains.

The material element, however, may not be in equilibrium with respect to plastic stretches. Consequently, the inequality (8) is reduced to

$$
(\sigma_1 - \sigma_3)\delta \log \lambda_1^p + (\sigma_2 - \sigma_3)\delta \log \lambda_2^p \geq 0
\tag{14}
$$

This thermodynamic inequality may be satisfied by many kinetic models—creep models that relate the rate of plastic strains to stresses. For simplicity, here we adopt a particular type of kinetic model: the model of time-independent plasticity.[28] A material is characterized by a yield strength. When the stress is below the yield strength, the rate of the plastic strain is taken to be so low that no additional plastic strain occurs. When the stress reaches the yield strength, the rate of plastic strain is taken to be so high that plastic strain increases instantaneously.

To calculate plastic deformation, we adopt the J_2 flow theory.[28] Recall that the plastic stretches preserve the volume, $\lambda_1^p\lambda_2^p\lambda_3^p = 1$. Consequently, (14) may be written in a form symmetric with respect to the three directions:

$$
(\sigma_1 - \sigma_m)\delta \log \lambda_1^p + (\sigma_2 - \sigma_m)\delta \log \lambda_2^p
$$
$$
+ (\sigma_3 - \sigma_m)\delta \log \lambda_3^p \geq 0
\tag{15}
$$

The J_2 flow theory is prescribed as

$$
\delta \log \lambda_1^p = \alpha(\sigma_1 - \sigma_m),
$$
$$
\delta \log \lambda_2^p = \alpha(\sigma_2 - \sigma_m),
\tag{16}
$$
$$
\delta \log \lambda_3^p = \alpha(\sigma_3 - \sigma_m)
$$

where α is a nonnegative scalar. This flow theory is symmetric with respect to the three directions, satisfies $\lambda_1^p\lambda_2^p\lambda_3^p = 1$, and is consistent with the thermodynamic inequality (15). In numerical calculations performed later, we will assume that the material is perfectly plastic. Let σ_Y be the yield strength measured when the material element is subject to uniaxial stressing. The yield strength is taken to be a function of concentration alone, $\sigma_Y(C)$. When the material element is under multiaxial stressing, the equivalent stress is defined by

$$
\sigma_e = \sqrt{\frac{3}{2}\left[(\sigma_1 - \sigma_m)^2 + (\sigma_2 - \sigma_m)^2 + (\sigma_3 - \sigma_m)^2\right]}
\tag{17}
$$

The material element yields under the von Mises condition: $\sigma_e = \sigma_Y(C)$. The value of α is specified by the following rules

$$
\begin{cases}
\alpha = 0, & \sigma_e < \sigma_Y \\
\alpha = 0, & \sigma_e = \sigma_Y, \quad \delta\sigma_e < \delta\sigma_Y \\
\alpha > 0, & \sigma_e = \sigma_Y, \quad \delta\sigma_e = \delta\sigma_Y
\end{cases}
\tag{18}
$$

III. A Spherical Particle of an Electrode

We now apply the theory to a spherical particle of an electrode material (Fig. 2). We have previously solved this problem using a theory of small plastic deformation.[22] Here we will allow large deformation. We model such an inelastic host of lithium by considering coupled lithium diffusion and large elastic–plastic deformation. In the following paragraphs, we specify the kinematics of large deformation, the kinetics of lithium diffusion, the flow rule of plasticity, and the thermodynamics of lithiation. The results of an amorphous silicon particle during charge and discharge are shown in Section IV. The detailed numerical procedure is included in Appendix A.

Before absorbing any lithium, the particle is of radius A, and is stress-free. This lithium-free particle is taken to be the reference configuration. At time t, the particle absorbs some lithium,

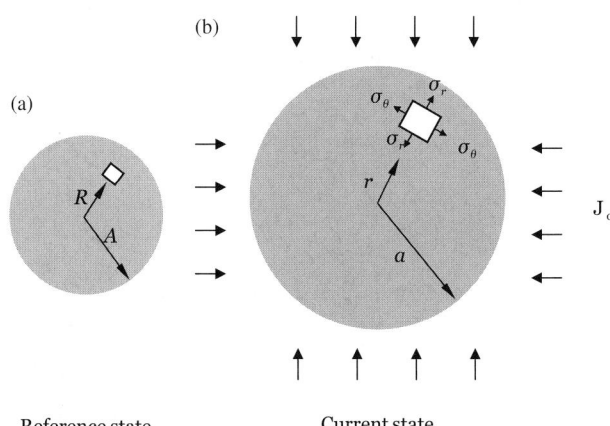

Fig. 2. (a) In the reference state, a spherical particle of an electrode is lithium-free and stress-free. (b) In the current sate, the particle is partially lithiated, and develops a field of stress.

the distribution of which may be inhomogeneous in the radial direction, but retains spherical symmetry. The inhomogeneous distribution of lithium induces in the particle a field of stress, and the particle swells to a radius a.

The kinematics of the large deformation is specified as follows. The spherical particle consists of a field of material elements. A material element a distance R from the center in the reference configuration moves, at time t, to a place a distance r from the center. The function $r(R, t)$ specifies the deformation of the particle. In representing a field, we may choose either r or R as an independent variable. One variable can be changed to the other by using the function $r(R, t)$. We will indicate our choice in each field explicitly when the distinction is important. The radial stretch is

$$\lambda_r = \frac{\partial r(R, t)}{\partial R} \tag{19}$$

The hoop stretch is

$$\lambda_\theta = \frac{r}{R} \tag{20}$$

The kinematics of lithium is specified as follows. Let C be the nominal concentration of lithium (i.e., the number of lithium atoms per unit volume of pure silicon in the reference state). The distribution of lithium in the particle is specified by the function $C(R, t)$. Because of the spherical symmetry, lithium diffuses in the radial direction. Let J be the nominal flux of lithium (i.e., the number of lithium atoms per unit reference area per unit time). The nominal flux is also a time-dependent field, $J(R, t)$. Conservation of the number of lithium atoms requires that

$$\frac{\partial C(R, t)}{\partial t} + \frac{\partial (R^2 J(R, t))}{R^2 \partial R} = 0 \tag{21}$$

Later we will also invoke the true concentration c (i.e., the number of lithium atoms per unit volume in the current configuration), and the true flux j (i.e., the number of lithium atoms per current area per time). These true quantities relate to their nominal counterparts by $J = j\lambda_\theta^2$ and $C = c\lambda_r\lambda_\theta^2$.

Write the radial stretch λ_r and the hoop stretch λ_θ in the form

$$\lambda_r = \lambda_r^e \lambda_r^p \Lambda^{1/3}, \quad \lambda_\theta = \lambda_\theta^e \lambda_\theta^p \Lambda^{1/3} \tag{22}$$

We will mainly consider high-capacity hosts that undergo large deformation by lithiation, and will neglect the volumetric change due to elasticity, setting $\lambda_r^e(\lambda_\theta^e)^2 = 1$. Consistent with this assumption, we set Poison's ratio $\nu = 1/2$, and set Young's mod-

ulus $E = 3G$. Recall that by definition the plastic stretches preserve volume, $\lambda_r^p(\lambda_\theta^p)^2 = 1$.

In the spherical particle, each material element is subject to a state of triaxial stresses, $(\sigma_r, \sigma_\theta, \sigma_\theta)$, where σ_r is the radial stress, and σ_θ the hoop stress. The state of elastic–plastic deformation is taken to be unaffected when a hydrostatic stress is superimposed on the element. In particular, as illustrated in Fig. 3(a), superimposing a hydrostatic stress $(-\sigma_\theta, -\sigma_\theta, -\sigma_\theta)$ to the state of triaxial stresses $(\sigma_r, \sigma_\theta, \sigma_\theta)$ results in a state of uniaxial stress $(\sigma_r - \sigma_\theta, 0, 0)$. The state of elastic–plastic deformation of the element subject to the triaxial stresses is the same as the state of plastic deformation of the element subject to a uniaxial stress $\sigma_r - \sigma_\theta$. We represent the uniaxial stress–stretch relation by the elastic and perfectly plastic model. Figure 3(b) sketches this stress–strain relation in terms of the stress $\sigma_r - \sigma_\theta$ and the elastic–plastic part of the true strain, $\log(\lambda_r^e\lambda_r^p) = \log(\lambda_r \Lambda^{-1/3})$. The yield strength in the state of uniaxial stress, σ_Y, is taken to be a constant independent of the plastic strain and concentration of lithium.

In the spherical particle the stresses are inhomogeneous, represented by functions $\sigma_r(R, t)$ and $\sigma_\theta(R, t)$. The balance of forces acting on a material element requires that

$$\frac{\partial \sigma_r(R, t)}{\lambda_r \partial R} + 2\frac{\sigma_r - \sigma_\theta}{\lambda_\theta R} = 0 \tag{23}$$

We specify a material model of transport as follows. We assume that each material element is in a state of local equilibrium with respect to the reaction between lithium atoms and host atoms, so that we can speak of the chemical potential of lithium in the material element. We further assume that the chemical potential of lithium in the material element takes the form:

$$\mu = \mu^0 + kT\log(\gamma c) - \Omega\sigma_m \tag{24}$$

where μ^0 is a reference value, γ the activity coefficient, and c the true concentration of lithium.

If the distribution of lithium in the particle is inhomogeneous, the chemical potential of lithium is a time-dependent field, $\mu(r, t)$, and the particle is not in diffusive equilibrium. The gradient of the chemical potential drives the flux of lithium. We adopt a linear kinetic model:

$$j = -\frac{cD}{kT}\frac{\partial \mu(r, t)}{\partial r} \tag{25}$$

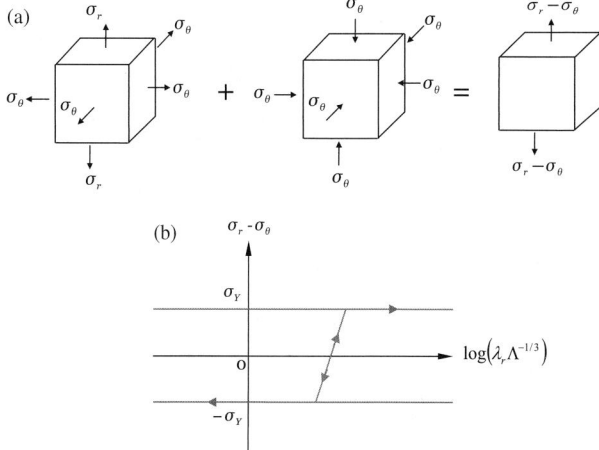

Fig. 3. (a) The state of elastic–plastic deformation of the element subject to the triaxial stresses $(\sigma_r, \sigma_\theta, \sigma_\theta)$ is the same as that of the element subject to a uniaxial stress $(\sigma_r - \sigma_\theta)$. (b) The uniaxial stress–strain relation in terms of the stress $(\sigma_r - \sigma_\theta)$ and the true elastic–plastic strain $\log(\lambda_r \Lambda^{-1/3})$.

This relation has been written in a conventional form, in terms of the true flux j and the true concentration c. Note that kT is the temperature in dimensions of energy, and that (25) may be regarded as a phenomenological definition of the diffusivity D. Recall that the flux relates to the drift velocity of lithium in the host by $j = c v_{\text{drift}}$. Thus, D/kT is the mobility of lithium in the host. The diffusivity may depend on concentration and stress.

The particle is subject to the following boundary conditions. Because of the symmetry, $r(0, t) = 0$ and $J(0, t) = 0$. On the surface of the particle, the radial stress vanishes at all time, $\sigma_r(A, t) = 0$. The particle is charged and discharged by prescribing on the surface of the particle a constant flux J_0, namely, $J(A, t) = \pm J_0$. The signs differ for charge and discharge.

IV. Numerical Results and Discussions

We solve the initial-boundary value problem numerically, as described in Appendix A. This section describes the numerical results, and discusses their implications. For simplicity, we set $\gamma = 1$ and assume that Ω, E, σ_Y, and D are constant. The initial-boundary value problem has three dimensionless parameters: $\Omega E/kT$, σ_Y/E, and $J_0 A \Omega/D$. For lithium in amorphous silicon, representative values are $\Omega = 1.36 \times 10^{-29}$ m^3, $E = 80$ GPa, $\sigma_Y = 1.75$ GPa,[20] and $D = 10^{-16}$ m^2/s,[29] giving $\Omega E/kT = 263$ and $\sigma_Y/E = 0.022$. The parameter $J_0 A \Omega/D$ is a dimensionless measure of the charging rate, and may be interpreted as follows. Let $C_{\max}$ be the maximum theoretical concentration of lithium. When the spherical particle of radius A is charged by a constant flux J_0, the nominal time τ needed to charge the particle to the theoretical maximum concentration is given by $4\pi A^2 J_0 \tau = (4\pi A^3/3)C_{\max}$. For silicon, the volume of fully lithiated state swells by about 300%, so that $\Omega C_{\max} \approx 3$. For a particle of radius $A = 1$ μm, $\tau = 1$ h corresponds to $J_0 A \Omega/D = 2.8$. In the following description, we use τ to represent the charge rate; for example, $\tau = 0.5$ h means that it needs 0.5 h to charge silicon to reach the theoretical maximum lithium concentration. Smaller values of τ represent a faster charge process. We normalize time as Dt/A^2. For a particle of radius $A = 1$ μm, the diffusion time scale is $A^2/D = 10^4$ s.

Figure 4 shows the evolution of the fields in the spherical particle charged at the rate of $\tau = 1$ h. The simulation is initiated when the particle is lithium-free, and is terminated when the concentration at the surface of the particle reaches the full capacity, $\Omega C_{\max} = 3$, and the interior of the particle is still much below the full capacity (Fig. 4(a)). At all time, $r(R, t) > R$, indicating all material elements in the particle move away from the center of the particle (Fig. 4(b)). The ratio λ_r/λ_θ measures the anisotropy of the deformation (Fig. 4(c)). The deformation is highly anisotropic near the surface of the particle, but is isotropic at the center of the particle. Plastic deformation occurs near the surface of the particle, but is absent at the center of the particle (Fig. 4(d)). The chemical potential of lithium in the particle is inhomogeneous, driving lithium to diffuse from the surface of the particle toward the center (Fig. 4(e)).

Figures 4(f)–(h) show the distributions of the radial, hoop and equivalent stresses. The traction-free boundary condition requires that the radial stress at the surface of the particle to vanish at all times. Because the distribution of lithium in the particle is inhomogeneous, the particle expands more near the surface than at the center, resulting in tensile radial stresses inside the particle. The hoop stress is compressive near the surface, and tensile near the center. For the spherical particle, the equivalent stress is $\sigma_e = |\sigma_\theta - \sigma_r|$, which is bounded in the interval $0 \leq \sigma_e \leq \sigma_Y$. By symmetry, the center of the sphere is under equal-triaxial tensile stresses. Because of the triaxial constraint at the center, the radial stress and hoop stress can exceed the yield strength.

The high level of tensile stresses at the center of the sphere may generate cavities. An experimentally measured value of the yield strength is $\sigma_Y = 1.75$ GPa.[20] Our calculation indicates that the stress at the center of the particle can be several times the yield strength. Let ρ be the radius of a flaw and γ_{surface} be the surface energy. For the flaw to expand, the stress needs to overcome the effect of the Laplace pressure, $2\gamma_{\text{surface}}/\rho$. Taking $\gamma_{\text{surface}} = 1$ J/m^2, we estimate that the critical radius $\rho = 1$ nm when the stress at the center of the particle is 2 GPa. Lithiation-induced cavitation have been observed in a recent experiment.[21]

To illustrate effects of yielding, Fig. 5 compares the two cases: $\sigma_Y = 1.75$ GPa and $\sigma_Y = \infty$ (no yielding). The fields are plotted at time $Dt/A^2 = 0.048$. As expected, yielding allows the particle to accommodate lithiation by greater anisotropic deformation, Fig. 5(b). Yielding also significantly reduces the magnitudes of the stresses, Figs. 5(c) and (d). This observation implies that fracture and cavitation may be avoided for electrodes with low yield strength.

As shown by (24), the chemical potential of lithium depends on the mean stress. This effect of stress on chemical potential has been neglected in some of the previous models. The representative values $\Omega = 1.36 \times 10^{-29}$ m^3 and $\sigma_Y = 1.75$ GPa give an estimate $\Omega \sigma_Y = 0.15$ eV, which is a value significant compared with the value of the term involving concentration in the expression of the chemical potential. Figure 6 compares results calculated by including or neglecting the term $\Omega \sigma_m$ in the expression of the chemical potential (24). The mean stress σ_m is compressive near the surface of the particle and tensile at the center. Consequently, the gradient of the mean stress also motivates lithium to migrate toward the center. Here we show the fields at the time when the surface of the particle attains the full capacity. The case with the mean stress included in the expression of the chemical potential absorbs more lithium at the center, Fig. 6(a). The stress gradient also decreases the chemical potential of lithium in the particle, Fig. 6(b). The contribution of the stress to the chemical potential helps to homogenize lithium distribution, and consequently reduces the stress.

Figure 7 compares the fields for three charging rates, $\tau = 0.5$ h, $\tau = 1$ h and $\tau = 2$ h. Each simulation is terminated when the surface of the particle attains the full capacity. The stress level is determined by the total amount of lithium inserted into the particle and the degree of inhomogeneity in the distribution of this amount of lithium. At a high charging rate, $\tau = 0.5$ h, the distribution of lithium is highly inhomogeneous, as in Fig. 7(a). However, the surface of the particle reaches its full capacity very rapidly. At this time, not much lithium has been inserted into the particle, so the stresses are fairly low and the deformation is anisotropic only near the surface of the particle. At such a fast charging rate, the particle will not store much lithium. At an intermediate charging rate, $\tau = 1$ h, lithium has some time to diffuse toward the center of the particle, but not enough time to fully homogenize the distribution of lithium. Consequently, relatively large stresses develop and the deformation is quite anisotropic. At a yet slower charging rate, $\tau = 2$ h, there is enough time for diffusion to nearly homogenize the distribution of lithium. At this slow charging rate, the particle is effective as an electrode in that it can store a large amount of lithium before the surface of the particle reaches the full capacity. This homogenization leads to relatively low stresses and nearly isotropic deformation.

Figure 8 shows the time evolution of the stress at the center of the particle for three charging rates. For a fast charging case ($\tau = 0.5$ h), the stress builds up until the outer surface reaches its full capacity. At an intermediate charging rate ($\tau = 1$ h), the stress builds up until it is large enough to significantly contribute to the chemical potential. At this point, both the stress gradient and concentration gradients tend to homogenize the concentration of lithium, and the stress decreases. However, the charging rate is still fairly fast relative to the time for diffusion. Thus, the concentration of lithium cannot be fully homogenized and the stress cannot be fully relaxed before the surface of the particle reaches the full capacity. By contrast, for the case of $\tau = 2$ h, the charging rate is slow enough for diffusion to homogenize the distribution of lithium before the surface of the particle reaches the full capacity, as shown in Fig. 7(a). As the distribution of

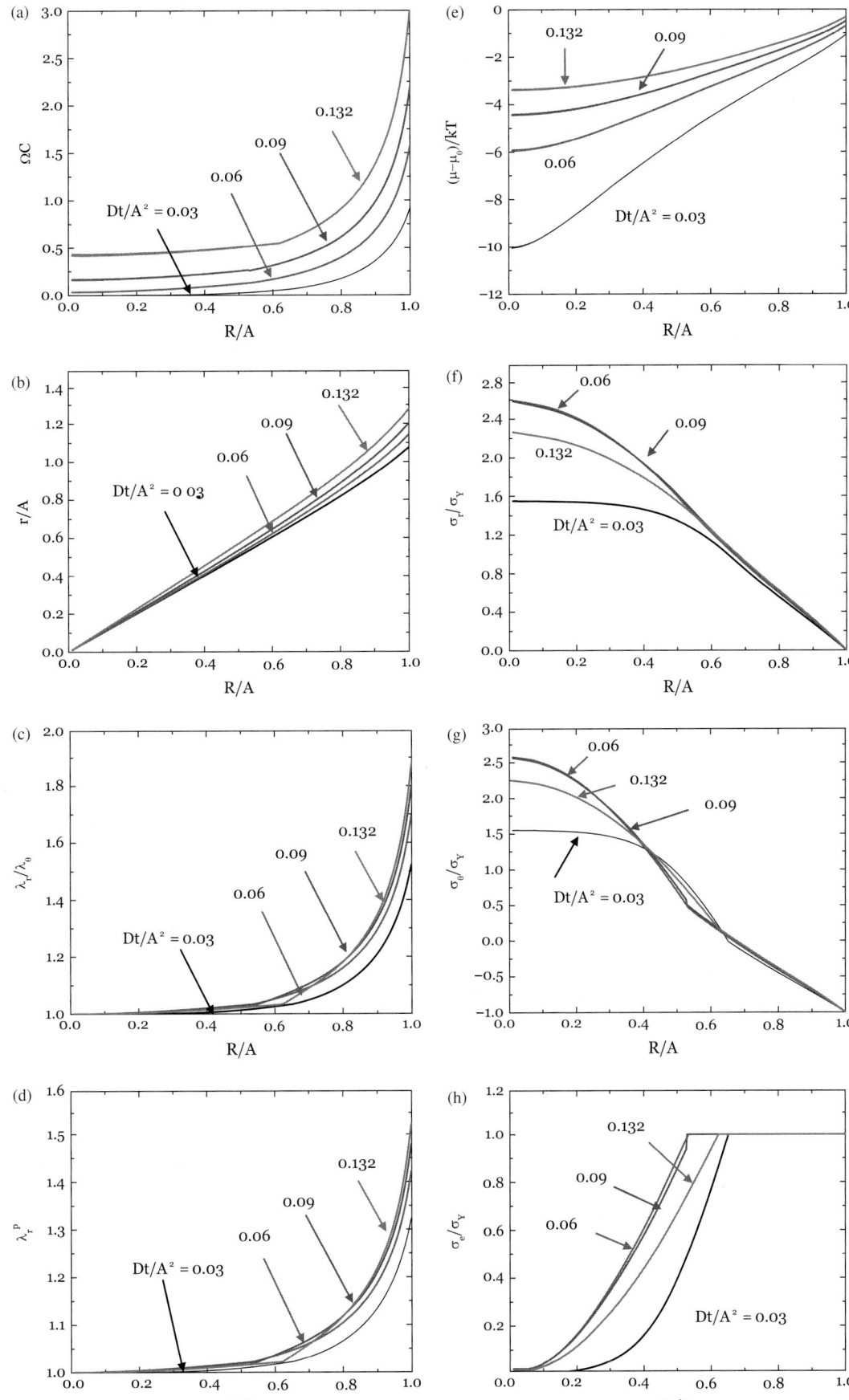

Fig. 4. As a spherical particle is being charged at the rate of $\tau = 1$ h, various fields evolve: (a) concentration of lithium, (b) deformation field, (c) ratio of the radial stretch to the hoop stretch, (d) plastic stretch in the radial direction, (e) chemical potential of lithium, (f) radial stress, (g) hoop stress, and (h) equivalent stress.

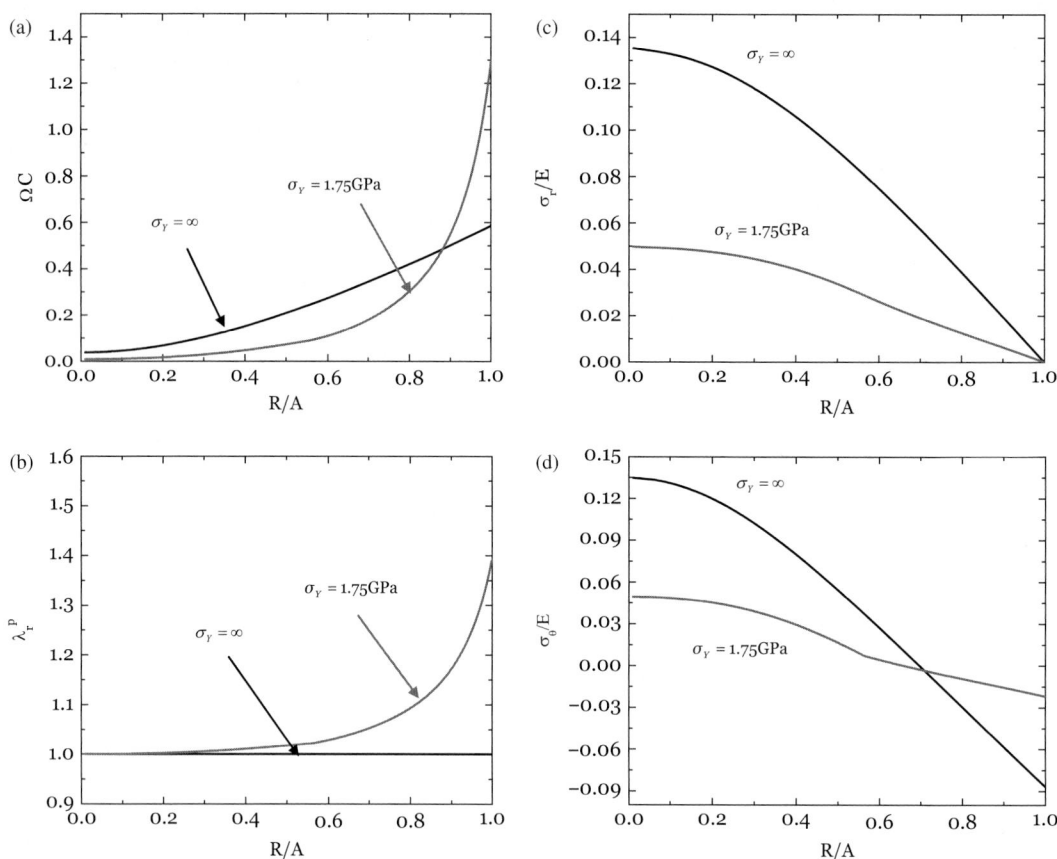

Fig. 5. The effect of plastic yield on various fields: (a) concentration of lithium, (b) plastic stretch in radial direction, (c) radial stress, and (d) hoop stress. The charging rate is $\tau = 1$ h, and the fields are given at time $Dt/A^2 = 0.048$.

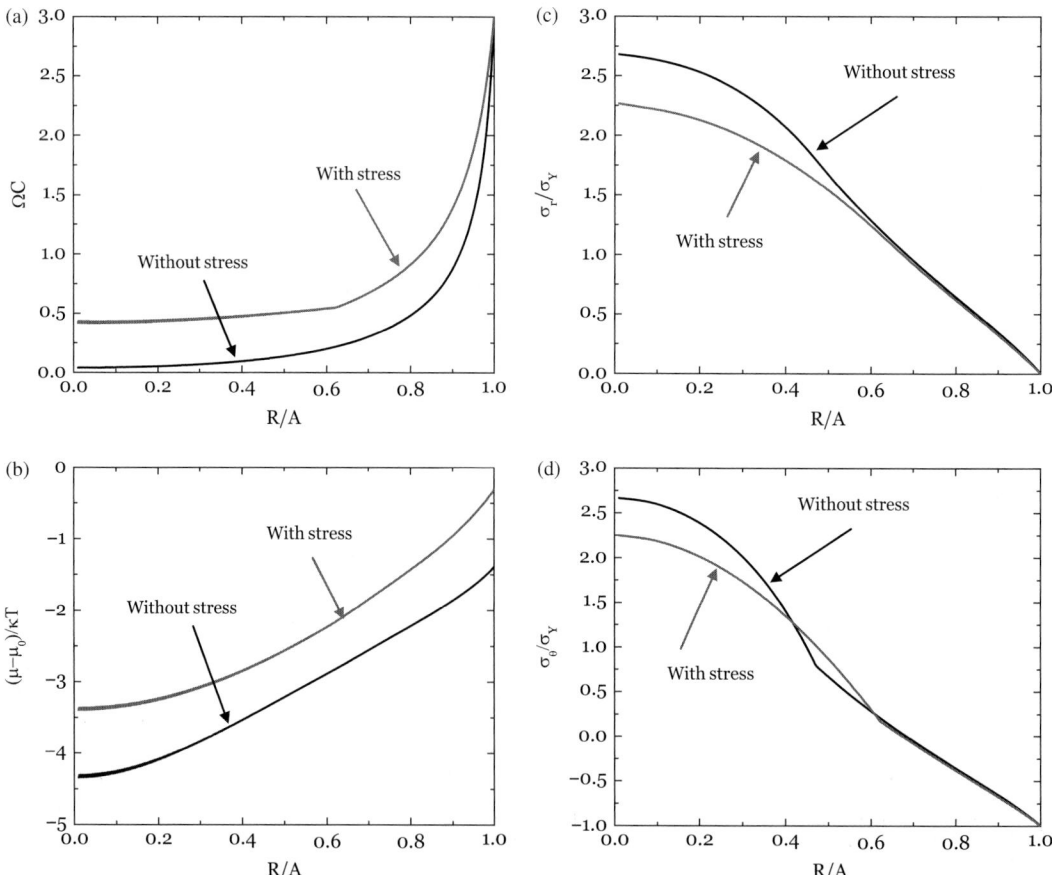

Fig. 6. Fields calculated with or without including the mean stress in the expression of the chemical potential of lithium are compared: (a) concentration of lithium, (b) chemical potential of lithium, (c) radial stress, and (d) hoop stress. The charging rate is $\tau = 1$ h, and both fields are given at the end of charge time, i.e., $Dt/A^2 = 0.132$ with stress calculation, $Dt/A^2 = 0.009$ without stress calculation.

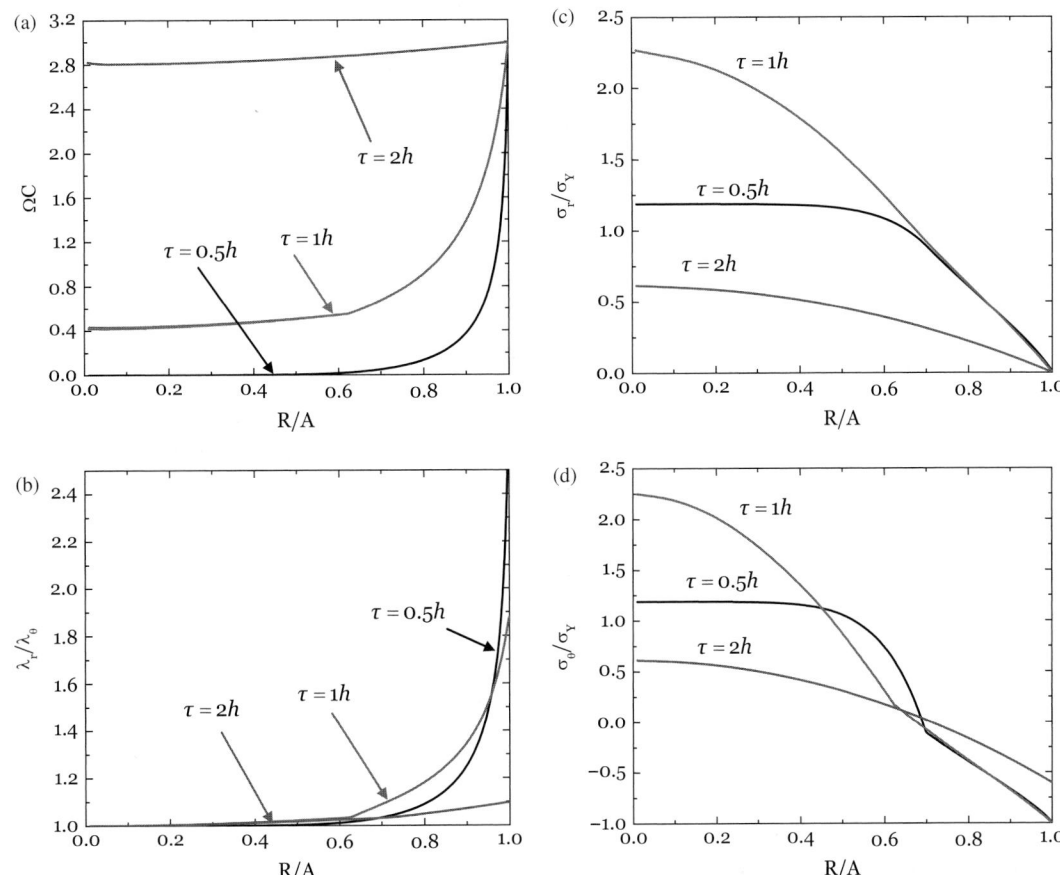

Fig. 7. The effect of the charging rate on various fields: (a) concentration of lithium, (b) ratio of radial stretch to hoop stretch, (c) radial stress, and (d) hoop stress.

lithium is homogenized, the stress is relaxed, as shown in Fig. 8. The effect of the charging rate on the triaxial tension at the center of the particle may be used to guide future experiments to study cavitation.

Figure 9 shows the time evolution of the lithium concentration and the stress fields as lithium desorbs from the spherical particle. The simulation begins when the particle is at full capacity of lithium and is stress-free, and is terminated when the concentration of lithium vanishes near the surface of the particle. As lithium desorbs, the concentration near the surface becomes lower than it is near the center of the particle, Fig. 9(a). This inhomogeneity causes the particle to contract more near the surface than at the center. Consequently, a compressive radial stress develops, Fig. 9(b). The hoop stress at the

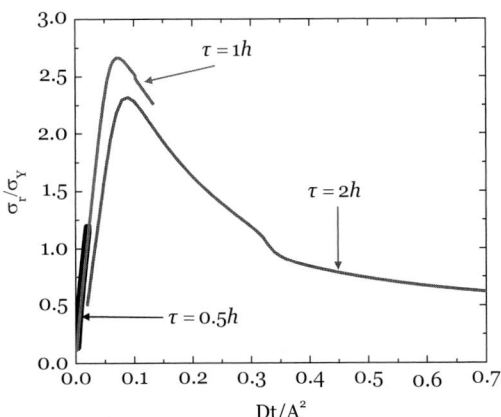

Fig. 8. The radial stress at the center of the sphere as a function of time at various charge rates.

surface becomes tensile with magnitude σ_Y, Fig. 9(c). This tensile stress may result in the propagation of surface flaws. Because the tensile stress is limited by the yield strength, fracture may be averted when the yield strength is low.

V. Conclusions

This paper formulates a theory that couples lithiation and large elastic–plastic deformation. The homogeneous lithiation and deformation in a small material element is analyzed using non-equilibrium thermodynamics. The material is assumed to undergo plastic deformation when the state of stress reaches the yield condition. A spherical particle subject to a constant rate of charge and discharge is analyzed by coupling diffusion and large plastic deformation. The effect of plastic yielding, stress on the chemical potential of lithium, and charging rates are studied. When the charging rate is low, the distribution of lithium in the particle is nearly homogeneous, and the stress is low. When the charging rate is high, the stress at the center of the particle can substantially exceed the yield strength. The developed stress gradient also greatly influences the diffusion of lithium, tending to homogenize the distribution of lithium in the particle. Plastic yielding can markedly reduce the magnitude of stress.

Appendix A: Notes on Numerical Procedure

We rewrite the governing equations in the form used in our numerical simulation. It has been assumed that neither elastic nor plastic deformation causes any volumetric change. It has been also assumed that Ω is a constant. Consequently, the volumetric change of a material element is

$$\lambda_r \lambda_\theta^2 = 1 + \Omega C \tag{A-1}$$

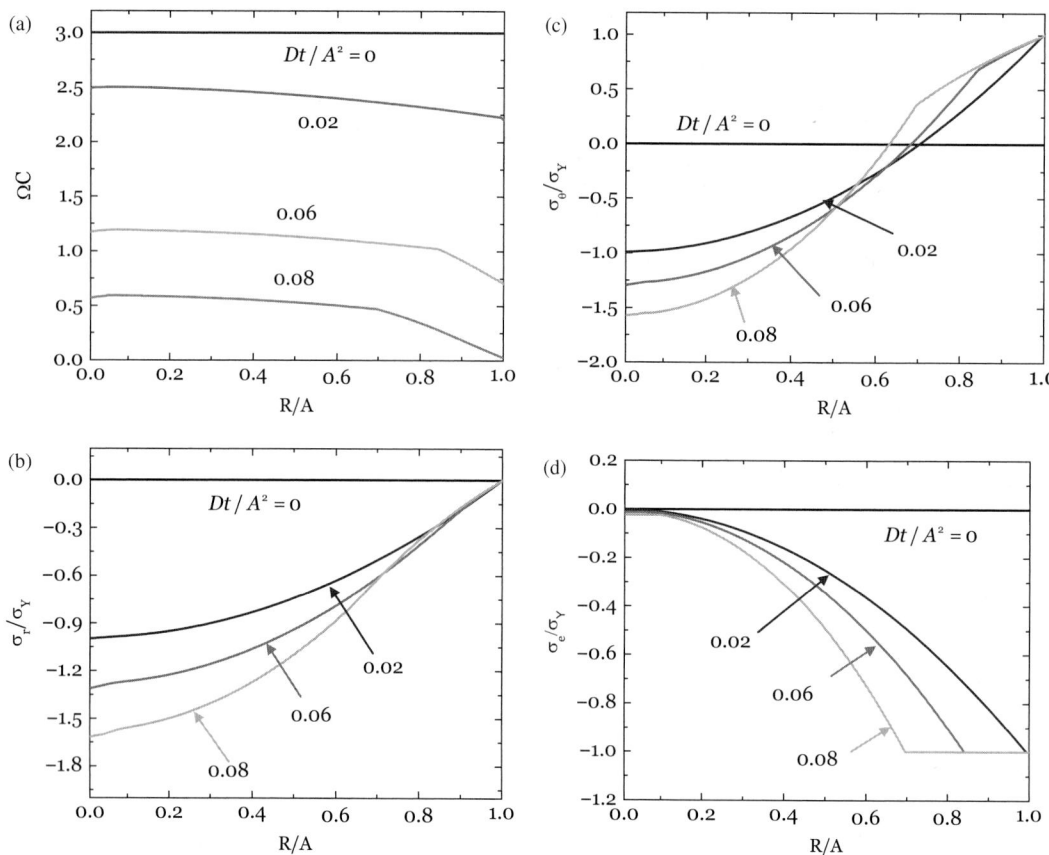

Fig. 9.   As a spherical particle is being discharged at the rate of $\tau = 0.5$ h, various fields evolve (a) concentration of lithium, (b) radial stress, (c) hoop stress, and (d) equivalent stress.

A combination of (A-1), (19) and (20) gives that

$$r(R, t) = \left[3 \int_0^R (1 + \Omega C) R^2 \mathrm{d}R \right]^{1/3} \qquad \text{(A-2)}$$

Here we have used a condition due to the spherical symmetry of the problem, $r(0, t) = 0$.

The kinetic model (25) can be written in terms of the nominal quantities:

$$J = -\frac{CD}{kT\lambda_r^2} \frac{\partial \mu(R, t)}{\partial R} \qquad \text{(A-3)}$$

A combination of (A-3) and (24) gives

$$J = -\frac{CD}{(1 + \Omega C)^2} \left(\frac{r}{R} \right)^4 \frac{\partial}{\partial R}$$
$$\times \left[\log \frac{\gamma C}{1 + \Omega C} - \frac{\Omega(\sigma_r + 2\sigma_\theta)}{3kT} \right] \qquad \text{(A-4)}$$

Recall that the flux satisfies the boundary conditions $J(0, t) = 0$ and $J(A, t) = \pm J_0$. The sign of the latter condition depends on whether the particle is being charged or discharged.

The stress $(\sigma_\theta - \sigma_r)$ and the plastic stretch evolve according to the ideal elastic–plastic model, Fig. 3(b). When $|\sigma_r - \sigma_\theta| < \sigma_Y$, the plastic stretch λ_r^p remains fixed, and the elastic stretch is given by $\log \lambda_r^e = (\sigma_r - \sigma_\theta)/E$, which is written as

$$\sigma_r - \sigma_\theta = E \log(\Lambda^{-1/3} \lambda_r / \lambda_r^p) \qquad \text{(A-5)}$$

when $\sigma_r - \sigma_\theta = \pm \sigma_Y$, the elastic stretch is fixed at $\log \lambda_r^e = \pm \sigma_Y/E$, and the plastic stretch adopts a value

$$\lambda_r^p = \lambda_r \Lambda^{-1/3} \exp\left(\mp \frac{\sigma_Y}{E} \right) \qquad \text{(A-6)}$$

Integrating (23), we obtain that

$$\sigma_r(R, t) = 2 \int_A^R \frac{(\sigma_r - \sigma_\theta)(1 + \Omega C) R^2}{r^3} \mathrm{d}R \qquad \text{(A-7)}$$

Here we have used the boundary condition $\sigma_r(A, t) = 0$.

We use the finite difference method, and divide the interval $0 \le R \le A$ into small elements. The initial condition $C(R, 0)$ is prescribed; for example, we set $C(R, 0) = 0$ to simulate the process of lithiation, and set $C(R, 0) = C_{max}$ to simulate the process of delithiation. The initial values of the function $J(R, 0)$ are set with $J(0, 0) = 0$ and $J(A, 0) = \pm J_0$ at the boundaries, and $J(R, 0) = 0$ at the interior points. We then evolve all functions with a time step Δt. At a given time t, the functions $C(R, t)$ and $J(R, t)$, along with the boundary conditions $J(0, t) = 0$ and $J(A, t) = \pm J_0$, are inserted into (21) to calculate $C(R, t+\Delta t)$. The result is then inserted into (A-2) to calculate $r(R, t+\Delta t)$. The field $\sigma_r(R, t+\Delta t)$ is calculated by the integrating (A-7), where $(\sigma_\theta - \sigma_r)$ is determined by the uniaxial stress–strain relation. We then calculate $J(R, t+\Delta t)$ by using (A-4). This procedure is repeated for the next time step to evolve the fields.

Acknowledgments

Matt Pharr acknowledges the support of the Department of Defense (DoD) through the National Defense Science & Engineering Graduate Fellowship (NDSEG) Program.

This paper is submitted to the *Journal of the American Ceramic Society* for a special issue dedicated to the memory of A. G. Evans.

References

[1]R. A. Huggins, *Advanced Batteries: Materials Science Aspects*. Springer, New York, 2009.

[2]J. Christensen and J. Newman, "Stress Generation and Fracture in Lithium Insertion Materials," *J. Solid State Electrochem.*, **10** [5] 293–319 (2006).

[3]X. C. Zhang, W. Shyy, and A. M. Sastry, "Numerical Simulation of Intercalation-Induced Stress in Li-Ion Battery Electrode Materials," *J. Electrochem. Soc.*, **154** [10] A910–6 (2007).

[4]S. Golmon, K. Maute, S. H. Lee, and M. L. Dunn, "Stress Generation in Silicon Particles During Lithium Insertion," *Appl. Phys. Lett.*, **97** [3] 033111, 3pp (2010).

[5]Y. T. Cheng and M. W. Verbrugge, "Evolution of Stress within a Spherical Insertion Electrode Particle Under Potentiostatic and Galvanostatic Operation," *J. Power Sources*, **190** [2] 453–60 (2009).

[6]Y. H. Hu, X. H. Zhao, and Z. G. Suo, "Averting Cracks Caused by Insertion Reaction in Lithium-Ion Batteries," *J. Mater. Res.*, **25** [6] 1007–10 (2010).

[7]R. A. Huggins and W. D. Nix, "Decrepitation Model for Capacity Loss During Cycling of Alloys in Rechargeable Electrochemical Systems," *Ionics*, **6**, 57–64 (2000).

[8]T. K. Bhandakkar and H. J. Gao, "Cohesive Modeling of Crack Nucleation Under Diffusion Induced Stresses in a Thin Strip: Implications on the Critical Size for Flaw Tolerant Battery Electrodes," *Int. J. Solids Struc.*, **47** [10] 1424–34 (2010).

[9]W. H. Woodford, Y. M. Chiang, and W. C. Carter, "Electrochemical Shock of Intercalation Electrodes: A Fracture Mechanics Analysis," *J. Electrochem. Soc.*, **157** [10] A1052–9 (2010).

[10]K. J. Zhao, M. Pharr, J. J. Vlassak, and Z. G. Suo, "Fracture of Electrodes in Lithium-Ion Batteries Caused by Fast Charging," *J. Appl. Phys.*, **108** [7] 073517, 6pp (2010).

[11]Y. Itou and Y. Ukyo, "Performance of LiNiCoO$_2$ Materials for Advanced Lithium-Ion Batteries," *J. Power Sources*, **146** [1–2] 39–44 (2005).

[12]P. Arora, R. E. White, and M. Doyle, "Capacity Fade Mechanisms and Side Reactions in Lithium-Ion Batteries," *J. Electrochem. Soc.*, **145** [10] 3647–67 (1998).

[13]D. Y. Wang, X. D. Wu, Z. X. Wang, and L. Q. Chen, "Cracking Causing Cyclic Instability of LiFePO$_4$ Cathode Material," *J. Power Sources*, **140** [1] 125–8 (2005).

[14]K. E. Aifantis and J. P. Dempsey, "Stable Crack Growth in Nanostructured Li-Batteries," *J. Power Sources*, **143** [1–2] 203–11 (2005).

[15]L. Y. Beaulieu, K. W. Eberman, R. L. Turner, L. J. Krause, and J. R. Dahn, "Colossal Reversible Volume Changes in Lithium Alloys," *Electrochem. Solid State Lett.*, **4** [9] A137–40 (2001).

[16]W. J. Zhang, "A Review of the Electrochemical Performance of Alloy Anodes for Lithium-Ion Batteries," *J. Power Sources*, **196** [1] 13–24 (2011).

[17]T. Takamura, S. Ohara, M. Uehara, J. Suzuki, and K. Sekine, "A Vacuum Deposited Si Film having a Li Extraction Capacity Over 2000 mAh/g with a Long Cycle Life," *J. Power Sources*, **129** [1] 96–100 (2004).

[18]C. K. Chan, H. L. Peng, G. Liu, K. McIlwrath, X. F. Zhang, R. A. Huggins, and Y. Cui, "High-Performance Lithium Battery Anodes Using Silicon Nanowires," *Nature Nanotechnol.*, **3** [1] 31–5 (2008).

[19]H. Kim, B. Han, J. Choo, and J. Cho, "Three-Dimensional Porous Silicon Particles for Use in High-Performance Lithium Secondary Batteries," *Angew. Chem.-Int. Ed.*, **47** [52] 10151–4 (2008).

[20]V. A. Sethuraman, M. J. Chon, M. Shimshak, V. Srinivasan, and P. R. Guduru, "In situ Measurements of Stress Evolution in Silicon Thin Films During Electrochemical Lithiation and Delithiation," *J. Power Sources*, **195** [15] 5062–6 (2010).

[21]J. W. Choi, J. McDonough, S. Jeong, J. S. Yoo, C. K. Chan, and Y. Cui, "Stepwise Nanopore Evolution in One-Dimensional Nanostructures," *Nano Lett.*, **10** [4] 1409–13 (2010).

[22]V. A. Sethuraman, V. Srinivasan, A. F. Bower, and P. R. Guduru, "In situ Measurements of Stress-Potential Coupling in Lithiated Silicon," *J. Electrochem. Soc.*, **157** [11] A1253–61 (2010).

[23]K. J. Zhao, M. Pharr, J. J. Vlassak, and Z. G. Suo, "Inelastic Hosts as Electrodes for High-Capacity Lithium-Ion Batteries," *J. Appl. Phys.*, **109** [1] 016110, 3pp (2011).

[24]E. H. Lee, "Elastic–Plastic Deformation at Finite Strains," *J. Appl. Mech.*, **36** [1] 1–8 (1969).

[25]M. N. Silberstein and M. C. Boyce, "Constitutive Modeling of the Rate, Temperature, and Hydration Dependent Deformation Response of Nafion to Monotonic and Cyclic Loading," *J. Power Sources*, **195** [17] 5692–706 (2010).

[26]M. Ben Amar and A. Goriely, "Growth and Instability in Elastic Tissues," *J. Mech. Phys. Solids*, **53** [10] 2284–319 (2005).

[27]T. Song, J. L. Xia, J. H. Lee, D. H. Lee, M. S. Kwon, J. M. Choi, J. Wu, S. K. Doo, H. Chang, W. Il Park, D. S. Zang, H. Kim, Y. G. Huang, K. C. Hwang, J. A. Rogers, and U. Paik, "Arrays of Sealed Silicon Nanotubes as Anodes for Lithium Ion Batteries," *Nano Lett.*, **10** [5] 1710–6 (2010).

[28]R. Hill, *The Mathematical Theory of Plasticity*. Oxford University Press, Oxford, 1950.

[29]N. Ding, J. Xu, Y. X. Yao, G. Wegner, X. Fang, C. H. Chen, and I. Lieberwirth, "Determination of the Diffusion Coefficient of Lithium Ions in Nano-Si," *Solid State Ionics*, **180** [2–3] 222–5 (2009). □

J. Am. Ceram. Soc., **94** [S1] S236–S240 (2011)
DOI: 10.1111/j.1551-2916.2011.04492.x
© 2011 The American Ceramic Society

Journal

Searching for a Biomimetic Method of Fabricating Network Structures

Brian Cox[†]

Teledyne Scientific Co. LLC, Thousand Oaks, California 91360

A recent theory has suggested that populations of migrating and proliferating cells can create network structures by responding to strain cues associated with the cell density variations that arise during network formation. Unlike prior theories of network formation, the strain-cue mechanism leads to nonfractal networks, consisting of closed loops rather than tree-like morphologies. In this paper, the possibility is suggested of developing a synthetic process for fabricating networks that mimic the mechanisms present in the theory. Such a biomimetic process should replicate three phenomena: (1) the extension of the existing network domain should be governed by the strain field that exists just outside the existing domain, rather than the strains within the domain; (2) the process should be stochastic; and (3) a relaxation mechanism must be present by which the strains that induce network extension at any location will fade with time. Simulations imply values for the key parameters of these mechanisms to achieve useful networks. Possible routes to realizing a biomimetic system are discussed.

I. Introduction

MULTIPHASE materials in which one phase constitutes a connected network can be desirable for a number of reasons. (1) A connected phase can provide pathways for thermal or electronic heat conduction. If the conducting phase is introduced as spatially random particles, conduction relies on achieving the percolation threshold and, if conductivity is required in three dimensions, percolation occurs at quite high volume fractions of the conducting phase. To reduce the volume fraction, some process is needed by which the particles can be assembled into line aggregates rather than being spatially random, which is not an easy feat. If conducting lines are achieved instead by using continuous fibers as the conducting phase, challenges arise with deploying the fibers in three directions—once again, isotropic conductivity is difficult to achieve—and with handling fibers of very small diameter. (2) Interconnected ligaments of a tough and strong material can provide mechanical bridging and therefore impart macroscopic toughness to an otherwise brittle material. Achieving isotropy is again a challenge; current methods based on chopped fibers or whiskers, for example, tend to produce approximately two-dimensional (2D) felt-like reinforcement (because of particle settling during processing), with relatively weak mechanical effect, for a given volume fraction of reinforcement, in the direction perpendicular to the layers of fibers or whiskers. An isotropic interconnected network of reinforcing ligaments could be significantly tougher than long-particle-reinforced materials for a given volume fraction. In a chopped fiber or whisker-reinforced material, a crack can run around the ends of the particles, which it cannot do if it propagates through a connected network of reinforcing ligaments. (3) If a connected network is formed in a solid body, simple

chemical processing (e.g., etching) can remove the matrix material, leaving the network as a free-standing structure. A free-standing three-dimensional (3D) network with open cell faces has numerous potential applications, including filtering, biological cell culture, catalysis, etc. The fact that it is a connected network gives it useable structural integrity, while control of the strut diameter and length and the network density would allow the surface-to-volume ratio and the permeability of the network to be tailored to a particular purpose.

While forming networks is a poorly developed art in materials engineering, nature forms a great variety of networks (especially nerve and vascular) during organogenesis with a wide range of forms and at spatial scales down to that of a single cell (~ 10 μm). Many of these networks are tree-like structures, often with fractal scaling properties.[1] But others are nonfractal, closed networks, characterized by roughly periodic repetitions of truss-like strut-and-node assemblies, a pattern that is more desirable for many engineering materials applications. Given the lack of processing routes to such networks in materials engineering, the questions of just how nature does it, and how nature's methods might be replicated, deserve some attention.

In a recent paper,[2] a theory was presented that can account for the formation of a 2D network by neural cells invading the gut of a mouse. The direction of migration of a neural cell invading the host population of the gut was assumed to be controlled by the magnitude of the strains in the host medium (cells plus extracellular matrix) that arise as the host medium deforms to accommodate the invader (Fig. 1). This single assumption of a strain cue that references strains external to the invader is sufficient to generate network structures. The strain induced by a line of invaders is greatest at the extremity of the line and thus the strain field breaks symmetry, stabilizing branch formation. The strain cue also triggers sprouting from existing branches, with no further model assumption. The characteristics of networks predicted by the theory depend primarily on the ratio of the rate of advance of the invaders to the rate of relaxation of the host cells after their initial deformation. A second model parameter arises from the assumption that cell advance is a stochastic process.

Prior models of branching structures and other shape instabilities that mark organogenesis have exploited concepts that originated in the physical sciences. Viscous fingering (a problem first identified in the oil industry and explained in Saffman and Taylor[3]) arises when a less viscous fluid is injected into a more viscous fluid in the confines of a porous medium. Repeated finger formation can result in fractal tree-like structures. While the formative shape instability depends on the viscosity ratio and the local pressure at branch tips, the driving force is ultimately the far-field pressure, whose transmission and diminution along the existing structure is important to the morphological outcome. Diffusion-limited aggregation (DLA) (first conceived to account for the aggregation of colloids[4]) generates shape instabilities in an expanding domain whose growth depends on the availability of some species that diffuses toward it from surrounding material. Enhanced particle collection by nascent protrusions favors their extension while shadowing the adjacent growth front, which therefore tends not to advance, resulting in fractal tree structures very similar to those caused by viscous fingering.

R. McMeeking—contributing editor

Manuscript No. 28613. Received September 21, 2010; approved February 11, 2011.
[†]Author to whom correspondence should be addressed. e-mail: bcox@teledyne.com

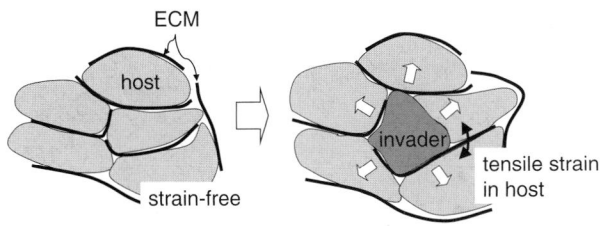

Fig. 1. When invaded, the host cells and their associated extracellular matrix must deform to accommodate the invader.

As pointed out by Halsey,[5] theories of viscous fingering and DLA are mathematically isomorphic, at least in simple renditions, in spite of the very different physics involved. Both can be framed as solutions to the Laplace equation, one using pressure as a dependent variable, the other using a probability field; in viscous fingering, the boundary between the two fluids moves in response to the gradient of the pressure across the boundary, while in DLA, the boundary of the accretion domain moves in response to the gradient of the probability variable across the boundary; and in viscous fingering, the boundary marks an isobar for the pressure variable, while in DLA, it marks an isocontour of the probability variable. Among other implications, this curious result issues a warning: finding a mechanism that replicates an observed morphology does not in itself prove that the mechanism is the operative one in nature. In the case of fractal tree-like networks, the observed morphology can be fitted equally well by two completely different physical phenomena. In the strain-cue theory of nonfractal network formation, a number of additional factors fall into agreement with experimental observations, which adds further credence to the strain-cue hypothesis[2]; but a strict test of the hypothesis remains for the future.

Nevertheless, with what belongs to the future left to the future, the strain-cue theory of Cox[2] does generate nonfractal networks and therefore one can pursue the following question: is it possible to mimic the phenomena identified in the theory by some invented laboratory process and thus replicate the method by which nature is hypothesized to form nonfractal networks with a view to applications in materials science? If so, this would constitute a significant advance in what is meant by the term "biomimetic." In the prior literature (to the author's knowledge), biomimicry has consisted of replicating morphologies from nature by whatever process can make them; in the present paper, the idea is proposed that the process itself might parallel the specific steps taken by living cells during natural morphogenesis. Thus, the entity being mimicked is not just the morphology, but the process by which a class of morphology is achieved in nature.

To open up a possible search for a biomimetic approach to fabricating nonfractal networks, this paper reviews the relevant details of the strain-cue theory of network formation and highlights the phenomena that must be replicated to achieve a biomimicking process.

II. The Strain-Cue Theory of Network Formation

The following is a summary of the theory presented by Cox.[2] The theory deals with the penetration of a 3D domain of host cells by an army of invading cells, which form a network. The invasion causes spatial variations in the density of cells as invaders intercalate themselves among the hosts. The density variations create spatially varying and time-dependent strain and stress fields. The fields are recomputed continually as the invasion progresses. The local values of strain components are used as cues to control the migration of invaders from domains that have already been invaded. The strain-cue rule is applied equally to all invaders on every iteration; all variations in migration patterns result solely from variations in the strain fields. No

explicit control over branch stability, sprouting, or any other network characteristic is introduced.

Exploiting intuition from the mechanics of inclusions, the strain cues needed to model network formation by cells are calculated as an eigenstrain problem. The problem is solved numerically on a simple cubic mesh with element size a, aligned with a Cartesian coordinate system (x, y, z). Each cubic element may be occupied by a combination of hosts and invaders, with number n_h and n_i, respectively, which need not be integers. An element is defined to be somewhat larger than a single cell. The cell number densities in any element are $\rho_h = n_h/a^3$ and $\rho_i = n_i/a^3$, respectively. The cell densities are smeared out to be uniform within each element, but they differ from element to element.

In the absence of mechanical stress, a single host and a single invader cell are assumed to have volumes V_h and V_i, respectively. The volume that the cells in a single element would occupy in the absence of stress is $V = n_h V_h + n_i V_i$. If this volume is larger than that of the occupied element, a mechanical stress will arise if the element is confined among other elements. The stress field is computed using analytical formulae derived for an "inclusion" in an elastic medium that contains a "stress-free eigenstrain."[6–8] The volume eigenstrain, $\bar{\varepsilon}$, measures the difference between the stress-free volume of the cells that occupy an element and the volume of the element. The system of cubic elements containing various eigenstrains generates spatially varying strain fields, which are computed by methods detailed in Cox.[2]

A critical element of the hypothesis for natural network formation is a strain-based criterion for cell migration. This criterion must appear in a successful biomimicking process. Of special interest is the quantity $\varepsilon_s = \varepsilon_1 + \varepsilon_2$, evaluated at a point P just outside any element that has already been invaded, where the directions 1 and 2 are the directions normal to the direction of a candidate migration step that passes through that point. The spatial variations in ε_s break symmetry, presenting a strain concentration when P lies at the extremity of any branch, which stabilizes branches. When a branch bends by a random variation in its direction to form an elbow, a strain enhancement also arises at the point of the elbow. This trend generates sprouting. These two effects together generate networks.

A second key element for which a biomimicking analogue must be sought is the assumed stochastic nature of cell migration (or cell proliferation) in nature. Networks only arise from a stochastic process. In the theory of Cox,[2] the stochastic process was implemented in the criterion for cell advance (migration). If element i has been invaded before the m-th time step, a strain-cue factor, S_{ij}, for migration into a contacting element j (i.e., an element sharing an entire face with the i-th element) during the m-th time step is defined to include a stochastic parameter that is determined by a pseudorandom number generator. The probabilistic nature of the migration criterion represents the effects of random variations in the local character of the host/invader system that are not represented by variations in the local strain fields. The random fluctuations have no spatial order; they introduce no spatial bias that might favor branch formation or stabilization, but they do influence network morphology and density.

The third key element in the theory of biological network formation is the introduction of strain relaxation. Where invading cells have migrated into regions already occupied by host cells, the local cell density rises and the host cells will tend to relax away from the invaders to restore uniformity. This motion is introduced into the simulations as a diffusion-like process in which only the host cells relax.

Relaxation reduces the stress and strain fields around an advancing branch of invaders and introduces a second rate process that competes with the rate of cell advance. The ratio of these two rates is critical to the outcome. If the branch does not advance during a time increment, further strain reduction will occur; and therefore, the longer a branch lingers at a given length, the lower the probability of its further extension. One possible outcome is that the branch will arrest permanently as the strains become evanescent. A network consisting of many branches can only continue to advance with undiminished density if new branches are formed from existing branches by sprouting at a rate per unit length of advance of a branch that at least equals the probability of arrest.

III. Characteristics of Networks Predicted by the Strain-Cue Theory

The simulations reported by Cox[2] are of the formation of planar networks; the invading cells are restricted to migrate within a single layer of elements whose centers lie on $z = 0$ in the Cartesian system (x, y, z). This restriction is appropriate for the formation of some nerve networks, such as in the gut and retina, and some instances of angiogenesis. For a biomimetic materials process, both 2D and 3D networks are of interest. While not addressed by Cox,[2] 3D networks would be created by the same mechanism if the constraint were removed that cells migrate in a single plane.

The following notes summarize events in network formation that should be part of a biomimetic process. They also highlight the local nature of the network forming mechanism.

(1) Networks Formed by Invasion from an Initial Band of Cells

In the simulations of Cox,[2] invasion begins from some initial set of invaders with an assigned spatial distribution. Consider first a relatively dense population of invaders distributed randomly in space, with statistically uniform density, over a narrow band $-5a < x < 0$. Given this starting population, the invasion tends to progress across a diffuse front toward positive large x, yielding networks exemplified by Fig. 2(a).

Figure 2(a) contains a number of features similar to those formed by neural crest cells invading the ileum and cecum of a mouse (Fig. 2(b)). The proportion of the width of a branch of the network to the width of typical voids are approximately equal in Figs. 2(a) and (b). The front of the invasion in Figs. 2(a) and (b) is comparable with the width of the larger voids (areas enclosed by branches) in the network. Comparison of other images from both nature and simulations shows some backfilling as the invasion proceeds, i.e., densification of the branch structure behind the leading extremity of the invasion. Branches occasionally appear to terminate permanently in the image of Fig. 2(b). Terminated branches in Fig. 2(a) occur at an approximately similar frequency, in a small percentage of enclosed void areas. Invading cells occasionally appear in clusters, which arise in simulations as fluctuations in the outcome of inserting the

local strain field values into the rule for advance. Where sprouting forms three-branch junctions in either Figs. 2(a) or 2(b), no two of the three branches involved are collinear. This is consistent with the prediction of the simulations that sprouting is triggered by the strain concentration that arises at an elbow when a branch's extension randomly changes direction.

(2) Network Development from a Single Invader

The characteristics of a predicted network do not depend on the initial spatial configuration assigned to the invaders. Figure 3 shows network development from a single centrally located invader cell. The problem definition in this simulation is similar to that in Fig. 2(a), the main difference being the initial distribution of invaders. Snapshots at increasing times show a network expanding in a roughly circular shape from the single initial invader, eventually filling the host domain. In the fully developed network, the branch density, sprout density, and distribution of enclosed void areas are statistically similar to those calculated from Fig. 2(a). Thus a biomimetic system can use various seeding or nucleation strategies to achieve a given network.

(3) Parameters that Control the Network

While a number of parameters were defined in the model of Cox,[2] the scaling properties of the governing equations together with statistical analysis show that most of these parameters do not affect network morphology independently. In fact, network morphology depends only on two parameters: one is the ratio of the rate of cell migration to the relaxation rate of the strains induced in the host by a migrating invader; the second is a parameter describing how sensitive the probability of cell migration is to fluctuations in the magnitudes of the strains around an invader.

(4) Physical Characteristics of the Mechanism of Network Formation

The key strain variations and the host relaxation mechanisms involve the interactions of 100 cells to order of magnitude; the influence on strain fields of cells beyond the nearest 100 cells in a branch falls rapidly with their distance. Thus, the network-forming mechanism has the robust characteristic of not depending on the maintenance of long-range order in the population.

Network formation requires that the strain stimulus mechanism be switched off at some point. If strain stimulus continues to operate indefinitely around a newly formed network branch, the branch disappears quite quickly due to the formation of an ever-increasing density of newly formed branches emerging from

Fig. 2. (a) A network grown from an initial band of invaders. Branches of invader cells (gray) are punctuated by occasional bifurcation or sprout points (black). Permanently arrested branches and clusters of invaders occur occasionally. (b) Nerve networks (white) formed in the mouse ileum and cecum, imaged at two different times during the invasion by neural crest cells (from Druckenbrod and Epstein[12]). The invasion is progressing from left to right in (a) and right to left in (b).

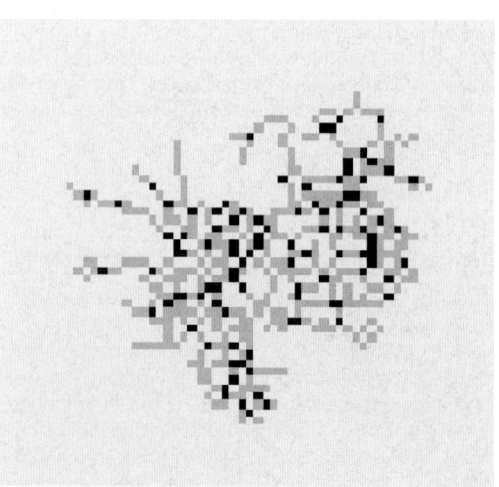

Fig. 3. A network grown from a single invader located near the center of the image. Branches of invader cells (gray) are punctuated by occasional bifurcation or sprout points (black). (For parameters similar to those used in Cox.[2])

it. The switching off is effected by relaxation due to host cell motions, which removes the signaling strains.

IV. An Inferred Template for Biomimetic Network Fabrication

(1) Strain Cues

The first key characteristic of the theory is that extension of the spatial domain occupied by the existing network must be triggered by values of a mechanical strain field (or stress field) at points exterior to, but immediately adjacent to, the existing domain. For successful mimicry, the strain field must first be generated and a mechanism then created for domain extension to be guided by the strain field. Following nature, the strain field can be generated if the process of network extension temporarily raises the local material density in the network domain.

Nature suggests several mechanisms by which domain extension can be guided by strain fields external to the existing network domain. Cells themselves can probe the space around them using filipodia and lamellipodia, which are thin protrusions of the cell membrane that can sense either chemical or strain signals. However, analogues of filipodia or lamellipodia in a nonanimate, biomimetic laboratory process require some imagination to conceive. Alternative candidate natural mechanisms whose mimicry may be more practicable include the formation of some species within the host domain under the stimulus of high strain levels, which can diffuse back to the network domain, where it could stimulate mass transport; and the transformation of host domain material into network domain material via a phase transformation triggered by strain.

(2) Stochastic Character

The strain-cue theory uses a stochastic algorithm to predict network extension. The stochastic character plays a pivotal role in introducing random sprouting from existing branches and in the occasional change of direction of an existing branch.

In nature, a number of mechanisms can be conjectured that would lead to the process of cell migration from the existing network domain being stochastic, including randomness in the locations of the host cells between which the invader must intercalate itself to effect migration. In the simulations described by Cox,[2] randomness was introduced by the Monte Carlo method using a pseudorandom number generator.

Random numbers were chosen each time migration from an already-invaded element was checked for advance in any direction. Thus, different cells with equal strain environments might differ in whether they advance or not on a given iteration. The fact that not all invaders subject to conducive strain conditions advance is one factor that controls the density of the network that is formed.

To assess prospective advance from the same already-invaded element on successive iterations, different random numbers were used. This can be interpreted as an expected result of the continuing motion of the host cells surrounding an invader, which changes the physical environment of the invader in a stochastic manner.

Thus, an analogue of a stochastic natural system would use a host medium with two features: random granularity (in correspondence with the discreteness of host cells) and continual motion, for example under thermal activation.

(3) Replenishment

The networks formed by migration from a planar reservoir are similar to those in the gut in general appearance and also in the fact that the entire network is formed by a cell population that is drawn initially from a reservoir on one side or end of the network (from the neural crest in nature). For a network to propagate indefinitely away from the reservoir of invaders, the invader population must be replenished. In nature, this occurs by a combination of new invaders migrating from the reservoir along existing network branches and the proliferation of invaders.

Replenishment by migration in a biomimetic system might be accomplished by, for example, ion diffusion. Replenishment by an analogue of proliferation could occur, for example, by a volume-enhancing phase transformation that draws on a continuously distributed source material.

(4) Relaxation Mechanism

In the strain-cue theory, relaxation of strains governs the probability of branch arrest and therefore the density of the network. Relaxation is represented as a mass transport process, occurring at a rate that is proportional to the spatial gradient of the cell density.

Any mass transport mechanism could serve as an analogue in a biomimetic system.

(5) Parameters Whose Values must be Maintained

Only two parameters control the outcome of simulations in the strain-cue theory: the ratio of the rates of the advance of the invasion front and the relaxation of induced strains, v_f/v_r, and the parameter, c_1, that defines the sensitivity of the probability of migration of an invader to deviations in its strain environment.

The value of c_1 should be selected to correspond to the expected variations in strain due to shape variations around a network branch. The mechanics of inclusions show that variations of 30% occur in the relative magnitude of the strain components computed in the strain-cue theory of Cox.[2] In nature, one can argue that cells are most likely to have evolved to sense strain variations of this magnitude, because that would make their action in forming networks most robust; if cells were substantially more sensitive to strain variations, slight aberrations in the spatial configuration of the cell population would cause spurious cell responses, while if they were substantially less sensitive, they would not sense the available strain cues. Consistently, the laboratory process used for network extension should occur at a rate that changes substantially when the guiding strain fields vary by approximately 30% in magnitude.

The acceptable range of values of v_r/v_f is determined fairly closely by the outcome of simulations. Depending on the value of c_1, v_r/v_f must lie approximately in the range of 0.03–0.06. Control of this ratio in nature appears to be affected mainly by the variation of v_f, the migration speed of the invading cells. For example, the relaxation velocity deduced for ameloblasts during the formation of human dental enamel is within 50% of the relaxation velocity that is required in simulations of the innervation of the mouse gut to obtain agreement with nature.[2] In contrast, the migration speeds of ameloblasts during human amelogenesis and neural cells during innervation of the mouse gut differ by three orders of magnitude (amelogenesis being very slow). One might surmise that the relaxation velocity is tied to rate processes in cells, which tend to occur at an invariant rate for different cell types under the stimulus of equal density gradients; whereas, a cell's migration speed might be varied more widely by switching its motility on or off. In a biomimetic system, either rate could feasibly be manipulated.

(6) Parameters Whose Values are Immaterial

The simulations are invariant with regard to the morphology of the predicted networks under changes of certain parameter values. The absolute migration and relaxation velocities are immaterial; only their ratio matters. The elasticity of the materials does not enter; the morphology is determined by the strain fields arising around a system of inclusions bearing eigenstrain, which are independent of the elasticity of the medium, provided the network and host domains have similar elasticity. The width of a network branch does not affect the network morphology; the network will simply scale in a geometrically similar manner. Neither does the choice of element size in the theory affect network morphology; the outcome is invariant under changes in spatial scale. Thus, biomimetic systems can target networks possessing any characteristic length scale (or characteristic area of the voids enclosed by branches).

(7)   Networks in Three Dimensions

If the restriction that migration can occur only on a plane is relaxed in simulations, the rules used by Cox,[2] without modification, will generate 3D networks. The strain concentrations around either branch tips or branch elbows are independent of permutations of the global directions (x, y, z). Therefore, both the sprouting rate and the probability of branch arrest per unit length of advance of a branch will be the same in 2D and 3D networks for given model parameters. The rate of branch coalescence will drop, implying a tendency toward open-tree rather than closed network structures.

V.   Some Material Concepts on which a Biomimetic System could be Based

Two material processes might offer developmental pathways toward biomimetic network formation. The first is strain-induced (or stress-induced) phase transformation in a solid material. Consider, for example, the phase transformation seen in partially stabilized zirconia, a phenomenon with a long presence in the ceramics literature. The image shown in Fig. 4 shows fingers or branches of transformed material (transformation of tetragonal zirconia to monoclinic zirconia at a critical value of local stress) in a specimen subject to applied mechanical load.[9] This phenomenon shows the potential of strain fields induced by existing transformed material to generate geometrically stable branch structures; the untransformed material around the tip of a nascent branch possesses spatial strain variations that favor the continued extension of the branch in its original direction of propagation,[10] in close analogy with the strain fields calculated in the simulations of Cox.[2] (The problem is also analogous to crack extension along a straight fracture path due to the strain concentration caused by the existing crack.) Presumably because the transforming material has a random microstructure, occasional bifurcation of transformation branches occurs, leading to the formation of closed void areas of Fig. 4, a key element of a network.

However, unlike the simulations of Cox,[2] no mechanism exists in the material of Fig. 4 to turn off the transformation process in the wake of the advancing front of branch tips. Therefore, the branches broaden indefinitely, merging and washing out any network morphology. This is analogous to results in Cox[2] for very low values of the ratio of rates, v_r/v_f. To achieve network morphology, a relaxation mechanism must be added to the transformation phenomenon. A simple method for achieving this would be to suspend transforming particles of zirconia (or other material) in a viscous matrix. The rate of branch extension can be controlled by varying the rate of the far-field load displacement in an experiment such as that of Fig. 4; the rate of strain relaxation around branches can be controlled by varying the viscosity of the matrix.

A second material phenomenon of possible interest is ion diffusion. Consider a material containing cavities into which ions might diffuse, e.g., dendrimer cages, selected to possess a chemical potential for the ions that varies when mechanical strain is applied to the cavities. If occupancy of the cavities by ions causes volume expansion, then occupied cavities will induce a strain field around themselves that will promote diffusion of ions into adjacent cavities that are not yet occupied. Mechanically, such a system is again analogous to the inclusion problem analyzed by Cox[2]; the strain fields will replicate the symmetry-breaking properties of the strains around invading cells in biological network formation. The domains of material in which cavities are occupied might therefore extend in geometrically stable branch structures.

VI.   Concluding Remarks

A laboratory process that mimics the way in which cell populations create networks must include a network extension mechanism cued by strains outside the existing network domain, a stochastic factor in network extension, and a relaxation process that causes the driving force for extension at any point to diminish in time. Given the presence of these phenomena, the key parameter that must be controlled is the ratio of the extension (migration) and relaxation velocities.

Acknowledgments

The author is indebted to Drs. Tim Bromage, Luisa Iruela-Arispe, Rodrigo Lacruz, Kerry Landman, Don Newgreen, C. E. Smith, and Mal Snead for conversations about biological aspects of networks and dental enamel. On aspects of physical and mechanical modeling, the author has accumulated much knowledge from old friends who worked at or visited the University of California, Santa Barbara, over the last 25 years, including Drs. John Hutchinson, Norman Fleck, Dave Marshall, Bob McMeeking, Zhigang Suo, and many others. He owes special debts to Dr. Fred Leckie, who, with characteristic charm, gently persuaded him of the value of seeking the simplest possible explanations of phenomena, even when the mathematics involved could easily run off into its own life; and Dr. Tony Evans, who brought us all together and made everything each one of us did seem an order of magnitude more important than we could otherwise have believed.

References

[1]V. Fleury, J.-F. Gouyet, and M. Leonetti (eds). *Branching in Nature.* Springer-Verlag, Berlin, 2001.
[2]B. N. Cox, "A Strain-Cue Hypothesis for Biological Network Formation," *J. R. Soc. Interface*, **8**, 377–94 (2011).
[3]P. G. Saffman and G. I. Taylor, "The Penetration of a Fluid into a Medium of Hele–Shaw Cell Containing a More Viscous Liquid," *Proc. R. Soc. London, A*, **245**, 312–29 (1958).
[4]T. A. Jr. Witten and L. M. Sander, "Diffusion-Limited Aggregation, a Kinetic Critical Phenomenon," *Phys. Rev. Lett.*, **47**, 1400–3 (1981).
[5]T. C. Halsey, "Diffusion-Limited Aggregation: A Model for Pattern Formation," *Physics Today*, (2000).
[6]T. Mura, *Micromechanics of Defects in Solids.* Martinus Nijhoff, The Hague, 1982.
[7]Q. Li and P. M. Anderson, "A Compact Solution for the Stress Field from a Cuboidal Region with a Uniform Transformation Strain," *J. Elasticity*, **64**, 237–45 (2001).
[8]J. D. Eshelby, "The Determination of the Elastic Field of an Ellipsoidal Inclusion, and Related Problems," *Proc. R. Soc. A*, **241**, 376–96 (1957).
[9]P. E. Reyes-Morel and I.-W. Chen, "Transformation Plasticity of CeO$_2$-Stabilized Tetragonal Zirconia Polycrystals: I, Stress Assistance and Autocatalysis," *J. Am. Ceram. Soc.*, **71** [5] 343–53 (1988).
[10]R. M. McMeeking and A. G. Evans, "Mechanics of Transformation-Toughening in Brittle Materials," *J. Am. Ceram. Soc.*, **65** [5] 242–6 (1982).
[11]B. N. Cox, "A Multi-Scale, Discrete-Cell Simulation of Organogenesis: Application to the Effects of Strain Stimulus on Collective Cell Behavior During Ameloblast Migration," *J. Theor. Biol.*, **262**, 58–72 (2010).
[12]N. R. Druckenbrod and M. L. Epstein, "The Pattern of Neural Crest Advance in the Cecum and Colon," *Dev. Biol.*, **287**, 125–33 (2005).

Fig. 4.   Strain-induced transformation in partially stabilized zirconia exhibits branch structure and elements of network formation (original photograph provided by Dr. I-Wei Chen).

Call for Papers Abstracts Due July 20, 2011

36TH INTERNATIONAL CONFERENCE AND EXPOSITION ON

ADVANCED CERAMICS
AND COMPOSITES

January 22-27, 2012
Hosted at Hilton Daytona Beach Resort and Ocean Center
Daytona Beach Florida, USA

www.ceramics.org/daytona2012

Organized by The American Ceramic Society and The American Ceramic Society's Engineering Ceramics Division

Author Services

Committed to providing the best possible service to journal authors

Visit the Author Services website at http://authorservices.wiley.com for:

- Online article tracking through production with optional e-alerts at key stages

- Information on how to **nominate up to 10 colleagues** to receive FREE online access to your article

- Author **guidelines** by journal

- **Resources**, FAQs and tips on article preparation, submission, artwork, copyright, offprints etc.

- **Free online access** to your article when it is published online

- **25% discount** on Wiley books

WILEY-BLACKWELL

http://authorservices.wiley.com